D0924121

Recent developments in molecular and statistical methods have made it possible to identify the genetic basis of any biological trait, and have given rise to spectacular advances in the study of human disease. This book provides an overview of the concepts and methods needed to understand the genetic basis of biological traits, including disease, in humans. Using examples of qualitative and quantitative phenotypes, Professor Weiss shows how genetic variation may be quantified, and how relationships between genotype and phenotype may be inferred.

This book will appeal to a wide range of biologists and biological anthropologists interested in the genetic basis of biological traits, as well as to epidemiologists, biomedical scientists, human geneticists, and molecular biologists.

Cambridge Studies in Biological Anthropology 11

Genetic variation and human disease

Principles and evolutionary approaches

Cambridge Studies in Biological Anthropology

Series Editors

Also in the series

G.W. Lasker *Surnames and Genetic Structure*

C.G.N. Mascie-Taylor and G.W. Lasker (editors) *Biological Aspects of Human Migration*

Barry Bogin *Patterns of Human Growth*

Julius A. Kieser *Human Adult Odontometrics – The study of variation in adult tooth size*

J.E. Lindsay Carter and Barbara Honeyman Heath *Somatotyping – Development and applications*

Roy J. Shephard *Body Composition in Biological Anthropology*

Ashley H. Robins *Biological Perspectives on Human Pigmentation*

C.G.N. Mascie-Taylor and G.W. Lasker (editors) *Applications of Biological Anthropology to Human Affairs*

Alex F. Roche *Growth, Maturation, and Body Composition – The Fels Longitudinal Study 1929–1991*

Eric J. Devor (editor) *Molecular Applications in Biological Anthropology*

Duane Quiatt and Vernon Reynolds *Primate Behaviour – Information, social knowledge, and the evolution of culture*

G.W. Lasker and C.G.N. Mascie-Taylor *Research Strategies in Biological Anthropology – Field and survey studies*

S.J. Ulijaszek and C.G.N. Mascie-Taylor *Anthropometry: the individual and the population*

Genetic variation and human disease

Principles and evolutionary approaches

KENNETH M. WEISS

Department of Anthropology,
Pennsylvania State University,
Pennsylvania, USA

PUBLISHED BY THE PRESS SYNDICATE OF THE UNIVERSITY OF CAMBRIDGE
The Pitt Building, Trumpington Street, Cambridge, United Kingdom

CAMBRIDGE UNIVERSITY PRESS
The Edinburgh Building, Cambridge CB2 2RU, UK www.cup.cam.ac.uk
40 West 20th Street, New York, NY 10011-4211, USA www.cup.org
10 Stamford Road, Oakleigh, Melbourne 3166, Australia
Ruiz do Alarcón 13, 28014 Madrid, Spain

First published 1993
First paperback edition (with corrections) 1995
Reprinted 1999

Printed in the United Kingdom at the University Press, Cambridge

Typeset in Times 10/12.5pt, in 3B2® [PN]

A catalogue record for this book is available from the British Library

Library of Congress Cataloguing in Publication data
Weiss, Kenneth M.
Genetic variation and human disease: principles and evolutionary approaches / Kenneth M. Weiss
p. cm. – (Cambridge studies in biological anthropology)
Includes bibliographical references and index.
ISBN 0-521-33421-7 (hc)
1. Medical genetics. 2. Human genetics. 3. Genetic epidemiology.
I. Title. II. Series.
[DNLM: 1. Evolution. 2. Genetics. 3. Genetics, Medical.
4. Variation (Genetics) QH 401 W431g]
RB155.W35 1993
573.2′1 – dc20 92-49275 CIP

ISBN 0 521 33241 7 hardback
ISBN 0 521 3366 0 paperback

This book is dedicated to my wife,
Anne,
and to my children,
Ellen,
Amie
and
Richard

Contents

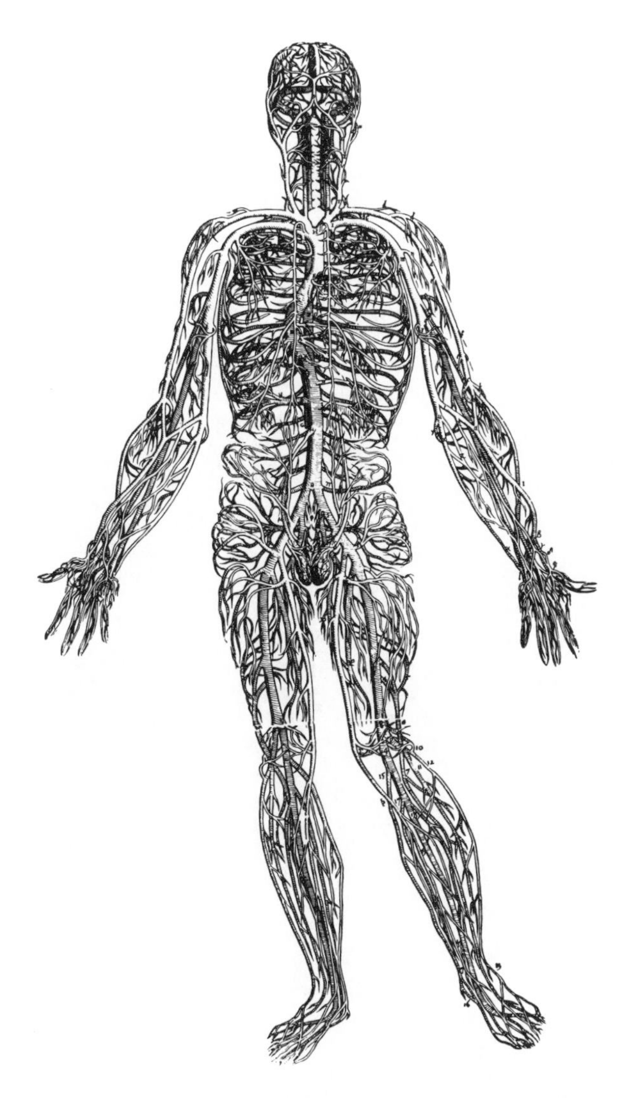

Frontispiece: This classic illustration of the human circulation by the great Belgian anatomist, Andreas Vesalius, shows the nested, hierarchical sets of branching vessels that provide nutrients to the body. While the larger arteries are specifically named, there are many more smaller arterioles and capillaries, providing multiple pathways to each area. Our understanding of the circulation depends on identifying the major individual vessels, but more fundamentally on knowing how the system works. The circulation serves here as a metaphor for similar nature of the genetic 'logic', or mode of control, of many complex phenotypes, one of the major themes of this book (see also Afterwords). (From Saunders and O'Malley, 1950.)

Preface: what is this book all about?

From the beginning of evolutionary population biology, studies of genetic disease have had an important place in the attempt to understand the relationship between the genes we inherit and the traits, or *phenotypes*, that we manifest during life. Diseases attract our deep personal interest. What causes them? Can they be inherited? Will *I* be affected? The very existence of genes that lead to traits that are deleterious to the individual also poses an apparent challenge to the idea of adaptive evolution by natural selection: how can such genes give rise to anything other than trivial frequencies?

This is an exciting time to be working in human genetics. A host of new statistical and molecular methods have become available in recent years; they enable us systematically to answer fundamental questions that previously we could only speculate about. Major new successes appear in the popular and scientific press every week, as genes related to one disease after another are identified. This has captured public imagination and its hopes. What is this all about? What does it tell us about basic biology?

In fact, the application of new methods to problems in genetic disease has, along with other developments in biology, led to a number of new generalizations about the genetic control of biological phenotypes that were unpredicted by classical evolutionary genetic theory, and which lead us to a revised view of genetic variation, how it works, and of the relationship between genotype and phenotype.

Among these generalizations are: (1) biological phenotypes are more heterogeneous and are affected by a more complex spectrum of genetic effects than had been thought; (2) these effects have characteristics that are generic across traits; (3) the genetic variants involved differ among populations, producing locally specific, cladistically structured ('tree-like') patterns of genetic variation that reflect the unique evolutionary history of each population; (4) the genes controlling a trait are often themselves cladistically structured, due to a deeper evolutionary history of duplication events, leading to further aspects of hierarchy and structure in the control of the phenotype; and (5) genetic variation can arise and be transmitted in several non-mendelian ways. These facts have greatly blurred the traditional distinction between qualitative traits (such as the presence or absence of cancer) and quantitative traits (such as cholesterol levels), and in a sense make biology more 'whole'.

This book is about the *logic* of the control of biological traits by genes and how we infer that control from data on variation in populations. The context is the genetics of disease, and I cover many of the recent discoveries, but my perspective is evolutionary and the subject is biology. The same principles that apply to disease also apply, generally without modification, to the genetics of normal biological variation. I hope this book is a useful introduction to *genetic epidemiology* for students of biomedical science, and that some will be led to pursue formal training in this subject. But I also hope to stimulate biologists, including my colleagues in anthropology, to apply these methods and ideas to the understanding of the evolution of normal variation.

One characteristic of human genetics is that we must rely on *observational* rather than experimental data. This is actually changing rapidly, as we become ever more able to rely on animal models for solving problems in human genetics, and as better gene identification techniques become available. But at present we are still forced to work largely with variation as it arises naturally in populations, and to rely on statistical *inference* from such data. This places a premium on our ability to find, and analyze, samples that are most revealing of underlying genetic mechanisms. Although a variety of such samples will be discussed, by far the most important are *families*. Genetically controlled traits vary in families in ways that follow evolutionarily determined rules. A specialized body of statistical methods, known generally as *segregation and linkage analysis*, has been developed to analyze family data to show whether genes control a trait and, if they do, to find the genes.

The parts of this book correspond to its major objectives. Part I summarizes the basic concepts of modern genetics and epidemiology needed to understand the relationships between genotypes and phenotypes in populations. Part II surveys the basic methods of genetic epidemiology. Part III considers how evolutionary processes generate the spectrum of genetic factors that cause variation in biological traits. Part IV discusses how the effects of the inherited genotype may be modified by the experiences of life and by various 'non-traditional' aspects of gene physiology, with important disease implications. Finally, in the Afterwords I draw these elements together into an overall synthesis of the genetic control of biological phenotypes, of which diseases are only the extremes of the natural pattern of variation.

I am not a statistician, and this is a conceptual survey rather than a full or formal statistical treatment. I assume that the reader has a basic general knowledge of statistics, but the mathematics in this book are descriptive rather than analytic, intended to show how things are described quantitatively, not to prove that those descriptions are correct nor to show in detail how the methods are implemented in practice. I hope my errors are few, and I provide extensive reference to the technical literature for readers who wish to undertake new studies of their own.

I discuss many diseases, and the book can be used as a text for graduate students and advanced undergraduates. However, my purpose is to search for a synthesis, and I have chosen examples preferentially from the recent literature to illustrate specific points, rather than providing a systematic survey of medical genetics. Some of these results may not stand the test of confirmation, but they give a better *feeling* of this area today than does the safer 'classic' literature. Every author also stresses what he/she is most familiar with and interested in. I can only apologize to authorities whose work I simply do not use or understand well enough to have included, such as path analysis for the study of quantitative phenotypes (e.g., Rao, 1985; Rao *et al.*, 1984; Rice, 1986), studies of twins (Cavalli-Sforza and Bodmer, 1971; Plomin *et al.*, 1990), or topical areas such as the study of behavioral and neuropsychiatric disorders (Kidd, 1992).

I have tried to make citation more efficient by using {curly brackets for general reviews that apply to a whole section or discussion}, [square brackets for references with technical details not discussed in the text], and (parentheses to document a specific point).

Acknowledgment

My understanding of this area has been greatly influenced over the years by interaction with many leading geneticists and epidemiologists, among whom are Fred Annegers, Ranajit Chakraborty, Aravinda Chakravarti, Robert E. Ferrell, Patricia P. Moll, James V. Neel, Peter E. Smouse, and William J. Schull. In writing this book I also had specific help from Eric Boerwinkle, John Blangero, Mike Boehnke, Robert Elston, and Sandra Hasstedt. I have tried not to use any of their ideas without attribution.

My graduate student Adam Connor developed the mathematical theory for the evolution quantitative allelic effects reported in Chapter 9, and the Gain discussion in the Afterwords, and post-doctoral researcher Steve Weeks did the computations for the fitness effects in Chapter 9 and the Afterwords.

I reserve special thanks for Charles F. Sing. Charlie and I have interacted for so long and to such an extent that it is probably impossible for me to be aware of all areas in which my ideas derive from his. As much as anyone in this area he has been struggling to understand the genetics of quantitative traits and the philosophy of complex causation. I also like his toy tractor collection.

I worked on this book as a summer guest in the Department of Biological Anthropology at Oxford University, and I thank Geoff Harrison and Tony Boyce for providing a stimulating environment and the infamous 'Black Hole' in which to work. I also thank Ken and Judy Kidd for the use of their basement apartment, where I wrote my frantic last draft while on sabbatical at Yale. I thank my students in Anthropology 473/573, who struggled with two early drafts, catching errors and confusion (nonetheless, your first line of defense against errors that remain is to check things for yourself). Sara Trevitt, Alan Crowden, and Sandi Irvine at Cambridge University Press and the series editors have graciously put up with my petulence.

My wife, Anne Buchanan, an epidemiologist in her own right, has discovered what it means to live with someone struggling to write a book, and has patiently helped me through incomprehensible early drafts, providing technical assistance of diverse kinds. My daughters Ellen and Amie also put up with a great deal, while son Richard grew up and left home in time to escape all of it.

Note to paperback reprinting. A few errors have been corrected in this reprinting. Since the first publication, advances have been made in many areas. These support rather than obsolesce the conceptual perspective of the book, but a few specifics are worth noting. Several diseases including Huntingdon's and Fragile X have now been shown to be due to mutational increases in the numbers of tandem trinucleotide repeats in or near the genes. Na–Li CNT seems less important to blood pressure variation, but other genes, notably ACE, have become good candidates. Microsatellite markers spaced 1 cM apart are now commercially available, and there is a drive for high-throughput typing of densely spaced single nucleotide polymorphisms (SNPs), for linkage and association mapping (Chs. 3 and 7), using hybridization arrays ('chips'). QTL and other forms of mapping have identified genes implicated in many complex diseases, including breast cancer (BRCA1, BRCA2), colorectal cancer, and several additional genes for Alzheimer's Disease. The web provides extensive documentation of these and many other facts (e.g., www.ncbi.nlm.nih.gov/, ariel.ucs.unimelb.edu.au:80/~cotton/dblist.htm, www.genethon.fr/genethon_en.html, www.uwcm.ac.uk/uwcm/mg/hgmd0.html). The pattern of complexity and variation that is at the core of this book is clear; how effectively we will be able to infer genetic causation frequent enough to be of public health importance for complex disease, remains less clear, and is a major challenge for the future.

Notational conventions used in this book

Following is a list of the main symbols and notational conventions used in this book.

For specific traits or phenotypes

I use the following conventions on human gene nomenclature.

1. A trait name is in Roman capitals (CF).
2. A locus name is in Roman italics (*CFTR*).
3. Alleles at a locus should be denoted by the locus name, an asterisk, and an allele designator (*CFTR**CF*); where clarity is served, I have used more traditional or mnemonic notation.

Main mathematical symbols used

b_{xy} regression coefficient of variable y on variable x

$C(r;s)$ combinatorial of number of ways to choose r things from s things

$\mathrm{Cov}(x,y)$ covariance between variables x and y

d dominance displacement (a measure of non-additive genetic effects)

$\mathrm{Det}(g|\phi)$ detectance of genotype g from phenotype ϕ

ELOD average LOD over all possible combinations of marker disease and locus

$f(x)$ normal (Gaussian) probability density for variable x

F kinship coefficient, inbreeding coefficient

g diploid genotype, single locus

G diploid genotype at an enumerated set of one or more loci (i.e., oligogenotype)

$Gain(H(b),b)$ Gain of function H over a specified set of genetic effects

$h(t)$ hazard function (usually age-specific risk, at age t)

h^2 heritability (fraction of variation due to genes)

H heterozygosity, or genotypic diversity

$\mathcal{L}$ likelihood
LOD natural logarithm of likelihood ratio ('Log Odds')
m admixture proportion
$m(t)$ age-specific fertility schedule
M mating type
N_e effective population size
NMF net maternity function = $S(t)m(t)$
Nor(μ,σ^2) designates a variable that is normally distributed, with mean μ and variance σ^2
PG multilocus aggregate polygenotype (not all loci enumerated)
PIC polymorphism information content, a measure of genetic diversity
$Pred(\phi|g)$ Predictance of phenotype ϕ from genotype g
$P_{o|f,m}$ Probability of offspring genotype given mother's and father's genotypes
q allele frequency at diallelic locus
p_i single-locus allele frequency, allele i (p,q for diallelic locus)
P_g genotype frequency for enumerated genotype
OR odds ratio
Pr probability
Prev prevalence of trait in a defined population
r coefficient of relationship
$r(t)$ risk of cell transformation with age
RR relative risk
$S(t)$ survivorship (probability of surviving some trait to age t)
SMR standardized morbidity (or mortality) ratio
t age in years or time in generations, depending on context
Var(x) variance of variable x, also denoted by σ^2
w coefficient of fitness
z standard normal distribution, Nor(0,1)
α (alpha) allelic values; mean effects of an allele on phenotype
β regression coefficient: effect of a risk factor variable on an outcome variable
γ (gamma) haploid (or gamete) genotype frequency
Γ (capital gamma) a specialized statistical function
δ (delta) (1) amount of linkage disequilibrium; (2) a 1/0 indicator variable

Δ (capital delta) 'change in'
θ (theta) recombination probability
Θ (capital theta) set of parameters of a statistical model
λ (lambda) likelihood ratio
μ (mu) mean; also used for mutation rates
ν_g (nu) genotypic effect: average deviation from population mean phenotype in individuals with genotype g
ξ (xi) standardized effect of a locus
$\rho(x,y)$ (rho) correlation coefficient between variables x and y
σ_x^2 (sigma) variance of variable x
τ (tau) genetic transmission parameters
ϕ (phi) phenotype
Φ (capital phi) cumulative normal distribution
χ^2 chi-squared distribution, used in statistical tests
ω (omega) segregation proportion
$\Omega_g(\phi)$ (capital omega) penetrance of phenotype ϕ given genotype g
$\in$ 'belongs to' or 'is a member of'

Estimates of parameter values are marked with a caret

Reference citations in the text

Curly brackets, general reviews; square brackets, technical details; parentheses, specific points (see p. xvii).

Abbreviations

Ab antibody
ACE angiotensin conversion enzyme
AD Alzheimer's disease
Ag antigen
apo apolipoprotein
APP amyloid precursor protein
AS ankylosing spondylitis
AT ataxia telangiecstasia
bp base-pair
BWS Beckwith–Wiedemann syndrome
C (1) cholesterol; (2) constant region of Ig heavy and light chains
CEPH Centre d'Etude du Polymorphisme Humaine
CETP cholesteryl ester transfer protein
CF cystic fibrosis
CFTR cystic fibrosis transmembrane conductance regulator
CGD chronic granulomatous disease
CHD coronary heart disease
CNS central nervous system
CR cholesterol ratio
DMD Duchenne muscular dystrophy
DS Down's syndrome
DZ dizygous (twins)
EGF epidermal growth factor
FAP familial adenomatous polyposis
FH familial hypercholesterolemia
GBCA gallbladder cancer
G6PDH glucose-6-phosphate dehydrogenase
HBLP hypobetalipoproteinemia
Hb hemoglobin
HD Huntingdon's disease
HDL high density lipoprotein
HexA β-hexosaminidase-A
HLA human leukocyte antigen
HPFH hereditary persistance of fetal hemoglobin
HWE Hardy–Weinberg equilibrium
i.b.d. identical by descent
i.b.s. identical by state
ICM inner cell mass
IDL intermediate density lipoprotein
IDDM insulin-dependent diabetes mellitus
Ig immunoglobulin
IVS intervening sequence
J joining region of V and C chains of immunoglobulins
LCAT lecithin/cholesterol acyl transferase
LDL low density lipoprotein
LDLR low density lipoprotein receptor
LINES long interspersed sequences
LOV loss of variation
MHC major histocompatability complex
MS multiple sclerosis
MZ monozygous (twins)
Na–Li CNT sodium–lithium countertransport
NFI neurofibromatosis I
NIDDM non-insulin-dependent diabetes mellitus
NTF non-transmitted factor
NTM normal transmitting males
OI osteogenesis imperfecta
PAH phenylalanine hydroxylase
PBD proliferative breast disease

PCR polymerase chain reaction
PKU phenylketonuria
PNG Papua New Guinea
PON paraoxinase
PWS Prader-Willi syndrome
PYr person-year(s)
QTL quantitative trait locus
RA rheumatoid arthritis
RB retinoblastoma
RFLP restriction fragment length polymorphism
SES socioeconomic status
SML single major locus
SSS stochastic self-similarity
STS sequence tagged sites
TB tuberculosis
TcR T-cell receptor
TG triglyceride
TSD Tay–Sachs disease
V variable region of Ig heavy and light chains
VLDL very low density lipoprotein
VNTR variable number of tandem repeats
YAC yeast artificial chromosomes

Part I *Genes and their expression*

1 *What is a gene?*

The usual idea of a gene is of a specific region of DNA that codes for a single protein or enzyme, and the position of a gene on a chromosome is known as its *locus*. Variants of the DNA sequence at this locus among individuals are known as *alleles*. This ‘one gene, one enzyme’ model has long been the basis for research by human geneticists trying to identify traits, or *phenotypes*, whose inheritance patterns are consistent with the action of individual genes.

However, recent advances in genetics have greatly revised our concept of what genes are and how they work, showing that the relationship between DNA sequence and phenotype is both more complex and more interesting than we had thought. Some functions of DNA do not even depend on its nucleotide sequence, and DNA sequence variation includes a variety of direct and indirect forms of feedback among various regions of the DNA within and between cells.

Human life begins with a fertilized egg that carries a set of chromosomes containing genetic instructions, plus enough basic compounds and nutrients to commence the cell cycle. This latter material includes messenger RNA (mRNA) coded for by the mother’s genes, which provides enough ‘information’ to direct the development of the new individual until its own genetic mechanism can be switched on. Thereafter, the organism depends entirely on its own genes, which contain the coding sequences and all the other signals needed for controlling embryological development, and the subsequent growth, tissue renewal, and physiology of the organism during its lifetime.

A differentiated organism results from the expression of different subsets of its genes in different tissues or developmental times. The regulation of gene expression is driven largely by proteins (themselves coded by genes) that work chemically, recognizing non-coding DNA regions located near the regions that do code for physiologically active proteins. To see how this works, we must first outline relevant aspects of DNA structure (Appendix 1.1) {Lewin, 1990}.

The arrangement of DNA in the cell

The full set of human genes, found in the nucleus of each cell, is called the *genome*. The genes themselves are arrayed linearly on *chromosomes*, long DNA molecules consisting of strings of chemical units called *nucleotides*. These are found in the cell nucleus (DNA is also found in energy-processing organelles known as *mitochondria*, in the cytoplasm outside the nucleus, a subject largely unrelated to our story here). There seem to be no prior practical chemical, or logical, limits as to how long a chromosome may be, nor to the sequence of the nucleotide bases of which it is composed.

The number and length of the chromosomes is characteristic of a species, but this is due to fortuitous evolutionary rather than functional reasons. Humans have 23 pairs of chromosomes, one set inherited from each parent, which are numbered in decreasing order of their length. Pairs 1 (the longest) to 22 (the shortest) are called *autosomes*. The final pair are the *sex chromosomes*, the only pair to have dimorphic forms, the *X* and the largely degenerate *Y* that are involved in sex determination. Each chromosome is packaged by end sequences known as *telomeres*, and is divided into two *arms* by a modified sequence, known as the *centromere*, that is used in cell division.

Under controlled laboratory conditions, chromosomes can be extracted from cells and chemically stained to reveal a series of lighter and darker *bands* under a light microscope. These bands, shown schematically in Figure 1.1, are characteristic in position and thickness, and have been coded for use as chromosomal landmarks. The set of stained chromosomes is known as the *karyotype*. The *short arm* of a chromosome is denoted *p* and the *long arm q*. For example, the first small band distal to the second major band from the centromere on the long arm of chromosome 1 is denoted 1*q*2.1.

The structure of single-gene coding regions of DNA

There is a high level of organization of individual genes at the micro as well as the karyotypic level. The general functional arrangement of a typical gene is shown in Figure 1.2. The details can vary, but this figure shows most of what we need to know to understand how genes control phenotypes.

By convention, DNA is labeled according to the carbon atoms in the sugar molecules by which adjacent nucleotides are attached to each other, drawing the 5′ (five-prime) end to the left and the 3′ end to the right. Genes are transcribed 5′ to 3′, and the code for the amino acids in the resultant protein is read in groups of three from left to right.

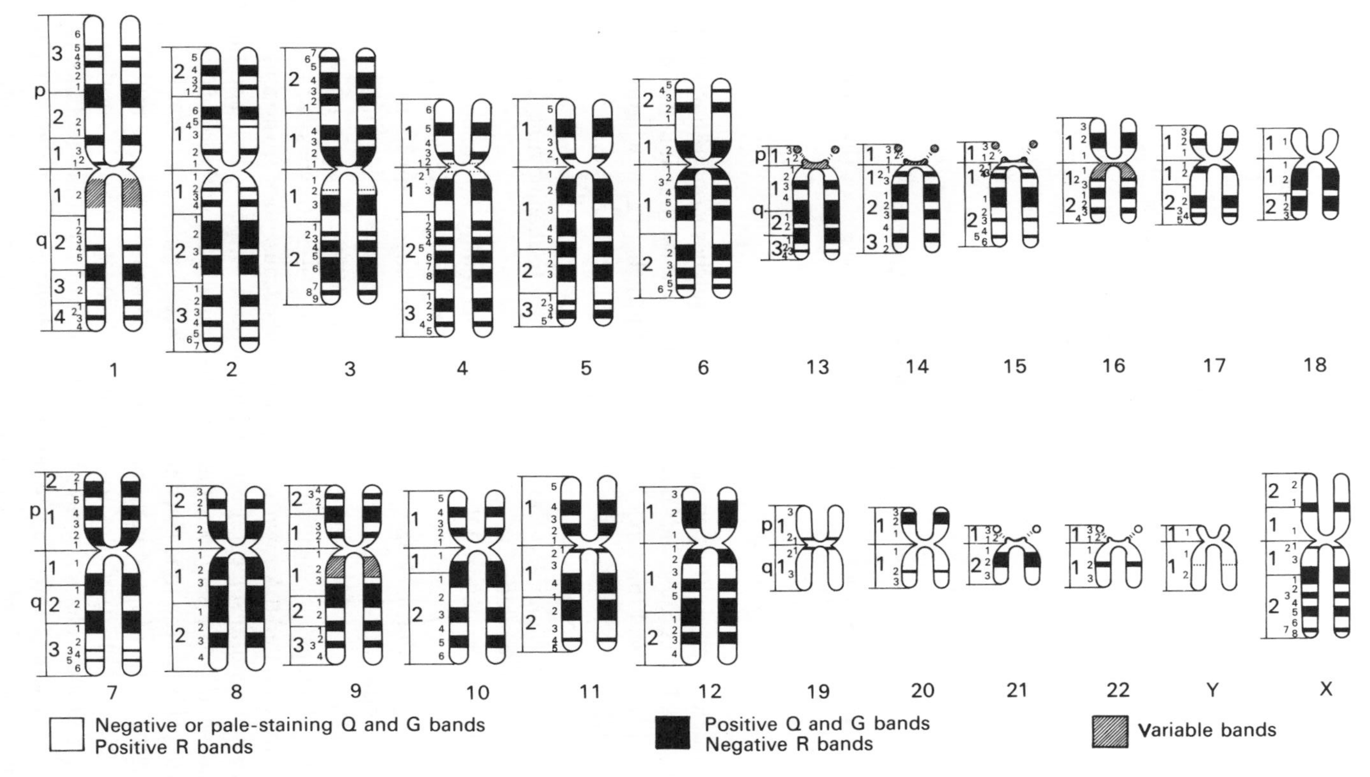

Figure 1.1. Human karyotype, showing a standard numbering system based on chemical staining of a chromosome preparation. Different staining methods yield different levels of detail that are used in refining the system (not shown). (From Thompson and Thompson, 1986.)

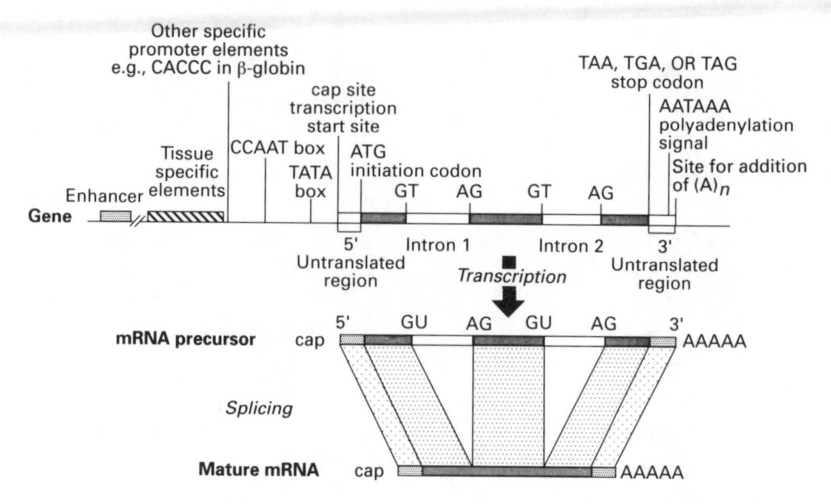

Figure 1.2. Schematic of structure and expression of an idealized human gene. (From Gelehrter and Collins, 1990.)

To the left and right of coding genes are stretches of non-functional DNA, so called because they have no currently known function; much of our total DNA is of this type. Precedent is of little help as to whether we are likely to find a function for these regions. Some clearly are the debris of millions of years of evolution and have no current function, but other non-coding DNA may serve as physical spacer or packaging functions or be involved in regulating gene expression.

The best examples of DNA with almost certainly no function are *pseudogenes* and *processed pseudogenes*. The former are simply old coding regions that lack the adjacent sequences needed for them to be actively coding protein; the latter are sequences complementary to former pieces of mRNA that were at some past time inserted into the genome, perhaps by retroviruses. The sequence of both types of 'dead' gene can be related to known active genes, but because they are no longer functional, natural selection is oblivious to such former genes, and the rate at which their sequences accumulate mutational variation can be used to calibrate the time since they were inactivated.

Another kind of non-functional DNA is segments known as repeats. These are stretches of tens to thousands of copies of some short sequences. Repeat elements have been strewn throughout the genome, in numbers ranging from a few to hundreds of thousands of copies in each genome. There are many types of repeat element. Some have sequences known from studies of viruses and other organisms to mediate the movement of genes and their insertion into chromosomes. These have no known physiological function. One such structure, the *Alu* repeat, is a sequence of about 300 nucleotides (a tandem pair of two 150 base-pair (bp) units) that can be copied and repeatedly reinserted more or less randomly all over the genome; there are hundreds of thousands of *Alu* copies in the human genome (Lewin, 1990). *LINES* (*long interspersed sequences*, or L1 sequences) are about 6500 bp of DNA whose genome copy number is in the thousands. These sequences are ubiquitous and approximately randomly distributed along the genome.

The recent publicity given to DNA fingerprinting refers largely to short sequences known as *VNTR*s (*variable number of tandem repeats*), whose copy number in a given locus varies so much that nearly every individual is a unique heterozygote. Similarly diverse are sets of *dinucleotide repeats*, such as $(CG)_n$, where n is the number of times the pair of nucleotides is consecutively repeated (sometimes this is denoted CpG because the nucleotides are connected via the sugar-phosphate backbone of a DNA molecule). These, too, are scattered all over the genome, and are so variable as to be useful in 'fingerprinting' individuals for forensic and genetic purposes.

Other regions of DNA shown in Figure 1.2 are transcription regulators. Several such regions are known, typically at roughly fixed distances upstream from the transcription initiation sequence, and usually with a nearly fixed nucleotide sequence of a few to a few hundred base-pairs. The farthest upstream from most coding genes is a region known from its base sequence as the *CAAT box*; closer to the initiation point is another regulatory sequence known as the *TATA box*. 'Islands' of C + G-rich repeated sequences are also found close upstream from coded genes and seem to be important in transcription. The chemical characteristics of these sequences are recognized by proteins, such as RNA polymerases, that bind to them and cause the downstream coding region to be transcribed into mRNA. Regulatory regions that enable transcription to take place are called *promotors*, and regulatory sequences that increase the number of copies or ease of transcription are called *enhancers*; the latter can be upstream or downstream from, or even in the introns (see below) of, a gene.

These 'generic' regulatory regions are similar among genes, but the control of individual gene expression requires that there also be locus-specific regulatory sequences or mechanisms of some kind. We are just learning how to recognize and characterize such regions, and this knowledge will be fundamental to understanding important biological traits in the future. Obviously, if a regulatory sequence is missing, or severely altered by mutation, the downstream gene cannot be transcribed and becomes a 'silent' pseudogene.

The *start sequence* (generally, ATG) is the place at which mRNA production begins, and moves forward through the coding region. The latter includes *exons*, whose number varies widely among genes and which contain the codes for the amino acids of the final protein. The exons are separated by regions labeled *introns*. These sequences are transcribed into immature message, but are then cleaved at their ends, or *splice junctions* (characterized by sequences GT and AG). The introns are then enzymatically removed from the mRNA molecule and the exons join together to form a single protein-coding sequence. At the end of the last exon is a stop codon that signals the end of mRNA transcription. The codes for starts, stops, and splice junctions are essentially common to all genes. With some further modifications (e.g., the addition of a long string of adenine residues called a *poly(A) tail*), the mature mRNA molecule can leave the nucleus to serve its *translation* function, to code for the amino acid sequence in a protein.

The functional constraints on the length or sequence of introns are not well understood. Some serve as regulatory or spacer sequences, others are actually coded in alternative coding frames in the same gene region. The exon–intron arrangement is typically conserved between species that have not shared a common ancestor for hundreds of millions of years, suggesting that mutations affecting the length or position of the introns have been selected against. However, there are examples of homologous genes (i.e. genes descended from a common ancestral gene) that function quite normally with introns missing or varying greatly in length. Bacteria have no introns. Probably introns are not inherently functional, but have occasionally become so due to the accidents of evolution.

Changes in DNA: mutations

The length of a chromosome is roughly stable, because chromosomes are replicated as units during cell division. However, various kinds of *mutation*, or change in the genetic material, may alter the chromosome length or arrangement. On rare occasions, chromosomes break into pieces which do not rejoin normally but may either form a *rearrangement* on the same chromosome, or may be *translocated* to join with another chromosome. These changes are known as *macro-mutations*, and can often be detected from the karyotype by ordinary light microscopy. Many types of repeat sequence (VNTRs, *Alus*, etc.) frequently change their copy number in a given region where they are located, through various mutational mechanisms.

Macro-mutations occurring in germ line cells (sperm or egg) can be inherited. This, though rare, is responsible for the karyotypic differences that accumulate between species over evolutionary time. Related species often have similar karyotypes. Indeed, the arrangements of genes on chromosome segments are conserved for tens of millions of years, to the extent that we can often identify genes in mouse chromosomes and then quickly find them in human chromosomes.

About once per one million to ten million meioses (productions of a gamete), a given nucleotide will be changed to some other nucleotide. This can occur if a chemical mutagen or ionizing or ultraviolet radiation has modified or deleted a nucleotide, or if DNA repair enzymes fail to replace the nucleotide correctly. Also, the DNA polymerase enzymes, which replicate DNA sequences, do not operate with perfect fidelity. Single nucleotide changes are called *point mutations*, or sometimes *micro-mutations*. If a point mutation results in a change in the amino acid sequence of the protein that the gene encodes, the phenotype of the individual may be affected. A well-known example is the point mutation that produces sickle cell hemoglobin. Unlike chromosomal mutations, point mutations cannot be visualized directly under a light microscope, and must be detected by molecular methods (see e.g., Appendix 1.2).

Point mutations appear to be by far the most common form of mutation. They include single-nucleotide *deletions* or *insertions*. These are known as *frameshift* mutations, because they change the triplet 'reading' (amino acid coding) frame, altering all downstream codons in the gene. Any nucleotide may be deleted, inserted, or changed for any other.

Many deletions of <20 bp have been associated with disease (or might cause other non-pathologic variation); many of these involve areas of short, direct repeated sequences (Krawczak and Cooper, 1991). Molecular mechanisms for the production of such mutations have been proposed.

Mutations can occur anywhere in the genome, but there appear to be some mutational hot-spots, regions in which chromosomal breaks or point mutations are more likely. For example, in CG pairs the C is often methylated and the deamination of methylcytosine leads to a C to T transition (Cooper and Youssoufian, 1988), and this is but one way that the local DNA environment may lead to mutability (Krawczak and Cooper, 1991). Of course, the *acceptability* of a mutation to natural selection depends on its functional impact, not its location or nature.

Phenotypic variation can be caused by mutations in any part of the system, except totally non-functional DNA, including the regulatory regions.

The regulatory mechanisms are themselves under genetic control, with their own regulatory mechanisms. A set of genes that have to be expressed simultaneously in a given cell may contain similar upstream regulatory sequences. The regulatory system is hierarchical and complex, sometimes closed in the sense that a gene's product can be auto-regulatory – i.e., can regulate the gene that coded for it.

APPENDIX 1.1: Essential aspects of DNA structure and function

The following is an outline of the biochemical nature of DNA and the mechanism by which it codes for the production of proteins {Lewin, 1990}.

DNA, a macromolecule with coding, replicating, and other functions

DNA (deoxyribonucleic acid) is the material of which the *chromosomes* are made. Each chromosome is a single, very large molecule, composed of millions of constituent atoms in a highly patterned way. As shown schematically in Figure A1.1, DNA consists of sequences of *nucleotides*, each of which contains a five-carbon sugar ring, a phosphate group, and one of four bases, *guanine*, *adenine*, *thymine*, and *cytosine*, denoted by their first letter (G, A, T, C). Two of these, A and G, are relatively long molecules called *purines*; the other two, T and C, are shorter and called *pyrimidines*.

A DNA molecule is double stranded, and is coiled into a helix. A helix is a structure with two 'backbones', the sugar-phosphate chains, connecting the series of nucleotides. The backbones maintain a fixed distance from each other, and twist by a fixed number of turns per unit length. To keep the strands parallel, it is necessary that the pairing between opposite bases maintains the same distance. This is accomplished by the specific *base-pair complementarity* of the purine–pyrimidine bases: chemically, A should pair only with T, and G with C. DNA is thus a long double helix comprising *base-pairs* (bp), A·T and C·G. Either strand can have any of the bases, and from a chemical point of view the pairs can be assembled in any order. Distances between specified positions on chromosomes are often measured in thousands of base-pairs, or *kilobase-pairs* (*kb* or *kbp*), or megabase-pairs (millions of base-pairs, Mbp).

Replicating function

The complementarity of the base-pairing mechanism is the key to the biological function of DNA. The essential feature is that DNA can replicate its sequence of base-pairs whenever a cell divides (*mitosis* in the case of somatic cell divisions, and *meiosis* in the formation of germ cells). The two strands of each chromosome (DNA molecule) are separated and enzymes known as *polymerases* bind to each strand and, selecting complementary nucleotides accessible in the surrounding fluid in the nucleus, pair them up one at a time in a sequence complementary to the existing strand to which the nucleotides are bound. In this way, we obtain two identical double-stranded DNA molecules.

guanine ['gwa:ni:n] 鸟嘌呤, adenine ['ædəni:n] 腺嘌呤, thymine ['θaimi:n] 胸腺嘧啶, cytosine [

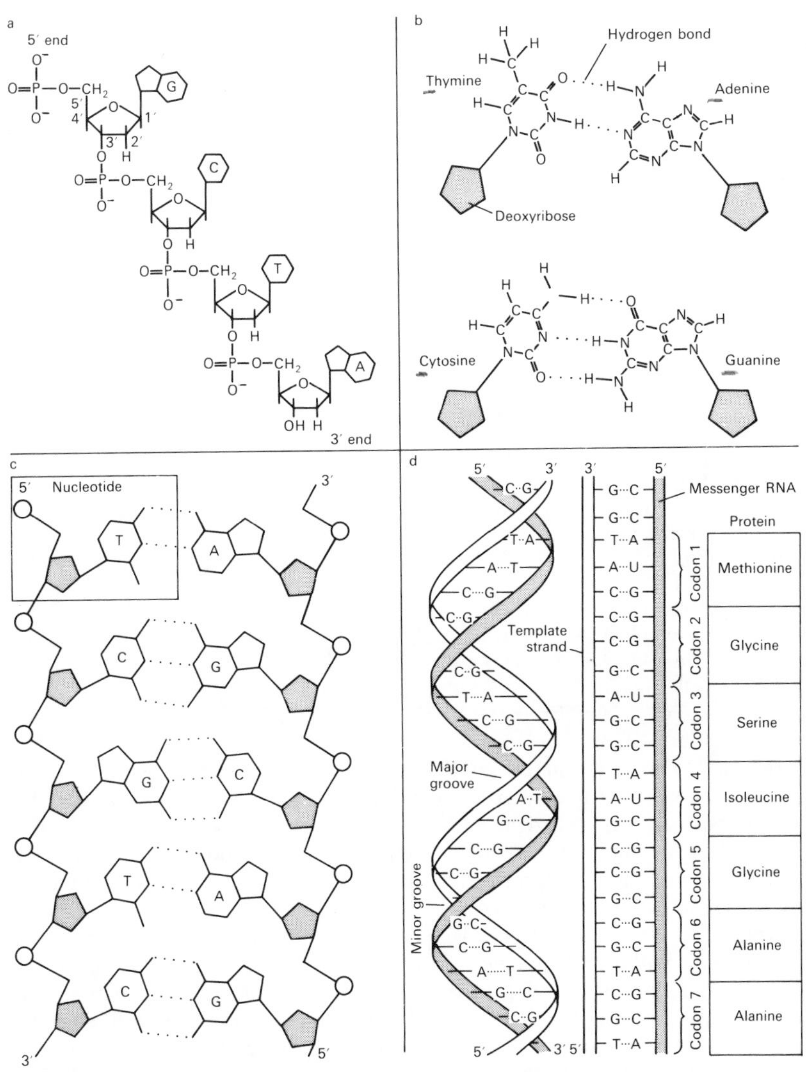

Figure A1.1. Basic structure of human chromosomes and DNA. (From Felsenfeld, 1985.)

Coding function: RNA production

Sequence complementarity is vital to the function of DNA as a codebook for the assembly of proteins. At specified chromosome regions, which depend on the particular cell involved, DNA-binding enzymes cause the two strands of a chromosome to separate, and one of the strands, known as the *template*, is *transcribed*. A single strand of nucleotides complementary to the template is

Table A1.1. *Primary amino acids and their three- and one-letter abbreviations*

Name	Three-letter abbreviation	One-letter abbreviation
Alanine	Ala	A
Arginine	Arg	R
Asparagine	Asn	N
Aspartic acid	Asp	D
Cysteine	Cys	C
Glutamic acid	Glu	E
Glutamine	Gln	Q
Glycine	Gly	G
Histidine	His	H
Isoleucine	Ile	I
Leucine	Leu	L
Lysine	Lys	K
Methionine	Met	M
Phenylalanine	Phe	F
Proline	Pro	P
Serine	Ser	S
Threonine	Thr	T
Tryptophan	Trp	W
Tyrosine	Tyr	Y
Valine	Val	V

assembled. This new single-stranded nucleic acid is called *messenger RNA*, or *mRNA*, and differs in structure from DNA in that (1) a ribose rather than deoxyribose sugar is used in the nucleotides, (2) a base known as *uracil* (U) is used rather than thymine, and (3) mRNA remains single stranded. The sequence of mRNA is similar to that of the *coding* strand, since it is complementary to that of the template strand, of the DNA.

There are other types of RNA. Messenger RNA functions to *translate* the DNA code into an *amino acid* sequence of a protein (see Table A1.2). *Ribosomal RNA* (*rRNA*) is a class of molecules that are constituents of intracellular organelles known as *ribosomes*. *Transfer RNA* (*tRNA*) is used to transport amino acids to ribosomes to be assembled into protein. A particular sequence of a given tRNA binds chemically to a specific amino acid. Because proteins are strings of amino acids, of which there are 20 types (Table A1.1), there are different tRNAs specific to each type of amino acid, and indeed to the redundant codons specifying a given amino acid. The genome contains multiple copies of the genes coding for these various tRNA types, and there are codon-usage biases, in coded genes, that reflect the relative abundance of these tRNAs that correspond to a given amino acid.

Self-regulating function

The DNA sequence itself can be modified in a way which is highly specific and is involved in regulating transcription. For example, G and C can be methylated by

the attachment of CH_3 to a specific carbon atom. DNA metl
enzymatic control, and methylated nucleotides are found in h
locations, often regions rich in CpG sequences. When DNA
methylation pattern is also enzymatically replicated, and persists i
until it is actively changed by existing physiological mechanisms, wh
sequence specific, or by mutational error.

Methylation affects the ability of regulatory proteins to recognize a
to specific regions of DNA. Generally, methylated regions cannot be
DNA transcription enzymes, so that methylation is a mechanism for t ..ning a gene off in a given cell. The inheritance of methylation patterns is known as *genomic imprinting*, a form of *epigenetic* inheritance that does not depend upon the DNA coding sequence (of course, it is dependent on the presence of the methylation sequences). Patterns of inheritance can be related to imprinting, as will be discussed in Chapter 14.

The DNA nucleotide sequence and the production of protein

The construction of a protein (Figure A1.2) is accomplished by the *translation* of the mRNA sequence into an *amino acid* sequence. After some modification in the nucleus, such as the splicing out of intron sequences (see the text) and the addition of a poly(A) (poly-adenine) tail, mRNA molecules migrate through pores in the nuclear membrane into the cytoplasm of the cell. There, they attach to, and move through, ribosomes as their sequence matches up, three nucleotides at a time, with exposed triplet sequences on tRNA molecules. A tRNA that binds a specific amino acid at one end always contains a nucleotide triplet, specific for that amino acid, at another part of its folded structure; the latter sequence binds to the mRNA as the attached amino acid links up with the protein that is forming. That is to say, DNA codes for amino acids via nucleotide triplets, known as *codons*. The code itself – the DNA – remains in the nucleus where it is sequestered from degradation, whereas mRNAs are often unstable and/or enzymatically degraded in the regulation of the metabolism of a given cell type.

The mRNA–tRNA matching begins at one end of the mRNA and works towards the poly(A) tail at the other end, as the mRNA moves through the ribosome. Each mRNA triplet brings a new amino acid to the vicinity of the ribosome, attached to its tRNA; subsequently, codon by codon in sequence, an enzymatically regulated process attaches the amino acids to each other to form an amino acid chain, the *polypeptide* or protein molecule. When the mRNA has moved through the ribosome, and all the coded amino acids are in place, the polypeptide is essentially completed.

A polypeptide can in principle have any length or amino acid sequence. It derives its physiological function from these characteristics, because, although all 20 amino acids share some chemical characteristics, they differ in the details of their physico-chemical properties. Some are larger than others. Some carry a net electrical charge, whereas others are electrically neutral. Two (cysteine and methionine) can form chemical bridges to each other, which, along with the other chemical variations among the amino acids, cause a polypeptide to form a particular three-dimensional shape. This *secondary* structure of the protein may later be modified into a higher structure if separate *chains*, or polypeptide subunits, are assembled into one large protein molecule. For example,

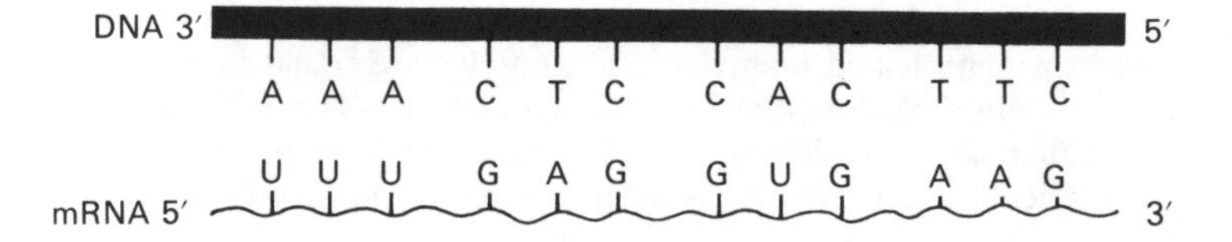

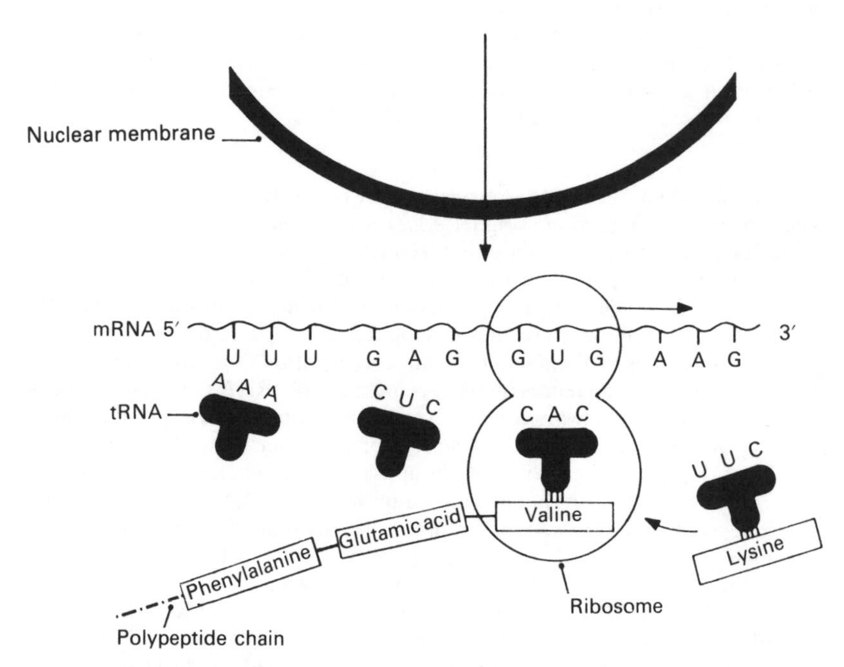

Figure A1.2. Schematic illustration of protein translation. (From *An Introduction to Recombinant DNA* by A. E. H. Emery, Copyright 1984 John Wiley & Sons Ltd. Reprinted by permission of John Wiley & Sons, Ltd.)

hemoglobin consists of one pair of each of two polypeptides (produced by two different DNA regions).

The final function of a protein depends on its shape and charge. Sometimes, as with antibody molecules, receptors for hormones, or enzymes, which speed up specific chemical reactions, the function depends on the ability of the protein to recognize the specific shape and charge of some other molecules with which it combines. On the basis of their final shape, size and charge, proteins also form the physical structures within cells, as well as those produced by cells. Mutations in DNA sequence can consequently lead to different physical characteristics in the coded proteins.

Aspects of the coding system

To choose among the 20 amino acids requires only 20 different codes, but a triplet of four different nucleotides can produce $4^3 = 64$ different codons. The DNA code is thus *redundant*, and there is typically more than one code for each amino acid (Table A1.2). Genes for the different forms of tRNA specific to each amino acid exist in the genome and are transcribed. Three codons, called *stop* codons, are used to signal the end of transcription.

The triplet code evolved from primitive single, and then doublet, coding systems {De Duve, 1991}. The way that this occurred has concentrated redundancy largely in the third, but with some in the second, nucleotide position in the codon. As a result, a substantial fraction of point mutations that arise in the third position are *synonymous*; that is, do not change the coded amino acid. Although there is some codon-usage preference, based on the abundance of the corresponding tRNAs, natural selection tolerates mutations in the third position much more readily than in the other positions {Li and Graur, 1990}. The redundant third position nucleotides are often of the same type; for example, the two codons for the amino acid phenylalanine, UUU and UUC, differ by U and C, both of which are purines. In addition to this DNA similarity between redundant codons, chemically similar amino acids (in terms of charge and size) often have similar codons.

These are the ubiquitous and unmistakable tracks of evolution, and lead to many ways in which mutations may have only minimal effects on the resulting protein. For example, some *non-synonymous* (amino acid changing) mutations substitute one amino acid for another with similar size or charge structure, and hence have little physiological effect.

Table A1.2. *The universal genetic code*

Codon	Amino acid	Codon	Amino acid	Codon	Amino acid	Codon	Amino acid
UUU	Phe	UCU	Ser	UAU	Tyr	UGU	Cys
UUC	Phe	UCC	Ser	UAC	Tyr	UGC	Cys
UUA	Leu	UCA	Ser	UAA	Stop	UGA	Stop
UUG	Leu	UCG	Ser	UAG	Stop	UGG	Trp
CUU	Leu	CCU	Pro	CAU	His	CGU	Arg
CUC	Leu	CCC	Pro	CAC	His	CGC	Arg
CUA	Leu	CCA	Pro	CAA	Gln	CGA	Arg
CUG	Leu	CCG	Pro	CAG	Gln	CGG	Arg
AUU	Ile	ACU	Thr	AAU	Asn	AGU	Ser
AUC	Ile	ACC	Thr	AAC	Asn	AGC	Ser
AUA	Ile	ACA	Thr	AAA	Lys	AGA	Arg
AUG	Met	ACG	Thr	AAG	Lys	AGG	Arg
GUU	Val	GCU	Ala	GAU	Asp	GGU	Gly
GUC	Val	GCC	Ala	GAC	Asp	GGC	Gly
GUA	Val	GCA	Ala	GAA	Glu	GGA	Gly
GUG	Val	GCG	Ala	GAG	Glu	GGG	Gly

APPENDIX 1.2: **Molecular methodology related to identifying genes**

This appendix outlines some of the current molecular biological methods for identifying genes and their characteristics {Weatherall, 1991}.

Restriction enzymes and DNA sequence patterns

Much of the revolution in molecular genetics, which has transformed all biomedical sciences in the past decade or so, derives from earlier discoveries that relate to the bacterial analogy of the human immune system. When a bacterium is invaded by an organism containing DNA, such as a virus, it can defend itself by producing one or more proteins known as *restriction endonucleases* (or restriction enzymes). These enzymes break DNA wherever (but only wherever) specific short strings of nucleotides are found. Each restriction enzyme is specific for a particular *recognition sequence*, sometimes called the cutting site or restriction site. For example, the enzyme, known as *Hpa*I cuts chromosomes everywhere the sequence GTTAAC occurs. Typically, the restriction site is a *palindrome*; that is, the sequence in one strand is the same as the sequence in the other (read in the other direction).

Restriction enzymes are generally named after the bacteria in which they are discovered (and occur naturally), plus a sequence number (e.g., *Eco*RI is the first restriction enzyme derived from *Escherichia coli*, the bacterium which inhabits our large intestine). Bacteria cannot simply destroy incoming DNA, or they would quickly digest their own genomes. Instead, they have enzymes that methylate the same sites that their restriction enzymes cut; most restriction enzymes do not attack methylated recognition sites.

When DNA is purified from human cells and digested by a particular restriction enzyme, it is cut into thousands of pieces whose number and length depend on the number of, and distance in kilobase-pairs between, the recognition sites. Enzymes that recognize longer sequences (e.g., six-base 'cutters') usually find fewer recognition sequences than do enzymes recognizing shorter sites. Individuals differ at thousands of nucleotides across their genome, and consequently produce different sets of *restriction fragments*. The resulting differences between individuals, in a given gene region, are known as *restriction fragment length polymorphisms*, or *RFLP*s.

RFLPs can be detected by separating the cut DNA pieces according to size (or in some methods, charge). This is done by exposing single stranded pieces of DNA to an electrical charge on an electrophoresis gel. DNA will try to move from the negative to the positive pole, but is restrained, according to its length, by the gel medium. Small pieces usually move further than longer pieces. After a suitable length of time, various methods can be used to visualize the fragments. In making a *Southern blot* (named after its developer), the DNA fragments are blotted (transferred) from the gel to a suitable filter or membrane material, and exposed to a DNA *probe* that contains sequences in a gene of interest. The probe, which has been synthesized with radioactive nucleotides, will *hybridize* to complementary sequences on the filter. The location of the bound probes, which reflects the length of the DNA that they recognize, can be seen as bands on X-ray film exposed to the filter. Alternatively a chemical DNA stain such as ethidium bromide, which fluoresces under ultraviolet light, can be applied and the pieces of DNA can be visualized in the positions along the gel to which they have migrated.

An RFLP exists when different individuals generate different numbers and lengths of bands in a region because their sequence in that region differs in the number and location of recognition sites. The presence of a restriction site is sometimes denoted by '1' or '+' and its absence by '0' or '−'. Individual chromosome haplotypes can then be scored for their pattern of a series of cut sites in the region (e.g., '++−+) or '1101' means sites 1, 2, and 4 were present, site 3 absent).

A powerful method of amplifying DNA is known as the *polymerase chain reaction* (*PCR*). DNA is mixed with short single-stranded sequences, known as *oligonucleotide primers*, plus nucleotides and DNA-replicating enzymes. Primers are synthesized so that their sequences are complementary to sequences that flank a short DNA region (one primer for each strand) whose polymorphisms are of interest (usually, only several hundred to a few thousand base-pairs). Under suitable conditions, the primers bind to their complementary sequences and the region between them is copied by the replication enzymes. DNA not 'primed' in this way does not replicate in this set-up. After repeating this replication process n times, the sequence between the primers is amplified 2^n times. Using this method a very small amount of DNA, such as from a hair bulb, mummified body, ancient preserved tissue, or even a single cell, can be amplified sufficiently so that its nucleotide sequence can be determined.

Because it is quick and efficient, many tricks have been developed to employ PCR to search for polymorphisms. When genes are newly identified it is now common practice to publish their sequence, or at least suitable primers spanning a part of the gene, so that other investigators can quickly study the same gene by PCR (i.e., without having to isolate and clone the gene itself). These are known as *sequence tagged sites* (*STS*).

It is now routine to determine the nucleotide sequence of any reasonable small length of DNA, and because we know what upstream regulatory regions, starts, stops, splice junctions, and exon codons look like, the nature of the gene can be analyzed, and in fact the coded amino acid sequence of the resulting protein can be inferred. Sequencing is generally done as follows. In four separate reactions, the DNA is replicated in an environment that contains a supply of all four nucleotides plus a fraction of one modified nucleotide (i.e., modified As in one reaction, Ts in another, etc.). These modified nucleotides chemically terminate the replication. The result in each reaction is an assortment of lengths corresponding to the lengths of DNA from the replication starting point to each occurrence of the modified nucleotide. For example, for the target sequence TTAGCAAG, the A reaction yields pieces TTA, TTAGCA, and TTAGCAA. These pieces can be sorted according to length in four parallel lanes on an electrophoresis gel; the lengths can be read off as the nucleotide sequence.

A selection of other strategies for identifying genes

Enzymes derived from retroviruses can be used to manufacture DNA that is complementary to mRNA. This is known as *complementary DNA* (*cDNA*). If mRNA is isolated from a given cell type, the cDNA made from that message represents all genes expressed in the cell. This can be useful for understanding gene expression, or characterizing differences among individuals in the physiology of a specific tissue type. cDNA can be used as probes to identify the chromosomal locus of the genes (even though the intron sequences are not present in cDNA). A *cDNA library* is a collection of cDNA isolated from a specific tissue. Such libraries are available for a wide variety of tissues.

The mRNA from a particular type of tissue can be isolated and separated by electrophoresis, transferred from the gel to a filter, and hybridized with DNA. For example, if one wishes to know whether a particular gene is being expressed in a particular tissue the mRNA can be separated in this way and exposed to a DNA probe for that gene. This is known as *Northern blotting*.

A relatively new genetic engineering approach is to create whole artificial chromosomes as cloning vectors for long fragments of DNA. *Yeast artificial chromosomes* (*YACs*) are made roughly as follows (Schlessinger, 1990). The human genome (or a selected large segment of it) is digested into long double-stranded fragments. Each of these is inserted into a piece of DNA that is engineered to contain yeast telomeres at each end, a yeast centromere, and a few other ingredients to enable this vector to be treated by yeast cells as a normal chromosome, and replicated. Sequences are also included for manipulating and isolating the inserted human DNA. A library of overlapping fragments in different YACs can be screened to see which ones contain a gene of interest (e.g., if there is a probe for that gene or its surrounding sequences). This locates the tested gene in the chromosomal region from which those specific YAC fragments came.

There are a variety of immunological methods that can be used to locate genes. Antibodies to a specific protein can be generated in experimental animals, such as rabbits. Human DNA can be digested into pieces, and individually inserted via special sequence 'vectors' engineered to be accepted, replicated, and expressed in bacteria. Bacterial clones (colonies) that express the gene of interest can be detected with the antibodies to the protein or in a variety of other ways, and the gene can then be excised from the transcription vector and sequenced, used as a probe, and so on.

These and other methods not described here form the large repertoire of techniques that make it possible routinely to analyze the gene structure and expression, molecular physiology, and genetic variation related to almost any human trait.

2 *The logic of the genome*

> We need not here consider how the bodies of some animals first became divided into a series of segments . . . for such questions are almost beyond investigation.
>
> C. Darwin, *The Origin of Species* (1859)

As the previous chapter indicated, a gene is much more, and more subtle, than simply a linear coding sequence. That sequence works only in the context of surrounding regulatory sequences and in the presence of the appropriate regulatory environment in the nucleus. Until recently, our knowledge of genetics was too rudimentary to enable us to make much headway in understanding how complex phenotypes result from genotypes. Darwin considered the problem essentially insoluble.

Phenotypes are also more complicated than can be explained simply by the linear coding sequence of a gene. Proteins have secondary and tertiary structures upon which their physiological functions depend. Even single proteins are only partial phenotypes, because their concentrations, induction, and tissue-specific environments and expression are also relevant to their action. Variation in regulatory and many other genes affects how the protein will function, just as variation in the sequence of the protein itself does.

An understanding of the relationships between genes and phenotypes requires an understanding of the *logic* by which life operates and has evolved. This is important because mutations can interrupt function at any point, and the pattern of pathology of a given genetic disease is often a direct reflection of where in the system the disruption occurs.

The organization of the genome

A second 'law' of evolution: duplication with variation

The basis of evolution is Charles Darwin's law or principle, of *descent with modification*, screened by natural selection. Modern evolutionary biology has confirmed evolution in exquisite detail. That model specifies, in principle, how the accumulation of changes in DNA sequence lead to accumulated differences in function between individuals or species. However, Darwin's model is not good at explaining how new function has arisen, or how complex organisms evolved from simpler antecedents.

In fact, new function arises by mechanisms quite different from the accumulation of incremental changes. Rather, from the beginning of life, a different but dramatically simple and very effective strategy has been adopted: to duplicate whole biological units, and allow these to undergo subsequent evolutionary diversification. This strategy has been so fundamental and pervasive to life that I believe it merits the status of a law of evolution unto itself: *duplication with variation* (Weiss, 1990a).

At each step in evolution, from the very first biological molecules, *metameric* (*segmented*) logic has been fundamental. Amino acids are variations on a common chemical theme and so, in a sense, are nucleotides. Thus, proteins are 'segmented' structures composed of strings of amino acids, mirrored by metameric structures (codons) in DNA and RNA {De Duve, 1991; Lewin, 1990; Stryer, 1988}. Duplication events are also of fundamental importance in the evolution of genes and the structures that they control, an idea first suggested as a general principle by Ohno (1970). Chapter 1 mentioned various repeat elements in the genome (e.g., *Alu* sequences) that apparently arose by duplication and rearrangement. The coding regions of a number of genes consist of repeated sequences that clearly arose by duplication; the genes for the structural proteins called collagens and those for the apolipoproteins involved in lipid transport are examples that will be discussed later. These are but the tip of an evolutionarily fundamental iceberg.

Exon 'shuffling': duplication and re-use of components within a gene

The division of genes into exons and introns reflects their basically segmented nature, because the functional domains of proteins often correspond closely to their exons. For example, when the insulin molecule is first translated it is called pre-pro-insulin, and contains a 'signal' peptide that prevents it from being secreted from the pancreatic islet cell in which it is produced. When this signal tag is cut off by an enzyme specific for that purpose, the pro-insulin molecule can undergo further pr[illegible]ssing, cleaving the 'pro' domain after mature insulin leaves the cell [illegible]ts job. Components such as this are frequently coded for by specific

[illegible] 2.1 shows the relationship between exons and various domains [illegible]man *leukocyte antigen* (*HLA*) proteins used by the immune

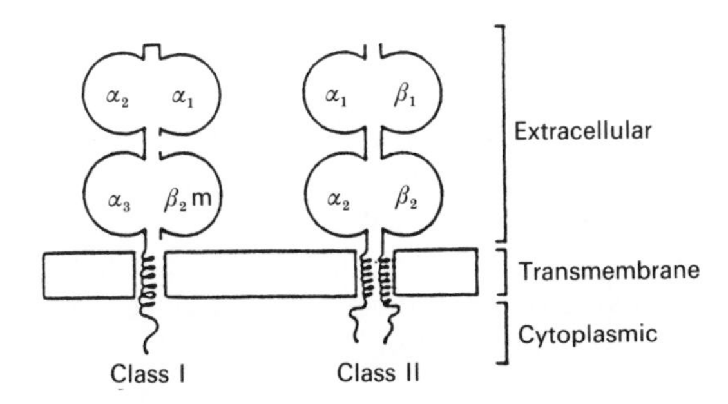

Figure 2.1. Relationship between exons and functional domains in HLA protein. (From Nei and Hughes, 1991.)

system to distinguish self from non-self. One end of the HLA molecule protrudes outside the cell, where it is involved in binding to foreign antigens (or to other molecules of the immune system), while part of the HLA molecule remains embedded in the cell membrane, and part remains inside the cell to signal when an antigen has been recognized.

DNA sequence analysis suggests that exons with similar function, even in diverse and essentially unrelated proteins, have arisen from some distant common precursor. That is to say, exons have occasionally been duplicated and used in the formation of new genes. Indeed, *exon shuffling* may be fundamental in the evolution of new protein functions (Dorit *et al.*, 1990; Gilbert, 1987). For example the first exon in apolipoprotein B, part of the lipid transport system, is found in the otherwise unrelated epidermal growth factor (EGF) receptor (Dorit *et al.*, 1990). Exons coding for the binding domain of the low density lipoprotein receptor (LDLR) gene (Appendix 2.1) are homologous to genes in the complement component of the immune system, and eight LDLR exons are homologous to exons in EGF (Sudhof *et al.*, 1985).

Duplication of entire genes

Occasionally, but often enough in evolution to be of fundamental importance, regions the size of a whole gene or more are duplicated. If the regulatory regions are intact this yields two copies of the gene(s), and with the accumulation of mutations over time, one of them may gradually diverge to take on a modified function. Today, we see examples of gene duplication in most if not all physiological systems that have been looked at in enough detail.

The minor changes in chromosome length occasioned by such duplication events do not preclude successful mitosis or meiosis. Sometimes, when copy number differs between two homologous chromosomes, mispairing at meiosis, followed by unequal crossing over, may cause copy numbers to change on the two chromosomes. Alternatively, misalignment of different copies which are slightly divergent in sequence may lead repair enzymes to correct one of the copies, so that within species sequences may evolve in a coordinated way. This is known as *gene conversion*. But duplicate genes often diverge in sequence and are occasionally translocated to other chromosomes where gene conversion and copy number changing no longer apply.

Sets of genes which have arisen by duplication from a common ancestral gene and retain similar function are known as *multigene families* (*supergene families* if they have different basic functions). Often, the chromsomal arrangement of related genes reflects their evolutionary ancestry. This can be seen in Figure 2.2 for the genes that code for the globin proteins involved in oxygen transport, including myoglobin, hemoglobin (a protein comprising sequences coded for by two related globin genes), and even leghemoglobin in plants (produced by a distantly related gene not shown in the figure). The bands in the pattern show schematically that the intron–exon structure has largely been conserved.

As a rule, the longer the time since the duplication event that separated two genes the more different their sequences will be (i.e., even among related genes within an individual). The sequences of members of a gene family can be arranged into an evolutionarily valid phylogenetic tree. For example, the gene sequence tree for the globin genes is shown in Figure 2.3. The genes themselves are located in two sets, the β-like genes on

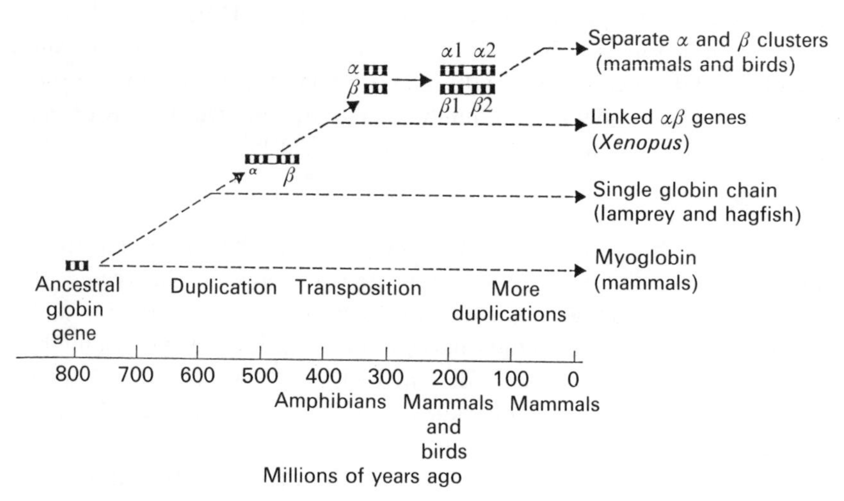

Figure 2.2. Evolution of globin genes by gene duplication. (From Lewin, 1990.)

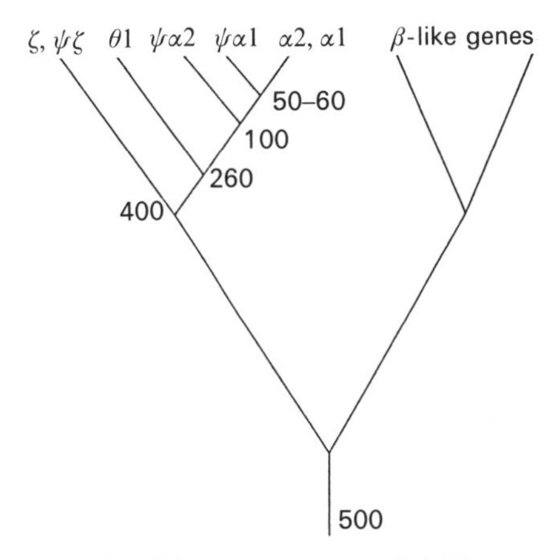

Figure 2.3. DNA sequence tree of globin genes. (From Higgs *et al.*, 1989.)

chromosome 11 and the α-globin region on chromosome 16. The figure shows the tree of relative sequence similarity among the genes in the α region (ψ denotes a pseudogene), and their collective relationship to the β-globin cluster, illustrating how sequence divergence corresponds to the history of gene duplication.

After billions of years of evolution, the genome has become essentially a huge complex of multigene families. Like the body, the genome is a metameric structure, composed of sets of repeated functional subunits, each modified in detail, which are distributed throughout the genome.

The metameric logic of gene expression

Higher organisms such as vertebrates are complex and differentiated, with hundreds of specialized cell and tissue types, and very sensitively timed developmental sequences required to turn a single cell – a fertilized egg – into a viable organism. This presents a multi-dimensional problem in gene expression that is solved largely through a hierarchically structured system of gene regulation, in which the activation of higher-level controlling genes causes the expression of sets of lower-level genes in a nested feedback sequence which we are only beginning to understand. How has evolution managed this seemingly hopelessly intricate problem without requiring a comparably large set of regulatory genes (and what will regulate *them*?)?

One key factor may be that related function involves related genes – and perhaps related regulatory sequences around them. Such genes often, though not always, remain contiguous after arising by duplication, a fact which can be used to coordinate the regulation of expression. For example, red blood cells must concurrently express a number of erythrocyte-specific genes, including of course the α- and β-globin chains that comprise hemoglobin. The latter must be controlled so that approximately equal amounts of each are produced. During embryogenesis and fetal development, the (related) β-globin region genes are expressed sequentially and in chromosomal order: first embryonic (ε) β-like globin is expressed, then fetal (γ) globin gene is used, and finally, the δ- and β-globin genes are expressed during postnatal life. Similar patterns pertain to the α-globin region genes.

There are a number of other examples of regulation by the expression of related genes sequentially in their chromosome order, although this is not the only means of gene regulation; however, members of gene families can be coordinately regulated even if they are not located together, if the duplication events which created them included their regulatory sequences. A good example may be the genes for color vision. These are all related to rhodopsin, the light–dark sensitive pigment. The rhodopsin gene is on chromosome 3; this gene is expressed in rod cells in the retina. The genes for color vision are expressed in the retinal cone cells; the blue-sensitive pigment gene is located on chromosome 7, but the genes for red and green vision are, in humans, located in a contiguous block on the X-chromosome. These genes are members of the same gene family, and all are expressed in retinal cells, but in no other cell type so far as is known.

Blocks of genes in the immune system are located contiguously and are expressed coordinately, also roughly in gene order. A highly intricate regulation system controls the expression of these genes, as is described in Chapter 13. In or near the cluster of HLA genes are genes whose coded proteins assist the HLA proteins to function.

A final important example are the *homeotic* genes; these are used in controlling the segmental development that is fundamental to embryogenesis. At least in part, this is accomplished by the expression of homologous genes in highly timed ways between regions and often in sequential gene order within a region. There are a variety of combinatorial systems by which a limited number of basic units, often derived from ancestral duplications, can specify the huge number of different structures, regulatory switches and so on that a complex organism requires {Gilbert, 1991; Shashikant *et al.*, 1991}.

Heritable variation may not be coding variation

In Chapter 1 we saw that around the typical eukaryote gene are a number of 'instructional' regions in which specific sequences were found which did not serve to code for amino acid sequences but rather as chemical information – e.g., binding sites – for regulatory processes. These sequences are not genes in the usual sense, but they are certainly related to phenotypes: mutation in such sites can seriously affect the coded protein, even inactivate a gene entirely. Such DNA sequences are replicated when cells divide, and are inherited. Methylation of DNA affects gene expression, is also inherited, and is sequence dependent (methylation sites), but not related to amino acid coding. Clearly we cannot study evolution by looking only at the amino acid structure of coded proteins.

The sequences discussed so far are represented in the germ line. The genes in the germ line connect us all directly back to the first life forms, and are passed down from parent to offspring. Individual germ cells can suffer mutations (called *germinal* mutations), or be non-functional, or can die. The genome can also change *within* the lifetime of individuals. We normally think of all cells in the body as being genetically alike, but there are important exceptions. An individual consists of billions of cells that are descended from a single fertilized egg, in a massive tree of cellular descent through embryogenesis and the subsequent cycles of cellular replacement and replenishment that take place in our blood, our skin, and the linings of all our organs, and so on. At each such cell division, *somatic* mutations can occur. These will be inherited by cells which subsequently descend from the first mutant cell. Somatic mutations are not heritable across generations, however, and die with the individual, but they can affect his/her phenotype in very important ways, as may be seen in Chapters 13 to 15.

Horizontal transfer of genes

It is sometimes possible for genes to be acquired during the lifetime of an individual. One example is infection by certain viruses, for example the *retroviruses*, which carry an enzyme known as reverse transcriptase; this causes the production of DNA copies of the viral RNA genome, which can then become integrated into the nuclear genome of the host's cell. Once incorporated, such viral genes become a normal part of the chromosome into which they have inserted. If this occurs in a somatic cell, then all somatic cells descended from that cell in the lifetime of the individual will inherit the inserted sequence. If the insertion occurs in a germ line cell, the insertion is heritable. If a viral gene is expressed in a human cell, aberrant cell behavior may occur, with pathologic consequences. Even smaller, transferable, genetically specified protein particles, called *prions*, may also be responsible for some diseases such as Jacob–Creuzfeld disease, a neurological degeneration.

On the basis of current knowledge, horizontal transfer of genes into the human germ line appears to be quite rare. However, a number of cancers are the result of such processes, and the widely dispersed sequence repeats, processed pseudogenes, and other insertion elements show that genes may move into and out of the genome more often than we have thought – at least somatically.

The organization of the body and its physiology

The result of evolution by gene duplication is an organism that has deep regularities in structure and physiology that correspond directly to regularities in the genome. Just as the genome is organized by sets of related genes that arose by periodic duplication, the metabolic pathways of most physiological systems and the structure of our tissues and organs can be arranged into trees of related units. We can take advantage of this fact in our effort to understand the genetic control of the phenotypes in which we are interested. Therefore, it is worth reviewing a few of these basic regularities.

Metameric physical logic: the second law again

One of the earliest metameric physical structures at a level higher than that of individual macromolecules is the cell itself {De Duve, 1991}. Although the cells in a mammal are highly differentiated in function, they share an ancient common ancestry in some single-celled organism, and much of the genetic mechanism used in creating and maintaining cell functions is repeated throughout the body. The basic sets of genes for cell maintenance, division, and the like are expressed generically in cells, although in some cases different types of cell may use related genes, from the same multigene family, to accomplish a function rather than using the exact same gene.

Tissues that are more complex than simple sheets of cells are usually metamerically structured. They involve the repeated construction of histological substructures, just as houses use bricks for walls and tiles for roofs. Clearly one strategy used by evolution to develop complex organs was to establish mechanisms for the periodic expression of sets of genes during development or tissue renewal. Examples are hair follicles in the skin, the finger-like villi in the intestine, papillae on the tongue, nephrons in the kidney, ovarian follicles, and so on. This also requires the evolution of coordinated regulatory mechanisms.

In these examples, the number of repeats is not programmed, and each unit represents a new replicate of the same basic structure, the expression of the same set of genes. However, at a higher morphological level, repeat elements are also found, that are constrained in number and differentiated in structure. Examples are the basic building blocks of the developing embryo, including the bones that form the limbs, the segmental structures called somitomeres and somites that are the precursors of gills and cranial nerves, and the vertebrae and their associated ribs and spinal nerves. In a sense, these are evolutionarily *homologous* rather than simply repeated subunits. They are coded for by homologous genes, the homeotic genes referred to earlier, whose chromosomal arrangement corresponds in orderly ways to the arrangement of the morphological segments themselves {Gilbert, 1991; Lewin, 1990; Shashikant *et al.*, 1991; Weiss, 1990a}.

Thus, at all phenotypic levels from that of physiological systems such as those using the genes for globin, immunoglobulin, and lipid transport, to histological structures, and to major morphological segments, evolution has produced complex structures by using gene duplication with subsequent mutational variation. Today, *we* use our understanding of this logical dimension of phenotypic control, which projects aspects of gene arrangement, organization, and control on to dimensionally complex phenotypes, to help us to unravel the geometry of gene control and evolution. For example, the function of a gene that has been newly identified is often quickly inferred by the researcher simply sitting at a computer and comparing the gene sequence to those of various gene families whose function has been identified. We search the genome for other genes with related sequence, intron/exon structure, or regulatory sequences. Regions containing one gene related to a trait are routinely sequenced to search for contiguous related genes under common regulatory control involved in the same trait.

Chemical compartmentalization: oil and water

Life appears to have begun in an aqueous environment, but for complex function to evolve it was necessary that its machinery be separated from the homogenizing effects of the surrounding environment. The evolution of cells achieved this end by using the immiscibility of polarized and non-polarized molecules. Water is a polarized molecule; that is, it has a natural electrical charge, forms multi-molecular crystals or structures, and tends to squeeze out molecules which, by having no charge, cannot be fitted into those structures. On the other hand, non-polarized molecules, such as oils and fats, do not carry a charge or form such structures, and tend to aggregate in the presence of water.

Cell membranes, around and within cells, consist mainly of two layers of lipid molecules. This effectively separates the aqueous environments on either side (e.g., the cytoplasm within and bloodstream without). Proteins or parts of proteins designed to work in an aqueous environment are said to be *hydrophilic* and contain a chemical charge, whereas those designed to work in a lipid environment are *hydrophobic* and uncharged.

The division of our bodies into intra- and extracellular domains is fundamental, because it enables the chemical environment to be different among different cells rather than being homogenized in the 'ocean' of blood or surrounding fluid. It is this fact above all others that enabled histologically differentiated multicellular organisms to evolve. But there is a price. If the cell membrane restricts the flow of molecules so that a cell can control its own business, the same restriction would prevent the acquisition of nutrients and release of waste by the cell, and the transfer of information among cells that have to function as a single organism.

To circumvent this problem, numerous proteins evolved whose structure allows them to reside stably in the cell membrane where they serve to regulate the exchange of molecules and information with the outside world. They may provide 'pores' through which certain molecules can pass, passively or actively. Exons coding for non-polar protein sequences of the appropriate length to reside in a bilipid membrane are common to such genes.

For example, there are receptors on the surface of cells that associate with circulating insulin and glucose molecules and bring the latter across the cell membrane into the cell, where they are used as a source of energy. Specific proteins bind to lipid molecules to transport them through the aqueous environment of the bloodstream, and specific cell-surface-bound *receptor* proteins recognize those lipid–protein complexes enabling the cell to obtain the lipids. Membrane-bound proteins pump sodium, calcium, and other ions across the membrane to maintain correct pH and chemical environments.

Intercellular communication: messengers and their receptors

Biological message transfer usually involves the complex of a receptor protein to some *ligand* or messenger molecule (e.g., hormone). This complex then triggers an appropriate physiological reaction. Some classes of biological messenger diffuse passively through cell membranes, where they are complexed with specific *binding proteins* to do their work. Examples are steroid hormones, such as estrogens, that are involved in reproduction, and retinoic acid, which plays a fundamental role in regulating embryogenesis.

Other biological messengers do not enter a cell, but rather are recognized specifically by cell-surface receptors coded for by genes that are only activated in cells that will be 'looking' for the specific extracellular message. The circulating messenger molecule binds to a receptor domain in the extracellular part of the receptor protein, and the presence of the receptor–messenger complex modifies intracellular parts of the receptor to trigger intracellular physiological responses. Examples of this receptor-mediated information exchange are the growth hormones, light-responding receptors in the retina, olfactory receptors, neurotransmitter receptors, and the immune system, which uses cell-surface receptors to recognize and respond to foreign substances circulating in the blood.

Mutations in receptor mechanisms are widely associated with disease. For example, aberrant antibody molecules can recognize molecules normally present in an individual and cause the destruction of cells in *autoimmune* disease. Persons who fail to synthesize the LDL receptor (Appendix 2.1) cannot remove cholesterol from the bloodstream and develop highly elevated cholesterol levels, and subsequent early cardiovascular disease. The absence of insulin, or of proper insulin recognition, leads to a failure of peripheral cells to take in blood glucose, and hence to diabetes.

Physiological compartmentalization and information transfer

In the same way that there are different chemical compartments in the body, biological traits often have a compartmental nature. Some are purely *local* in nature; that is, affect one specific tissue. Other traits involve circulating molecules, and affect the body as a whole; these are *systemic* traits. The distinction may blur at the edges, but it is important in understanding the nature of phenotypes, especially diseases. Mutations have effects that correspond to the cells in which they are expressed or in which the function they affect is manifest. Cancer of the stomach, broken bones, and cataracts are examples of local pathologies. Even when cancer metastasizes (spreads) to other tissues, it is the result of a purely local original clone of cells. A heart attack is a local phenotype, due to the occluding of a specific artery, but usually reflects high levels of circulating fats that can be deposited across the vascular system.

When we ask what the risk factors for a disease might be, it is wise to consider the nature of the disease. Smoking is a risk factor for cancer in tissues exposed to smoke (local effects), but, because it stimulates the nervous system in ways which can lead to cardiovascular disease, it also has systemic effects. When we ask if dietary constituents can be responsible for a disease, such as diabetes, we should consider what constituents would be involved and what tissues or systems they might affect.

Conclusion

The body and its physiology have similar levels of order, and use similar logic, which we can increasingly map on to corresponding structures in the genome. Therefore, understanding phenotypes themselves may help us to infer aspects of the underlying genotypes, and vice versa. This leads to an important general principle. Many biological traits are complex and have evolved by the piecing together of various genetic mechanisms that were already present and could be duplicated and modified for new function. As is discussed in Chapter 9 and thereafter, this inefficient, *ad hoc*, and 'noisy' evolutionary process results in traits affected by a spectrum of genetic effects. These genetic effects can be interrupted or modified in many ways by the many types of mutation that can occur in the genes involved, from modifications of the amino acid code to effects on regulatory sequences.

If the Golden Rule of society is that order is maintained by a mutual agreement to do good, there is what I call the *Rusty Rule* of disorder in genetics: whatever *can* go wrong *will* go wrong, somewhere, sometime, in some individuals. We will see that this is more than a platitude: it is a fact of life.

APPENDIX 2.1: **A brief summary of the lipid system**

On the banks of the great, grey-green, greasy Lipoprotein River.

Post-translational modification of R. Kipling, *Just So Stories* (1902)

Lipids (fats), including triglycerides and cholesterol, are important sources of energy, precursors of biologically active molecules (e.g., cholesterol for vitamin D and steroid hormones), and are necessary constituents of cell membranes and other structures. The system by which lipids are acquired, transported, utilized, and disposed of is complex, but reasonably well understood due to intensive research motivated by its relevance to the cardiovascular diseases that result from the deposit of fats 'on the banks' (interior walls) of the great, and especially the lesser, arteries {Humphries, 1988; Lusis, 1988; Sing and Boerwinkle, 1987; Sing and Moll, 1989}.

The lipid system also exemplifies the logic, genetic structure, and evolution of complex phenotypes that concern much of this book. Because of this, and because many of the examples to be used are from the lipid system, this Appendix provides a cursory overview of the lipid system.

A network of interconnected, interacting, biochemical pathways . . .

The general plan of the lipid system is illustrated schematically in Figure A2.1. The ports of origin of lipids in the body are the diet and the cells, especially liver and some peripheral cells that manufacture triglycerides and cholesterol. These substances are transported to their main sites of storage (liver, adipocytes in fat tissue) or usage (muscle, other cells needing energy). Because lipids are largely non-polar and not water soluble, they must be actively transported in the

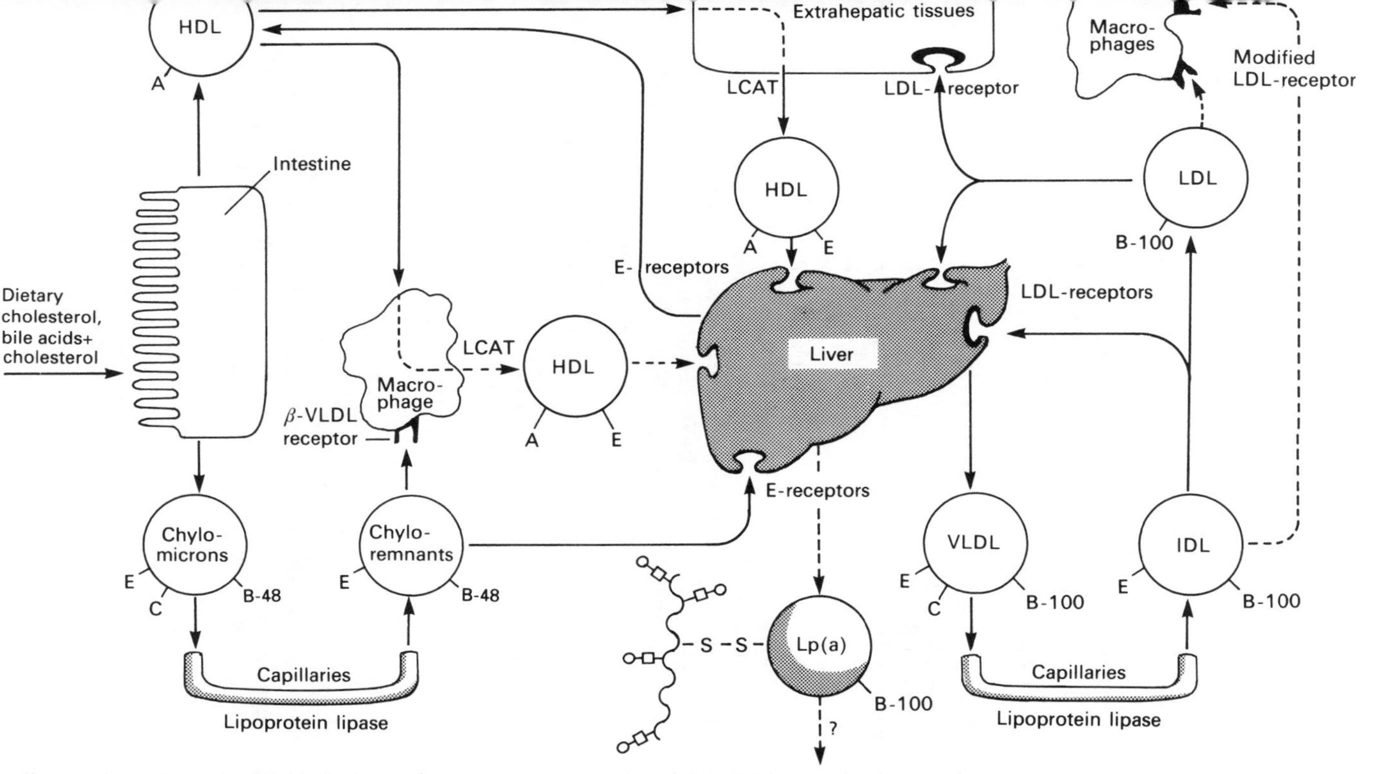

Figure A2.1. Schematic of lipid physiology (for explanation see the text). HDL, high density lipoprotein; VLDL, very low density lipoprotein; LDL, low density lipoprotein; LCAT, lecithin/cholesterol acyl transferase; IDL, intermediate density lipoprotein; Lp(a), lipoprotein A. (From Utermann, G., Apolipoproteins, quantitative lipoprotein traits and multifactorial hyperlipidemia. In *Molecular Approaches to Human Polygenic Diseases*. Copyright 1987 John Wiley & Sons Ltd. Reprinted by permission of John Wiley & Sons Ltd.)

bloodstream in packages of specific composition. The transport is from the points of origin through the bloodstream to the peripheral tissues, and on to the liver. The ports of exit are thus decomposition by metabolism and excretion via the liver.

Lipids are transported in particles, known as *lipoproteins*, composed of cholesterol esters, triglycerides in the form of phospholipids, and *apolipoproteins*. The almost unpronounceable (and for me, untypable) latter are the coded protein products of specific genes that are denoted by abbreviations, apo-AI, apo-B, and so on. Depending on their relative concentration of triglycerides, cholesterol, and apolipoproteins, different lipoproteins have different density as determined by centrifugation; the higher the proportion of triglycerides the lower is the density. The structure and composition of these lipoproteins are enzymatically regulated and depend upon the structure of the apolipoproteins they contain.

As indicated in Figure A2.1, dietary lipids are absorbed through the intestinal mucosa into the lymphatic system in the form of *chylomicrons*, high in triglycerides and carrying apo-B48, and apo-CII and apo-CIII on their surface. In the capillaries (near peripheral tissues that can utilize the energy of triglycerides or store them as adipose tissue), the apo-CII molecules on the chylomicrons enable the enzyme *lipoprotein lipase* (*LPL*) to hydrolyze the triglycerides in the particle, releasing free fatty acids for absorption by nearby tissues. LPL is attached to the endothelial cells lining blood vessels. The apo-Cs are lost in the process, and the resulting particles, known as *chylomicron remnants*, take on apo-E and apo-B48 on their surface, converting them into smaller particles that are actively recognized and bound, and absorbed, by apo-B–E receptors on cells in the liver.

Triglycerides and cholesterol are synthesized *de novo* in the liver and secreted, thus becoming available to peripheral tissues. This endogenous source supplements exogenous sources in a system that is concentration monitored. Initially, triglyceride-rich *very low density lipoproteins* (*VLDL*s) are secreted. These also carry some cholesterol as well as apo-B100, apo-E, and apo-C proteins on their surface; the first of these proteins is needed for VLDL synthesis and secretion, and the latter prevents the remnant apo-E receptors on the liver from reabsorbing the particle too quickly.

As with the chylomicrons, LPL hydrolyzes VLDL triglycerides in capillaries, releasing free fatty acids to peripheral tissue. The VLDL is converted into *intermediate density lipoproteins* (*IDL*s) and these change to *low density lipoproteins* (*LDL*s), which carry only the apo-B100 lipoprotein. Peripheral tissues, and the liver, bear LDL (apo-B100) receptors which bind these particles and bring their constituents into the cells. LDL receptor-mediated access to cholesterol regulates endogenous synthesis of cholesterol by cells, by down-regulating the production of the LDL receptors and other enzymes in the cholesterol synthetic system.

LDL particles are thus involved in transport of lipids to peripheral cells where they can be used for physiological purposes (i.e., as a source for metabolic energy) or stored. The concentration of LDL-bound cholesterol (LDL-C) is positively associated with risk of *coronary heart disease* (*CHD*), mainly heart attacks, atherosclerosis, and other diseases associated with arterial function. The reason is that these circulating lipids can be deposited as fatty atherosclerotic plaques on the interior of the arteries, eventually closing the arteries and/or leading to roughened surfaces or other characteristics that can trigger blood clotting reactions that clog the arteries.

There is also a system of *reverse transport* that removes excess triglycerides and cholesterol from the body. These lipids are transported by *high density lipoproteins* (*HDL*) to the liver where they are absorbed, and either resecreted later when needed or excreted into the gallbladder. The liver can do the latter directly, or by first converting cholesterol into *bile acids*. These compounds exit the liver to be stored in the gallbladder, and then secreted into the upper small intestine when the gallbladder is stimulated to do so by the ingestion of a fatty meal. Bile acids serve to emulsify fats so that they can be absorbed as chylomicrons, thus closing the cycle.

HDL particles are constructed by macrophages (using chylomicron remnants), the liver, and intestinal tissues, as disk-like structures of phospholipids plus apo-AI. These disks contact and absorb cholesterol from cell membranes, and the cholesterol is kept in the nascent HDL particle by esterification by the enzyme *lecithin/cholesterol acyl transferase* (*LCAT*), which transfers one of the fatty acids of the triglyceride phospholipid (lecithin) to a cholesterol molecule; the new non-polar cholesterol ester then migrates to the inside of the particle. Further lipids continue to be added onto the HDL particle by the breaking down of chylomicrons and VLDL, and apo-C and apo-E are transferred to the maturing HDL from these sources.

At the same time, another enzyme, *cholesteryl ester transfer protein* (*CETP*), can transfer cholesterol from the HDL to convert IDL and VLDL to LDL (not shown in Figure A2.1). Thus, there is complex and multiple interaction among these pathways, much of it enzyme regulated. Eventually, the HDL cholesterol is absorbed by the liver. For this reason, HDL is involved in depriving the body of cholesterol, and the level of circulating HDL-cholesterol is associated with decreased risk of heart disease.

These steps are dependent on gene products – the apolipoproteins, enzymes, and cell-surface receptors referred to in the above outline. Other particles may also be important in disease risk; for example, another lipoprotein known as *Lp*(*a*) forms a molecular covering for some LDL particles and may make them particularly relevant to heart disease.

Risks of different diseases are associated with different levels and aspects of these lipoprotein particles. But because one or more particle serves as the substrate for the production of another, the system is interactive, homeostatic, and has many feedback loops. Clearly, it is of importance to know the effect of: (1) structural variation in the genes discussed here, which may affect their physiological efficiency; (2) variation in the concentrations of their gene products, which may affect or be affected by the various lipid concentrations; and (3) variation in the concentrations of the lipoprotein particles themselves, which depends on (1) and (2) as well as on the availability of cholesterol and the triglycerides. These are the many points of control in a system that has important phenotypic relevance, as we see in later chapters. To paraphrase ideas by C. F. Sing (personal communication), there are networks of strong and weak correlations among the various lipoprotein components. Genetic variation may be able to alter either of these, but most variants appear to affect the weaker correlations;

the stronger correlations may represent biologically necessary relationships that are difficult to modify very much.

. . . Controlled by networks of related genes

The importance of gene duplication in the evolution of complex systems is typified in the lipid system. The apolipoproteins that determine the composition of lipoprotein particles and transport lipids in various physiological directions are coded for by a series of homologous members of a multigene family {Li *et al.*, 1988; Rajavashisth *et al.*, 1990; Taylor *et al.*, 1987}. The extensive DNA sequence similarities among the apolipoprotein genes can be used to construct their gene tree, as shown in Figure A2.2 (a few recently characterized apolipoprotein genes are not shown).

The apolipoprotein genes are also clustered in a way that reflects their gene duplication origin: apo-E, -CI, and -CII are together on chromosome 19, and apo-AI/CIII/AIV comprise a cluster on chromsome 11. Apo-AII is now alone on chromosome 1. Apo-B is on chromosome 2, but may not be part of this family. This arrangement can be explained by a series of duplication and translocation events that are consistent with the sequence tree (Luo *et al.*, 1986). Sequence similarities further suggest that the apolipoproteins may be distantly related to the gene for LCAT, closely linked to that for CETP on chromosome 16.

The exons vary in length among apolipoprotein genes, but mark functionally conserved domains that the genes have in common (Figure A2.3). Each gene contains a 5′-untranslated region and a signal peptide that correspond to the first two exons (first exon in apo-AIV). The signal peptide enables the protein to pass through the cell's lipid membrane but is then cleaved to activate the protein, whose physiological role is determined largely by the last two exons.

These genes share basic physiological characteristics. They all code for proteins that take the shape of an *amphipathic* α-helix; that is, they have a coiled structure that contains a non-polar face, which can be inserted between the non-polar fatty acid chains of the phospholipid, and a polar face, which can interact with the charged portion of the phospholipid as well as the bloodstream. This keeps the apolipoproteins on the surface of the lipoproteins where they can interact with

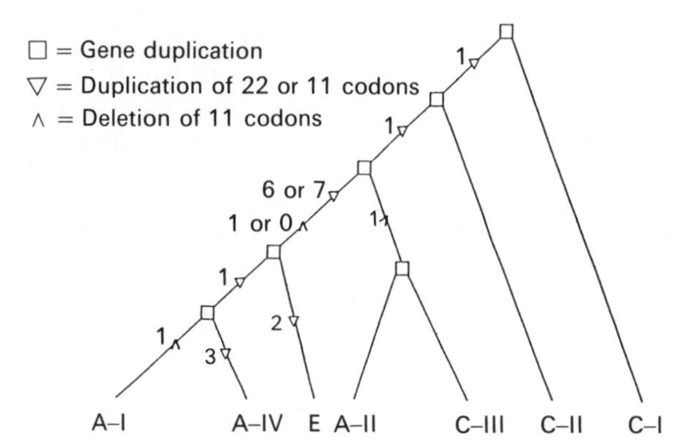

Figure A2.2. Apolipoprotein DNA sequence tree. (From Li *et al.*, 1988.)

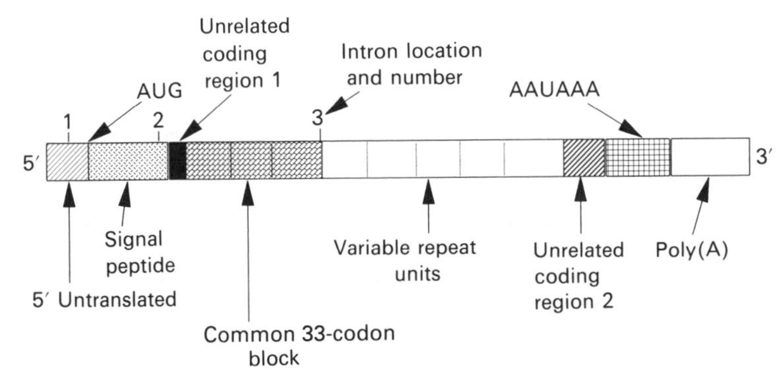

Figure A2.3. Schematic of apolipoprotein mRNA structure. (From Li *et al.*, 1988.)

cell-surface receptors or lipid related enzymes, as well as with the non-polar cholesterol ester and triglycerides that comprise the core lipids of the lipoprotein products.

The mechanism of evolution of these genes for their particular function shows some other ramifications of gene duplication. One reason for the difficulty in constructing a sequence-based tree of evolutionary relationships among the apolipoprotein genes is that they have high amounts of internal repeat structure (Li *et al.* (1988) used exon–intron boundaries for alignment). The apolipoproteins have a periodicity of 22 amino acid residues, each unit of which is itself a tandem repeat of 11 residues. These arose by duplication of an original 11 residue unit. The resulting structure has amino acid periodicity that fits the functional constraints of the amphipathic protein. Charged and uncharged amino acids retain their position even after millions of years of mutational modification of this repeated metameric coding structure. In some of the genes there are higher-order periodicities of 33 and 66 nucleotides as tandem or three-way duplications of the primary 11-nucleotide unit. The sequence structure thus has nested subunits and sequence similarity among the related genes.

These genes may have originated from an original sequence of 11 nucleotides (TCGGACGAGGC) that was at some point tandemly repeated several fold. Eleven is not an even multiple of three, so the sequence does not make a simple set of codons; but a string of repeated copies of such a sequence could be read at any starting point and would quickly get into 'phase' so that eventually, downstream, the same amino acid sequence would be coded for, buffering the gene against the effects of point mutation (Ohno, 1985). The amino acid sequence coded by strings of the core 11-mer has the physiological properties desired in a primitive lipid-related protein. For example, the thymines (Ts) always occur in the middle of a codon, thereby always encoding a hydrophobic amino acid. In the helically shaped protein structure that results from this code the amino acid residues are spaced in such a way that enhances the lipid interaction characteristics of the protein. The functional details of the different apolipoproteins follow from mutational modifications in the basic repeated core string.

The apo-B gene exemplifies another phenomenon: *differential transcription* of the same gene region to code for different proteins {Lewin, 1990}, in this case, apo-B48 and apo-B100. This is accomplished by a tissue-specific DNA modification process not yet understood, but the result of which is that cells in the small intestine produce apo-B48, used in chylomicrons, whereas hepatic (liver) cells transcribe the longer apo-B100 sequence. These serve the different functions outlined above.

It is not clear whether apo-B is related to the other apolipoprotein genes (Li *et al.*, 1988). Apo-B is one of the largest monomeric proteins known, with 29 exons, very different from the other apolipoprotein genes. However, like them, apo-B has a cyclical amphipathic repeat structure, perhaps with 22-meric periodicity, and may produce binding sites with sequence homology to the apo-E binding sequences.

3 *Concepts of frequency and association in populations*

How common is a phenotype in a population? How often does it arise? This chapter introduces some basic concepts of the impact or 'frequency' of phenotypes and of the relationship between that and the frequency of genes which may help to produce them.

Basic concepts of trait frequency and risk

Static measures: prevalence and its use as a probability

We can denote by ϕ the phenotype that we observe on an individual. The prevalence, or frequency, $\Pr(\phi)$, of a specific phenotype at any given time is the fraction of individuals in a specified population who have the trait, equal to the *probability* that a randomly sampled individual has that phenotype[1]. A phenotype can usually take on a range of possible values, and over this range $\Pr(\phi)$ gives the *phenotype distribution*. Since every person must have exactly one phenotype, the prevalences over the acceptable range must sum to unity.

A phenotype can be a continuous measure such as the level of blood glucose, for which we can write the prevalence of a given glucose level, for example, as Pr(100 mg/dl). Or, a phenotype can be a qualitative trait such as the presence or absence of a disease, which we can denote $\phi = 1$ (presence) and $\phi = 0$ (absence). The term prevalence usually refers to the fraction of individuals who have the trait and is denoted $\Pr(\phi = 1)$, or when the context is unambiguous just by Prev.

Prevalence is usually *estimated* from samples of a whole population, by the fraction of the sample who are affected: $\acute{\Pr}(\phi) = n_\phi/N$, where N is the sample size. The uncertainty in an estimate depends on the sample size and the true prevalence, and is often described by such measures as the confidence interval of the estimate (in this case, of a proportion). Such uncertainty statements require a number of assumptions, which are discussed in biostatistics texts {Armitage and Berry, 1987; Sokal and Rohlf, 1981}.

An important point is that prevalences refer only to some *population of inference*; the latter can be a species, national population, geographical area, village, group specified as to sex, age, socioeconomic status (SES), or exposure to some variable, including genotype. It is important in thinking about genetic traits and genetic variation to be aware not only of the population of inference but also whether a sample reflects that population properly. This is important when the population is structured or some particular sample is taken. For example, the prevalence of disease may vary with age or sex. The prevalence of a random sample of the population may not be very informative of the prevalence in certain age or sex classes. In fact, the prevalence of a disease can differ between two populations, even if the risks of disease for each age are the same in both populations, if the populations have different age distributions.

In the case of genetic disease studied by a sample of families, the age and sex distribution of the families may not be representative of the population, in which case the prevalence structure in the sample may be a poor reflection of that in the population. For example, if nuclear families are *ascertained*[2] via individuals affected with a late-onset disease, such as cancer or diabetes, the families may overrepresent older people in the population.

Samples can be structured in such a way as to represent various segments of a population, by *stratifying* or ensuring that each segment is represented by a certain number of observations. If the fraction of the segments in the general population is known, then an estimate of the population prevalence can be derived from the stratified (and hence non-representative) sample. If i represents the stratum, and the frequency of this stratum in the population is f_i, then $\widehat{\mathrm{Prev}} = \Sigma_i f_i \widehat{\mathrm{Pr}}_i(\phi)$. An important stratification for us is by genotype.

Stratification is often important for various reasons in a study design, for example to remove the effects of *nuisance* variables, such as age or sex, that are causally related to a trait but are not the causes of primary interest. The nuances of sampling strategies and estimation are described by Cochran {1976} or in other texts on sampling methods. The important point is to understand the relationship between the sample and the population of inference.

Dynamic measures: incidence and the concept of a rate

Prevalence refers to the proportion of existing instances of a particular phenotype. The occurrence of *new* cases is measured by the *incidence*. This is defined as the absolute count of new occurrences of the trait over some span of observation, usually a time interval – specific, like prevalence, to some population of inference. The incidence of non-fatal cases of disease is the *morbidity* incidence, while that for fatal conditions is the *mortality* incidence.

The *incidence rate* is the number of new occurrences *per unit of observation*, for example time measured in *person-years* (*PYr*). Incidence is measured in whole numbers of cases, but because one might observe any amount of units of observation, the incidence rate is an infinitely divisible, or *continuous*, variable. For example, if there are 7 new cases for every thousand PYr there will be 0.7 cases per 100 PYr.

The number of PYr of risk exposure is often obtained from census or other public sources. For many biological purposes, however, what is relevant is the rate of new occurrences per *at-risk* individual. For some traits an individual can be multiply affected (e.g., breast cancer in the contralateral breast, second heart attacks), and in such cases it is appropriate to include affecteds. For traits such as diabetes that can occur only once and then persist, the denominator should include only those persons in the population who, by virtue of being unaffected, are still at risk.

The age-specific incidence or mortality rate: the hazard function

Incidence is typically strongly affected by age, and it is often important to account for this. The age-specific incidence or mortality rate is known as the *hazard* rate, or function, $h(t)$, at age t. The hazard is a continuous measure but is usually estimated from data grouped into discrete age classes, such as 5–9, 10–14, . . . If some continuous biological process is responsible for these age-specific rates, they can often be fitted to a parametric mathematical model of the process {Elandt-Johnson and Johnson, 1980}. Human hazard functions vary greatly among traits, and there is no general rule for their shape; however, the overall mortality hazard function is generally as shown in Figure 3.1: A rapid decline of overall mortality risk from birth to age 10, followed by an accelerating increase. We presume that overall infant mortality declines because: (1) there is differential elimination of very vulnerable infants with congenital problems; and (2) the newborn is immature with regard to its immune and other systems and gradually matures, becoming hardier. Some specific causes, e.g., some cancers, do increase in risk with age during childhood. Most chronic diseases have exponentially increasing risks in adult ages.

An important quantity determined entirely from the hazard function is the probability of surviving the risk involved over some time period, such

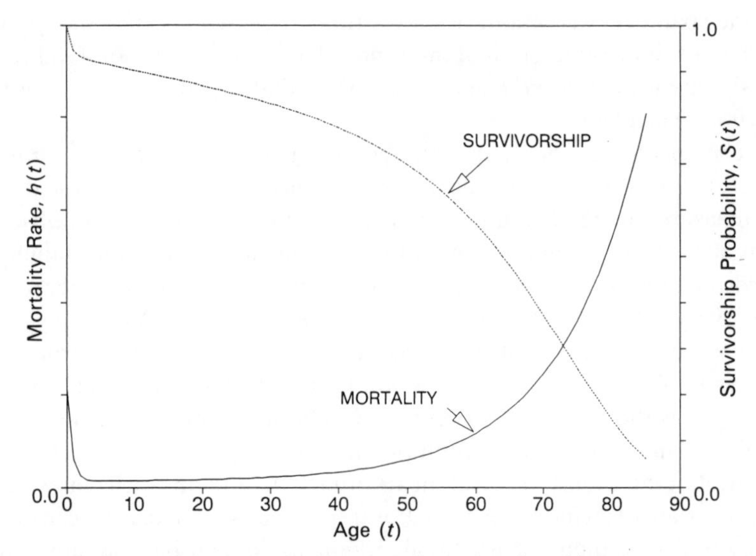

Figure 3.1. Schematic of human mortality and survivorship curves.

as from birth to age t. This is known as the *survivorship function*, defined as $S(0) = 1.0$, at birth, and

$$\Pr(\text{Surviving to age } t) =$$

$$S(t) = \exp\left[-\int_0^t h(s)\,\mathrm{d}s\right] \approx \prod_{i=0}^{t-1}(1 - h_i) \approx \prod_{i=1}^{t/n}(1 - nh_i). \qquad [3.1]$$

The notation 'exp' means exponentiation to the power e, the base of natural logarithms. The first form of [3.1] is used for a continuous hazard function, and summing over a continuous variable is denoted with the integral sign from calculus and the accompanying term 'ds' denotes small intervals on this scale. The second version of [3.1] is a discrete approximation for hazard rates estimated from 1-year age groups (i = age) that is reasonable when h is small; the third version is a similar approximation for discrete n-year age classes (i indexes the classes). Figure 3.1 shows both mortality and its associated survivorship.

Though expressed as continuous rates, most biologically relevant hazards are at least partially probabilistic in nature: one has a *chance* of dying this year from the cause, but Nature rarely insists in advance that any specific individual *will* die. In genetics, survivorship is usually treated as a probability. $S(t)$ is the probability that death or some condition of interest has not arisen by a specified time.

The complement of survivorship is the probability of disease arising *by* age t; a related measure fashionable in epidemiology is the *cumulative incidence*, C-I(t), defined as

$$\text{C-I}(t) \approx \int_0^t h(s)\,\mathrm{d}s \approx \sum_{i=0}^{t} nh_i \approx 1 - S(t). \qquad [3.2]$$

If there is no differential migration, death, or loss to follow-up of affected individuals, the cumulative incidence is approximately equal to the prevalence at a given age.

These measures of age-specific risk have another important characteristic. Formally, they are derived as the experience of a *cohort* of individuals undergoing the same risk schedule, $h(t)$ (or, equivalent to the probabilistic life experience of a single individual exposed to that hazard). But often the available data are *cross-sectional*; that is, come from observations made at a given point in time on individuals of different ages. The hazard estimated from such data applies to a cohort only if the risks are not changing over time; if they are, then the risk to today's n-year olds when they are, say, m years old, will not be the same as that to today's m-year olds. This is a very important point, especially when we are dealing with traits caused by genes interacting with environments that have changed over the generations represented by a sample of family members.

Sometimes, for example in analyzing family data for genetic risk, it is necessary to determine the probability that a given individual has a specific phenotype (e.g., occurrence of some disease), when this is a function of age. This can be specified in terms of survival functions. Typically, an individual is followed prospectively (even if data are ascertained retrospectively) from birth until he/she dies from some other cause, is lost to follow-up (disease-free) for any other reason, or experiences onset. The first two conditions are known as *censoring* relative to the cause of interest, since presumably the individual might have manifested the disease had he/she remained alive and under observation. Typically, the 'age' assigned to such a person is that at censoring or onset. The probability of observing a person at age t defined in this way is

$$\left.\begin{array}{ll} \prod_{i=1}^{t-1} (1 - h_i), & \text{if lost to follow-up, cause-free} \\ & \text{alive or dead, at age } t \\ S(t-1)h_t, & \text{if first manifests cause at age } t, \end{array}\right\} \qquad [3.3]$$

where $S(t)$ refers to surviving the occurrence of the cause in question, and $h(t)$ refers to the hazard only for that cause. This assumes that the causes of censoring are unrelated to the trait in question.

Model-free measures of association

Perhaps the basic question in science is whether a given cause has a given effect. In genetics, we are interested in whether genetic variation is causally related to phenotypic variation. Standard statistical tests of association can be used to gain a preliminary answer to such questions. If we do not or cannot specify the nature of the genes that might be involved, such tests are called *model-free*, in contrast to formal, *model-based*, studies used in more advanced stages in the investigation, when the specific characteristics of genetic risk are taken into account. Model-free studies in genetic epidemiology use the same general methods as do other epidemiological studies {Armitage and Berry, 1987; Breslow and Day, 1980, 1987; Sokal and Rohlf, 1981}. But there are some important conceptual differences in the genetic context.

The following is a brief summary of the major model-free strategies used to demonstrate association between *risk factors*, such as genes or environmental exposures, and outcome phenotypes. Somewhat different methods are typically used for quantitative phenotypes and for discrete traits.

Genes as risk factors for qualitative traits

Prospective study designs: cohort studies

Probably the simplest question is to ask whether a specific genotype is associated with risk of a specific qualitative disease. The most analytically simple way to answer such a question is with a *cohort* study, one of the classical approaches of epidemiology. Cohorts of individuals, usually contemporaries, are identified who are exposed, or not exposed, to specific genes; they are followed prospectively to determine whether their exposure affects their risk of disease or some other phenotype. The data, in the form of counts, are arrayed as a *contingency table*, as shown in Table 3.1.

The analysis is straightforward. The question is: what fraction of the exposed and unexposed individuals have developed the disease? These two values are $R_1 = a/(a + b) = a/N_1$ and $R_2 = c/(c + d) = c/N_2$, respectively. The ratio of $R_1/R_2 = [a/(a + b)]/[c/(c + d)] = a(c + d)/c(a + b)$ is known as the *relative risk*, *RR*. As given here, it is a ratio of proportions, or cumulative incidences (or prevalences) after some follow-up period.

Table 3.1. *Contingency table for simple cohort study*

	Disease	No disease	Totals
Exposed	a	b	N_1
Not exposed	c	d	N_2
Totals	M_1	M_2	T

Note: N, row total; M, column total; T, grand total.

Under the null hypothesis, the risk factor makes no difference to the occurrence of disease and the *RR* should equal 1. This is unlikely to be observed exactly in any sample, so a statistical test must be applied to determine whether the observed *RR* differs significantly from 1; one test is

$$\chi^2_{1\text{d.f.}} = (ad - bc)^2 T/(M_1 M_2 N_1 N_2), \qquad [3.4]$$

an approximate chi-squared distribution with 1 degree of freedom (d.f.). The data can be stratified by age or other variables, or adjusted (see below).

Sometimes it is not possible to follow an unexposed control group, or it is preferred to use risks in the population as a whole. For example, it may be of interest to estimate the number of cases expected in the general population, for the number of PYr of risk observed in one's sample of exposed individuals. If N is the number of PYr observed and h the population hazard, then $E = hN$ is the expected number of cases; this can be summed over all observed ages, with appropriate age-specific hazards and PYr. The ratio of the *observed* (O) to expected number of cases is the *standardized morbidity (mortality) ratio*, $SMR = O/E$. Under the null hypothesis the *SMR*, like the *RR*, has an expected value of 1. If the absolute risks are small, the variance of the *SMR* is about O/E^2, and an approximate chi-squared test is $\chi^2_1 \approx (O/E)^2/O$.

Because the follow-up period is usually determined by the study itself, rather than by some inherent process, *RR* and *SMR* can be somewhat arbitrary measures. If the exposure is supposed to work by altering the hazard function for the trait, one might wish to compute the *rate ratio*, $h_1(t)/h_2(t)$, instead of a cumulative measure of proportions of cases. The rate ratio can be constant for all ages after exposure or it may take a more complicated form that may be informative about the biological processes involved {Breslow and Day, 1987; Kalbfleisch and Prentice, 1980}.

Table 3.2. *Familial association for rheumatoid arthritis*

	Disease	No disease	Totals
First-degree relatives of cases (exposed to case's genes)	21	475	496
Spouses of cases (unexposed)	12	661	673
Totals	33	1136	1169

Source: Del Junco *et al.*, 1984.

Family data: retrospective–prospective designs

Most studies of the impact of inherited risk on family members are *retrospective cohort* studies; individuals are ascertained at the time of the study, but their experience is followed statistically from the time of their birth forward. A common model-free approach to genetic disease is to ask if individuals exposed to the genes of a relative known to be affected have a higher risk than do unrelated individuals, such as spouses of the affecteds, who are not exposed to those genes. Environments may or may not be shared among family members, but genes always are; a positive test result suggests genetic risk.

Table 3.2, presents a simplified example from a study of rheumatoid arthritis, an inflammatory joint disease known to be associated with specific HLA genotypes (see below). The $RR = (21/496)/(12/673) = 2.37$. The significance test $\chi^2 = 6.35$, highly significant at the 5% level, indicating that risk for rheumatoid arthritis is familial.

Retrospective designs: case-control studies

Another common model-free design for detecting association is the cost-efficient *case-control* study, also of retrospective design. Whereas a cohort study identifies and classifies individuals initially by their exposure status, and tests for subsequent disease, a case-control study identifies subjects by disease status and tests retrospectively for exposure. Case-control data are also arrayed in contingency table form, as in Table 3.3, but this is turned around from the layout of the cohort study.

Here, the investigator chooses in advance how many cases and controls to sample, so the risk that an exposed person will become a case cannot be estimated directly. However, the odds that a case has been exposed to a risk factor, a/b, can be compared to the odds for a control, c/d. The ratio of these is ad/bc, often called the cross-product ratio or *odds ratio*[3], OR.

Table 3.3. *Contingency table for simple case-control design*

	Exposed	Unexposed	Totals
Case (disease)	a	b	N_1
Control (no disease)	c	d	N_2
Totals	M_1	M_2	T

Note: N, row total; M, column total; T, grand total.

If the disease is rare, the OR can be close in value to the RR. Under the null hypothesis, we expect OR to equal 1, and the significance of the deviation of the observed OR from 1 can be tested as before. Contingency tables can accommodate a variety of data stratifications (e.g., by age and sex) and other complications {Breslow and Day, 1980; Fleiss, 1981}.

Epidemiologists sometimes refer to designs such as that in Table 3.2 as case-control studies, because relatives are ascertained from cases and unaffected controls. Note, however, that the real risk is not in them, but in the persons exposed to their genes (by inheritance), so such studies are really cohort studies (Weiss *et al.*, 1982).

Genetic markers and allelic association in populations

In the designs just described no genes are identified. Disease associations among family members may indicate genetic risk factors, but families share environments and the study itself tells us nothing about what those environments may be. We can often be more specific. Exposure status can be tested relative to some known genes for which the sampled individuals can be typed. Thus, a case-control study can see whether cases have a higher frequency of a given genotype than controls do, or a prospective study can compare disease in individuals with or without a specific test genotype. The test genotypes can be *genetic markers* or *candidate* genes. A genetic marker is a polymorphic locus identified independently of any physiological causal function related to disease, often just randomly chosen from the genome. A candidate locus is one that on prior physiological grounds we suspect may be involved in the trait.

For example, Amerindians seem generally to be susceptible to non-insulin-dependent *diabetes mellitus* (*NIDDM*). Samples of Arizona Pima and Papago Amerindians were typed for variants in immunoglobulin (antibody) genes known as *Gm*, to determine whether there was an

Table 3.4. *Gm markers and diabetes in Pima and Papago Indians*

	Diabetes		
$Gm^{3;5,13,14}$	Present	Absent	Totals
Absent	1343	3284	4627
Present	23	270	293
Totals	1366	3554	4920

Source: Knowler *et al.*, 1988.

association with the occurrence of NIDDM (Knowler *et al.*, 1988). *Gm* variants represent different *haplotypes* (alleles at several closely linked immunoglobulin genes), which are highly variable between human populations. The haplotype denoted as $Gm^{3;5,13,15}$ is found in Europeans but not pure-blood Amerindians. If Amerindians are genetically susceptible to NIDDM, will this be reflected by a negative association between NIDDM and the European-derived *Gm* type? Table 3.4 shows this laid out as a cohort design. The *RR* from this table for the absence of the *Gm* European marker type is 3.63. The chi-squared test is $\chi^2_{1\text{d.f.}} = 55.27$, a highly significant negative association between marker and disease; surprisingly, this does not mean *Gm* types are causally related to diabetes (Chapter 10).

Statistical associations never prove causation and often seem difficult to believe. Tables 3.5A and B provide two sets of examples. In both, many different diseases were reported to be associated with alleles at the *same* locus. This raised many questions, but above all, could all the associations be true?

The ABO blood groups were the first human genetic system to be typed on a mass scale. Because blood was easy to type, and typing is important in blood transfusions, ABO typings have been collected routinely from patients, and even from the general population. A great number of diverse diseases are associated with ABO genotypes. This includes both infectious and non-infectious diseases (Table 3.5A) affecting different organ systems. How can this be so? We might suspect various artifacts or biases, but many of these results have been confirmed by multiple studies, and not all tested diseases showed positive associations.

Many of these diseases involve tissues such as lung and the digestive tract, in which the ABO substances are produced and displayed on cell

Table 3.5A. *Associations between ABO blood groups and non-infectious disease*

Disease	Comparison types	No. of studies	Cases	Controls
Cancers				
Stomach	A:O	101	55 434	1 852 288
Colon and rectum	A:O	17	7435	183 286
Cervix of uterus	A:O	19	11 927	197 577
Corpus of uterus	A:O	14	2598	160 602
Ovary	A:O	17	2326	243 914
Breast	A:O	24	9503	355 281
Multiple primaries	A:O	2	433	7823
Benign salivary tumors	A:O	2	12 968	12 968
Other internal diseases				
Duodenal ulcers	O:Not-O	44	26 039	407 518
Gastric ulcers	O:Not-O	41	22 052	448 354
Rheumatic diseases	Not-O:O	17	6589	179 385
Pernicious anemia	A:O	13	2077	119 989
Diabetes mellitus	Not-O:A	20	15 778	612 819
Ischemic heart disease	Not-O:O	12	2763	218 727
Gallstones	A:O	10	5950	112 928

Source: Summarized from Vogel and Motulsky, 1986.

surfaces, during development or in wound healing. These tissues are also exposed to various environmental agents, such as viruses and pollen grains, that may be antigenically similar to the ABO substances. On the basis of current knowledge of the genes, their mutations, and the epidemiology of the associated diseases, immune explanations seem plausible (see e.g., Vogel and Motulsky, 1986).

The other example, in Table 3.5B, the HLA system, was discovered because of its role in tissue-graft rejection, and is involved in the self/non-self recognition and protection functions of the immune system. The positive HLA association with rheumatoid arthritis was referred to above. At the time the reports of associations appeared, the HLA-associated diseases seemed to have little if anything in common, nor even any obvious self/non-self autoimmune attributes. Again it seemed unlikely that all the diverse associations could be true; again it has turned out that immune explanations may apply in some cases, although none has yet been fully worked out, as is discussed in Chapter 13. Other HLA associations appear to implicate the HLA chromosomal region rather than specific genes, as is discussed in Chapter 7.

Table 3.5B. *HLA and disease*

Condition	HLA type	Frequency (%) Cases	Frequency (%) Controls	Odds ratio
Hodgkin's disease	A1	40	32.0	1.4
Idiopathic hemochromatosis	A3	76	28.2	8.2
	B14	16	3.8	4.7
Behçet's disease	B5	41	10.1	6.3
Congenital adrenal hyperplasia	Bw47	9	0.6	15.4
Ankylosing spondylitis	B27	90	9.4	87.4
Reiter's disease	B27	79	9.4	37.0
Acute anterior uveitis	B27	52	9.4	10.4
Subacute thyroiditis	Bw35	70	14.6	13.7
Psoriasis vulgaris	Cw6	87	33.1	13.3
Dermatitis herpetiformis	D/DR3	85	26.3	15.4
Celiac disease	D/DR3	79	26.3	10.8
	D/DR7	Also increased		
Sicca syndrome	D/DR3	78	26.3	9.7
Idiopathic Addison's disease	D/DR3	69	26.3	6.3
Graves' disease	D/DR3	56	26.3	3.7
Insulin-dependent diabetes	D/DR3	56	28.2	3.3
	D/DR4	75	32.2	6.4
	D/DR2	10	30.5	0.2
Myasthenia gravis	D/DR3	50	28.2	2.5
	B8	47	24.6	2.7
Systemic lupus erythematosus (SLE)	D/DR3	70	28.2	5.8
Idiopathic membranous nephropathy	D/DR3	75	20.0	12.0
Multiple sclerosis	D/DR2	59	25.8	4.1
Optic neuritis	D/DR2	46	25.8	2.4
Goodpasture's syndrome	D/DR2	88	32.0	15.9
Rheumatoid arthritis	D/DR4	50	19.4	4.2
Pemphigus (Jews)	D/DR4	87	32.1	14.4
IgA nephropathy	D/DR4	49	19.5	4.0
Hydralazine-induced SLE	D/DR4	73	32.7	5.6
Hashimoto's thyroiditis	D/DR5	19	6.9	3.2
Pernicious anemia	D/DR5	25	5.8	5.4
Juvenile rheumatoid arthritis				
Pauciarticular	D/DR5	50	16.2	5.2
All cases	D/DRw8	23	7.5	3.6

Source: Vogel and Motulsky, 1986.

Genes as risk factors for quantitative traits

Disease-related phenotypes such as cholesterol level or weight are quantitative. Genes may affect such phenotypes, and so may other exposures such as age, diet and the like. A risk factor cannot cause a quantitative trait like weight, but can *affect* its value, and these effects are of interest.

Generally, we identify a set of risk factors, $\boldsymbol{x} = \{x_i\}, i, = 1, 2, 3, \ldots, n$, for which levels of exposure are *measured* (i.e., observed) on each individual (writing $\boldsymbol{x}$ denotes a vector, or ordered list, of measures). The variables can be quantitative risk factors, such as calories consumed, or qualitative ones, such as specific genotypes. We can model the effect of 'dose' x_{ij} to individual j of risk factor i on his/her phenotype, ϕ_j, by a regression coefficient, β_i, through an equation such as the following:

$$\phi_j = \alpha + \sum_i \beta_i x_{ij}. \qquad [3.5]$$

Here α is some baseline value (which can be zero). An example might be

$$\text{Height} = \alpha + \beta_1 \times \text{age} + \beta_2 \times \text{sex} + \beta_3 \times \text{calories/day} \\ + \beta_4 \times (\text{genotype } AA),$$

dropping the js for convenience. Non-linear effects can also be included, for example by adding terms such as $\beta_5 \times \text{age}^2$. Qualitative risk factors, such as genotype in [3.5], are modeled with *dummy variables*, set to 1 if the person has the genotype, and to zero otherwise.

These are standard multiple regression methods. Statistical computer packages estimate the values of the parameters (α, βs) that provide the best prediction of an individual's observed phenotypes given the values of his/her risk factors, based on the available data. If a given risk factor, x_i, is important enough for its effects to be detected in the sample collected, its regression coefficient ($\hat{\beta}_i$) will be significantly different from zero and provides our best estimate of the dose–response effect of that risk factor: for every unit increase in an individual's exposure to risk factor x_i, the individual's trait value changes by $\hat{\beta}_i$ units.

In the simplest genetic case, known as a *measured genotype* approach, non-genetic risk factors may be nuisance variables, and the regression is used to *adjust* the data for these variables: each person's phenotype is expressed as a deviation from the average person with his/her age, sex, etc. The deviation that remains is what we hope to explain by genotype, which now represents a categorization of the data, and which can be analyzed equivalently by analysis of variance (see e.g., Sokal and Rohlf, 1981). In some cases there is *genotype by environment* ($G \times E$) interaction; that is, the environmental exposures have different phenotypic effects for different genotypes, and we need a separate regression such as [3.5] for each genotype. These situations are considered in Chapter 12.

A couple of special cases

It is worth mentioning briefly how special risk-factor situations are modeled, because the methods are used later in the book. There are occasions in which we have to model the effects of quantitative risk factors, e.g., calories or cholesterol, on qualitative outcomes such as the probability of disease. The most common approach is to use *logistic regression*. We express the probability that individual j has a disease, given exposures levels $\mathbf{x}_j$, by

$$\Pr(\phi_j = 1|\mathbf{x}_j) = 1/\{1 + \exp[-(\alpha + \Sigma\beta_i x_{ij})]\}. \quad [3.6]$$

This may seem awkward, but it allows quantitative variables that can take on any value to be translated into probability effects; that is, risks that are between 0 and 1. Conveniently, the natural logarithm of the odds ratio, based on [3.6], is

$$\ln[\Pr(1)/(1 - \Pr(1))] = \alpha + \Sigma\beta_i x_{ij}, \quad [3.7]$$

which in this scale is an ordinary linear regression {Breslow and Day, 1987}.

Another important problem is how to model risk factors, including genes, that work strictly by affecting the age-specific hazard for a given qualitative outcome. To do this, we usually use another related regression method, known as *proportional hazards*. The risk factor is assumed to act in a multiplicative way on some baseline hazard:

$$h_j(t|\mathbf{x}_j) = h_0(t)\exp(\alpha + \Sigma\beta_i x_{ij}) \quad [3.8]$$

{Breslow and Day, 1987; Kalbfleisch and Prentice, 1980}. Explanations on how to develop and use regression models to analyze multiple risk-factor data are provided by the references cited in this section. For simple situations, standard commercially available statistical computing packages can be used directly. Genetic analysis often requires models not available in these packages, but there are several genetic analysis packages that can accommodate some of these situations (see Chapter 6, note (2), Software). Unfortunately, the investigator must sometimes write his/her own program for the details specific to some genetic analysis problems.

Conclusion

This brief survey of basic concepts of frequency and association has been given to illustrate some of the more common general approaches to the study of genetic epidemiology. The measures given here are statistical. They involve uncertainty as reflected, for example, in the standard errors of the estimates of prevalence, regression coefficients, and association measures. The subjects of significance testing, confidence intervals, statistical power to detect effects etc. are too extensive to discuss here, but are fundamental to understanding the nature of genetic epidemiological evidence, i.e., evidence on genetic control derived from population data. Effects that we detect are inherently related to the nature and size of the samples we choose to collect. We never know the true state of Nature with certainty. When our knowledge rests on sampled data, its strength is inherently dependent on the nature of the samples.

Notes

1. Generally in this book, the notation $\Pr(x)$ will refer to the probability of event x. For simplicity of notation I use the same symbol, such as ϕ for phenotype, for a variable and a specific realization of it, and I use the same symbol for a parameter, or characteristic, of a population or model, and an estimate of that parameter from a sample. Estimates are denoted, for example, by a caret (e.g., $\hat{p}$), or the subscript MLE (Chapter 4).
2. In epidemiology, the term ascertainment is often used for the acquisition of data *about* subjects who have already been chosen for study, whereas in genetics the term refers to the identification of those subjects for study (epidemiologists call this 'selection').
3. Geneticists usually use the term odds in a way different from the epidemiologists' usage just given. Rather than the ratio of probabilities exposed/unexposed in a given group, geneticists have used 'odds' to refer to the ratio of likelihoods of the data under competing hypotheses about gene location (Chapter 7).

4 *Genes and phenotypes in populations*

In short, in matters vegetable and animal,
the very model of a modern Major-Gene . . .

Mutated from W. S. Gilbert, *Pirates of Penzance*

How can the almost unimaginable amount of genetic variation, arising in an almost unimaginably variable environmental context, be related to specific phenotypes? This chapter extends the concepts of frequency and association developed in the previous chapter, to provide the basic concepts needed for a *genetic model* of how genes may affect a trait and how they act. Such models take advantage of the special constraints that billions of years of evolution have placed on traits controlled by genes.

Frequency concepts for genetic traits

The prevalence of a genetically related trait in a population depends on the amount of genetic variation that affects it.

Allele and genotype frequencies

The most fundamental quantitative variable in population genetics is the *allele frequency* (often carelessly called the 'gene' frequency), a prevalence measure. If, among the $2N$ copies of a given gene in a population of N diploid individuals, n_i are allele i, then the frequency of that allele is defined as $p_i = n_i/2N$. There is no theoretical restriction on the number of alleles that can exist at a locus, but their frequencies must sum to 1. Thus, if one allele is very common, others must be correspondingly rare. We estimate the allele frequency from the same formula, using n and N for the individuals and alleles in a sample, and the standard error of the estimate is that of a binomial sample, $\sqrt{[p(1-p)/2N]}$. When a locus has only two alleles, we use a shorthand form for denoting their frequencies: p and $q = 1 - p$.

The analogous frequencies of the different *genotypes*, g, at a locus, among the N diploid individuals, are defined as $P_g = n_g/N$. Again, to account for all genotypes, $\Sigma_g P_g = 1$. Here, however, the number of genotypes *is* constrained, and equals $m(m + 1)/2$ if there are m alleles at the locus: m homozygotes, and $m(m - 1)/2$ heterozygotes, representing every possible pairing of alleles. A diallelic locus with alleles A and a has three genotypes, AA, Aa, aa. However, not all possible genotypes need be found in any population or sample.

When all genotypes are identifiable, the system is said to be *co-dominant*, and the allele frequencies can be determined simply by counting. The frequency of the i-th allele, p_i, is estimated by its frequency in the sample (here using ns and N to refer to the sample):

$$\hat{p}_i = \left(2n_{ii} + \sum_{i \neq j} n_{ij}\right) \Big/ 2N, \qquad [4.1]$$

where the ns are the counts of the subscripted genotypes. By [4.1] it can be shown that $\hat{p}_A = \hat{P}_{AA} + \frac{1}{2}\hat{P}_{Aa}$ in a diallelic locus. Weir (1990) gave details on estimating allele and genotype frequencies.

Some characteristics of genotype frequencies

Because of sexual reproduction, there is a relationship between the genotypes in one generation and the mating pattern of the parents of that generation. If mating is random with regard to genotypes at a locus, the genotype frequencies are strictly determined by the allele frequencies. The probability of an i homozygote is $P_{ii} = p_i \times p_i = p_i^2$, and of an ij heterozygote $P_{ij} = p_i p_j + p_j p_i = 2p_i p_j$ (since the 'order' of alleles is biologically unimportant). Populations that satisfy this relationship are said to be in *Hardy–Weinberg equilibrium* (*HWE*). For a diallelic system the genotypes have frequency $P_{AA} = p^2$, $P_{Aa} = 2pq$, $P_{aa} = q^2$.

Despite the enormous amount of psychic energy and resources that we invest in carefully choosing our mates, genotypes for most human loci are in HWE, to within the limits of statistical variation (tests for HWE are given in Appendix 4.1). This is because the criteria for mate choice, e.g., behavior and physical appearance, involve only a fraction of all loci. It is important that this is so, because we must often make some assumption about genotype frequencies for unseen loci that we are trying to understand. However, mates *are* sometimes chosen non-randomly with respect to at least some genetically determined characteristics (e.g., visible morphology), and inbreeding, social subdivision of a population and other factors can cause deviation from HWE that may be important;

Table 4.1. *Two-locus genotype frequencies under Hardy–Weinberg equilibrium*

		Father			
	Gamete:	AB	Ab	aB	ab
	Prob:	pr	ps	qr	qs
Mother	$AB\ pr$	r^2p^2	p^2rs	pqr^2	$pqrs$
	$Ab\ ps$	p^2rs	p^2s^2	$pqrs$	pqs^2
	$aB\ qr$	pqr^2	$pqrs$	q^2r^2	q^2rs
	$ab\ qs$	$pqrs$	pqs^2	q^2rs	q^2s^2

Note: Table gives frequencies in offspring of genotypes corresponding to the gametes defining the cell. Two loci with allele frequencies p,q, and r,s, respectively.

methods exist for taking these deviations into account (see e.g., Hartl and Clark, 1989; Hedrick, 1985).

Multilocus genotypes

Multilocus genotype frequencies are determined by the allele frequencies of the loci involved. We can denote this as $G = \{g_i\}$, a *set* of genotypes at specific loci, g_1, g_2, etc. If these loci are in HWE, the frequency of G is the product of the genotype frequencies for the individual loci: $P_G = \Pi_{g \in G} P_g$, indexed over all single-locus genotypes, g, being considered. These must sum to unity over all the genotypes. Table 4.1 provides genotype frequencies for two diallelic loci in HWE.

Linkage disequilibrium in populations

Not all sets of loci are independent; that is, in HWE with respect to each other. Consider two diallelic loci: one locus disease related, with alleles D and d, and allele frequency p_D, and another with alleles M, m with allele frequency p_M. Under HWE there should be no association between alleles at the two loci. But several factors can induce dependence among loci; these include recent admixture between genetically disparate populations, genetic drift, natural selection, or recent mutation. However, the loci may by *syntenic*, or physically *linked*; that is located on the same chromosome.

Such association is easiest to describe in terms of *haplotypes*; that is, alleles found in the same gamete. With no association, the probability of observing a given D allele and M allele in a gamete should simply be the product of the allele probabilities at each of the loci. But there may be a difference between this and the haplotype frequencies, γ_{DM}, that we actually observe. Let δ represent this discrepancy, known as *linkage disequilibrium* (gametic disequilibrium, allelic association). We can write

$$\gamma_{DM} = p_D p_M + \delta. \qquad [4.2]$$

There are four possible haplotypes (DM, Dm, dM, dm). The value of δ is the same for one complementary pair of haplotypes (e.g., DM, dm), and $-\delta$ for the other (Dm, dM), and δ can range between -0.25 and $+0.25$ for two diallelic loci. Several authors give details of the statistical analysis of linkage disequilibrium {Hedrick, 1985; Weir, 1990}.

The probabilities of diploid genotypes can be worked out in terms of the haplotype frequencies under random mating, since under these circumstances the genotype frequencies are simply the product of the gamete frequencies. For example,

$$\Pr(DDMM) = \gamma_{DM}\gamma_{DM} = (p_D p_M + \delta)^2. \qquad [4.3]$$

A full set of such frequencies is shown in Table 8.2, p. 141.

Linkage disequilibrium is often used as evidence for a causally important association between genetic markers and hypothesized disease loci and/or to help to refine the chromosomal location of a gene of interest, and may also be informative about gene history.

Measures of genetic diversity in populations

The amount of genetic variability at a locus depends on the frequencies of the alleles at the locus. Two measures of genotypic diversity in a population are often used. One is the *heterozygosity*

$$H = 1 - \Sigma_g p_g^2 = \sum_{i \neq j}^{n} p_i p_j \qquad [4.4]$$

for a locus with n alleles. In HWE, H equals the probability that a random individual is a heterozygote. When all alleles have equal frequency ($= 1/n$), heterozygosity is maximized at $H = 1 - (1/n)$.

The second measure, used especially in genetic counseling and gene mapping through the use of genetic markers, is the *polymorphism information content* (PIC) of a locus, defined as

$$PIC = 1 - \sum_{i=1}^{n} p_i^2 - \sum_{i=1}^{n-1} \sum_{j=i+1}^{n} 2p_i^2 p_j^2 = 2\Sigma\,\Sigma p_i p_j (1 - p_i p_j), \qquad [4.5]$$

which, for reasons whose importance will be seen later, is the probability that one of two random individuals (e.g., parents in a mating) is a heterozygote and the other is a different genotype (Ott, 1991). The maximum $PIC = (n - 1)^2(n + 1)/n^3$, when all alleles have equal frequency.

Frequency relationships between genotype and phenotype

The frequency of a genetically related disease depends upon the relevant allele and genotype frequencies.

The concept of penetrance

A given genotype does not always produce the same phenotype. The association between the two is known as the *penetrance*. Individuals with a given genotype will have some *distribution* of phenotypes; the *penetrance function* Ω_g specifies the probability that an individual with genotype g has phenotype ϕ:

$$\Omega_g(\phi) = \Pr(\phi|g). \qquad [4.6]$$

The penetrance function is defined over the entire range of biologically possible values for the phenotype. Since this is a probability distribution, it must sum to unity over this range.

The penetrance probability ranges from 0 (phenotype impossible) to 1 (phenotype inevitable). In principle every genotype has a penetrance probability relative to every phenotype; that is, there is some probability that a given phenotype would arise in any genotype. Of course, this may be zero for some phenotypes.

Often, only one phenotype is of particular interest, for example the presence of disease (say, $\phi = 1$). In such cases we say that the penetrance of genotype g is the probability that someone with that genotype has the disease, and when the context is clear we can write $\Omega_g(1)$, or even just Ω_g for this. Since 'trait absent' ($\phi = 0$) is also a phenotype, $\Omega_g = 1 - \Omega_g(0)$.

We often want to relate genotypes at a diallelic locus to the presence of a disease. If the trait is always present in, say, AAs and Aas, and absent in aas, then $\Omega_{AA} = \Omega_{Aa} = 1.0$, $\Omega_{aa} = 0.0$, and the A allele is said to be *dominant*, and a is *recessive*. If a disease is dominant, then 'normality' is recessive, and vice versa.

For a continuous quantitative phenotype, such as height or levels of cholesterol, the penetrance function for a given genotype is usually expressed as a mathematical probability density function, from which the probability of observing a phenotype within any range on the scale can be computed for each genotype. For many quantitative biological traits there is some measurement scale on which the phenotypes are approximately *normally* distributed; that is, follow a normal (or *Gaussian*) distribution in the population. Such a distribution is specified by its mean (μ) and variance (σ^2), and the phenotype distribution has the standard formula for the normal density function:

$$\Omega_g(\phi) = f(\phi|\mu_g, \sigma_g^2) = \frac{1}{\sigma_g\sqrt{(2\pi)}} \exp\left[-(\phi - \mu_g)^2/2\sigma_g^2\right] \qquad [4.7]$$

Here, the index g indicates that this is specific to a given genotype. We denote a normally distributed variable, say ϕ, by $\phi = \text{Nor}(\mu, \sigma^2)$, and its probability density by $f(\ )$. The normal distribution ranges from $\phi = -\infty$ to $+\infty$; while this is not literally possible for a biological trait, it is often a reasonable approximation. The penetrance function of several genotypes can be identical, for example if there is dominance.

Penetrance is a statistical, population-specific association between genotype and phenotype, not a biological *explanation* of such a relationship. Many factors may affect the expression of a given genotype. These can include other genes, including those that regulate the tested genotype, environmental factors, errors in measurement or classification, sampling error, and so on. Such factors can generate a quantitative distribution of phenotypes associated with a discrete set of genotypes, for example the distribution of plasma cholesterol levels among individuals with a given genotype at the apo-E locus.

A population concept of causality for genetic inference

The real objective is to understand how genotypes are 'causally' related to some phenotype. There are at least two relevant meanings of the concept of 'causation' in biology. One is physiological, and basically mechanistic (often carelessly equated with 'deterministic'). It refers to the direct mechanism of action of the locus in question. Globin genes code for hemoglobin molecules and hence are causally related to respiratory physiology. This is the mechanistic realm of causation.

Often, however, our knowledge is insufficient to enable us to ascribe such effects to a particular genotype, and we must turn to the other, epidemiological, type of causation. This is the strictly probabilistic *association* between the occurrence of alleles and phenotypes in a population or sample, as specified by penetrance functions. Other things being constant, we can say that a system of genotypes G is causally related to phenotype ϕ if, and only if, for at least two genotypes, i and j in the system,

$$\Omega_i(\phi) \neq \Omega_j(\phi), \qquad [4.8]$$

for at least some phenotype value ϕ. 'Other things being constant' is an escape clause in case some factor, e.g., an environmental exposure, is differentially associated with different genotypes. Epidemiological causation is a strictly population-specific concept, and is sample size dependent (since tests of the equality of two penetrance distributions are statistical), and does not imply physiological causation.

Occurrences of a disease that are not attributable to a specific genotype are called *sporadic* (or *phenocopies*). Often, we use this term to refer to cases arising in the lowest-risk of a set of tested genotypes – as if those cases reflect the effects of purely environmental agents. However, protection is simply the anti-world of susceptibility and it seems better to avoid any mysticism about causation, and say merely that each genotype has some risk. We can worry about cause in the mechanistic sense separately.

Interpreting causation from penetrance functions

Figure 4.1 shows schematically various aspects of the penetrance relationship between genotypes and phenotypes. Suppose that on some

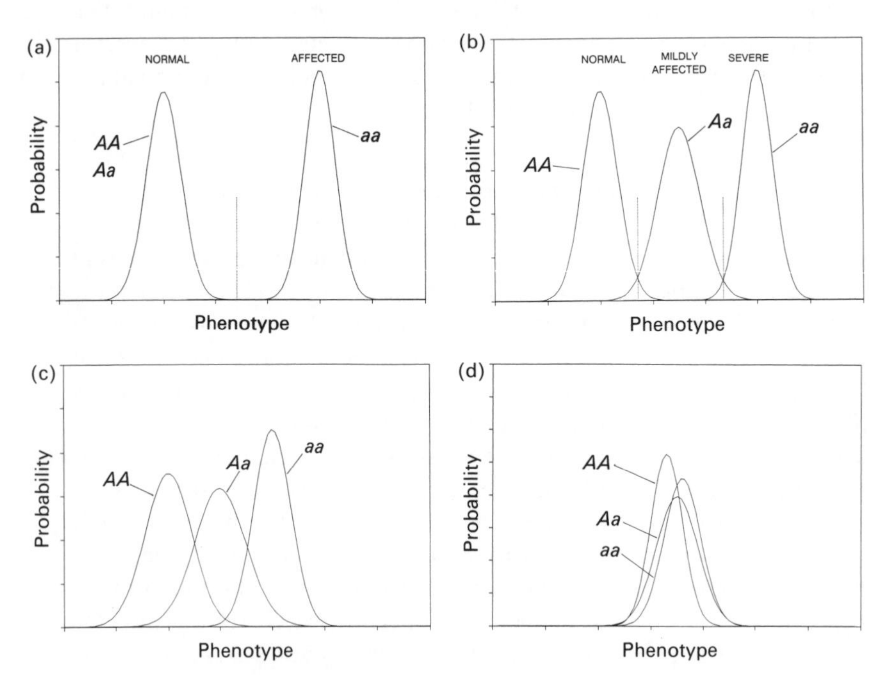

Figure 4.1. (a) to (d) Schematic of penetrance functions for specific genotypes, on an arbitrary phenotype scale. *AA*, *Aa* and *aa* could be the penetrance functions for three hypothetical genotypes at a locus. See the text for details.

scale a set of genotypes is associated with quantitative phenotypic variation, due to the effects of measurement error, environmental factors, and so on.

Figure 4.1 shows the penetrance functions for genotypes at a single diallelic locus. The shape of the functions is arbitrary, although, because they are distributions, the area under each curve is 1 (the peaks differ because of the shape differences). In (a) there is complete dominance and two of the penetrance functions are identical, and have essentially no overlap with the third; here, a phenotype can be attributed either to the dominant or recessive genotypes, with minimal misclassification. Indeed, in such cases we may often classify the population into two discrete phenotypes, 'normal' and 'affected', ignoring the quantitative variability within each category. This is the case for the disease phenylketonuria (PKU), whose underlying quantitative phenotype is the activity level of the enzyme phenylalanine hydroxylase.

In Figure 4.1(b) three penetrance distributions overlap slightly. Some phenotypes are unambiguously associated with specific genotypes, but there are phenotypes for which the penetrance probability is the same for two different genotypes (i.e., where the lines intersect in the figure). Overall, the overlap is sufficiently little that the population may still be naturally divisible into discrete categories, such as normal, mildly affected, and severe cases of some disease, with some individuals difficult to classify. An analysis of variance could show, with adequate sample sizes, that these distributions differ.

In Figure 4.1(c) the penetrance distributions have substantial overlap, and a high fraction of individuals would be of ambiguous classification. Figure 4.1(d) takes this to an extreme, in which there is very little difference among the three penetrances. It would require very large samples to demonstrate, or even to imply, these differences, and even so we would hesitate to attribute much 'causal' effect for such a locus.

Some genotypes may have disjoint penetrance functions (e.g. the sum of the left and right modes in Figure 4.1(c)). This is rarely considered and could lead to confusion if it occurs, but there are documented occurrences such as cholesterol levels and apo-E genotypes.

The causal spectrum and the concept of a genetic model

Figure 4.1 show *conditional* penetrance distributions; that is, each applies to a collection of individuals all with the same known genotype. Until we can classify individuals in this way, we may have to work with the phenotype distribution of the general population. In the population, the size of the distribution associated with each genotype is equal not to 1.0

but to P_g, the genotype probability. The population distribution is the sum of these, and has area equal to 1.

This is shown in Figure 4.2, corresponding to panels from Figure 4.1. Note that now the 'visibility' – and hence detectability – of the different genotypic effects depends on their frequencies in the population, can vary among populations, and may depend heavily on sample size.

In Figure 4.2(a), the affecteds are rarer than normals, but the population distribution is frankly bimodal and the discrete dichotomization of the two phenotypes is obvious. In (b) the rarer upper mode can verge on undetectability except with large samples. The distribution of cholesterol levels in the population is like this, with upper modes due to rare genotypes at the LDL receptor locus. In (c), the overlap problem makes the detection of multiple genotypic effects quite problematic except with large samples. The dotted line shows the population phenotype distribution, the sum of the three components. This could easily be mistaken for the kind of right-skewed distribution often found for biological traits

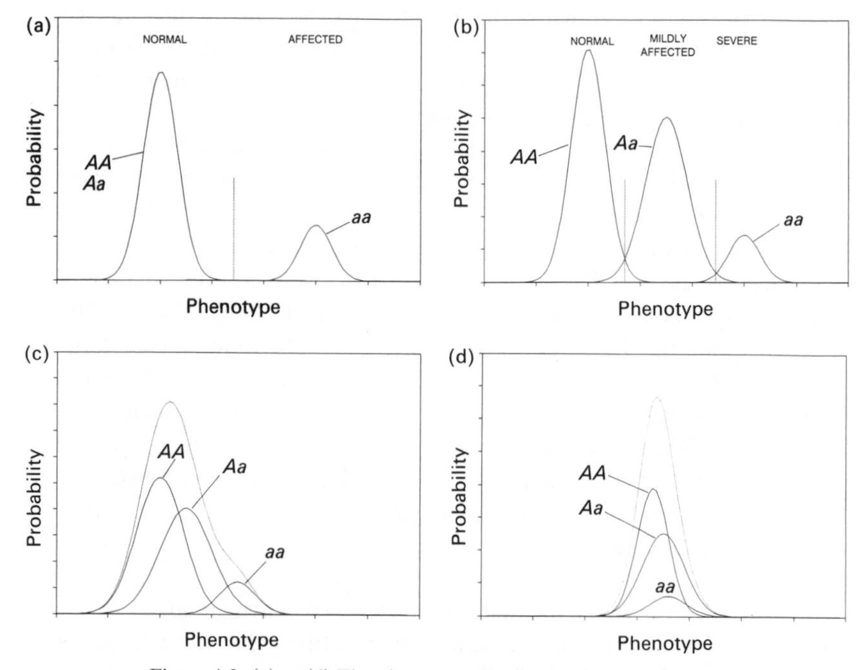

Figure 4.2. (a) to (d) The phenotype distribution in a population as expressed in terms of components due to identified genotypes. Panels correspond to those of Figure 4.1, except that here the relative area under each curve is proportional to its respective genotype frequency in the population. See the text for details.

(Wright, 1968); e.g. distributions of triglyceride levels associated with apolipoprotein gene variants are sometimes skewed. Finally, (d) provides a situation essentially unresolvable from population data; most alleles that affect quantitative traits probably have such minor effects (Chapter 9).

Figure 4.2 identifies variation at a single locus, but for traits affected by many loci, every individual has genotypes at *all* the loci, and the sum of all of these distributions can generate a very smooth phenotype distribution in the population, masking the discrete, genotype-specific nature of the underlying genetic causes.

The concept of a genetic (epidemiological) model

Three main factors determine the genetic causation of a given trait: (1) the distribution of the *allelic effects* on the trait, at loci that affect it; (2) the frequency distribution of those alleles and hence of the genotypes involved; (3) the relationships among the different loci, and between them and the environment, in their causal effects. These factors comprise the genetic *causal spectrum*, or *genetic architecture*, of the phenotype.

A generic formula for such a spectrum is

$$\Pr(\phi) = \Sigma_G P_G \Omega_G(\phi). \qquad [4.9]$$

The prevalence, or probability, of a given phenotype ϕ in a population is the weighted sum, over all relevant multilocus genotypes, of the genotype probability times the penetrance, or probability of the phenotype given the genotype. Environmental effects and interactions are latent in the Ωs in [4.9].

Much of the remainder of this book deals with the nature of the causal spectrum, how it is brought about by evolution, and how we infer its characteristics. However, the full spectrum for a given phenotype is usually an inaccessible abstraction. Rather, we must be satisfied with specifying a simplified subset of the spectrum that we can realistically hope to understand. Such a specification is known as a *genetic model*. A genetic model is a hypothesis about, or specification of, relevant components of statistical association that would be predicted by phenomena specific to the behavior of genes.

Although a genetic model of this type begins to bridge the gap between epidemiological and physiological causation, such a model is developed for application to epidemiological, not physiological data, and is often designed for a specific kind of sample. Its objectives are: (1) to infer that genetic factors are involved; (2) to describe their frequency, variation, and mode of action; and (3) to locate and identify the genes, so that the phenotype becomes as predictable as possible and physiological mechanisms can be determined. The steps in using a genetic model to do this are (1) to collect appropriate *samples*, (2) to *estimate* the parameters of the model from the data, and (3) to *test* the hypothesis of the model in some statistically meaningful way.

The current repertoire of genetic models with which we can effectively work is quite limited and very simplified. The models usually provide for (1) one or at most a few loci with very large effects on penetrance compared to those of any other loci, and/or (2) a large number of loci with individually very small effects typically treated in aggregate, and/or (3) environmental effects.

The simplest example is a disease caused by a single diallelic locus in HWE. The prevalence of the disease following the definition of [4.9] is

$$\text{Prev} = \Sigma P_g \Omega_g = P_{AA}\Omega_{AA} + P_{Aa}\Omega_{Aa} + P_{aa}\Omega_{aa}. \quad [4.10]$$

This has three independent parameters, the allele frequency and two of the penetrance probabilities. If the disease is dominant, this is simply p^2+2pq.

Conclusion

This chapter illustrates basic ways in which genetic variation is expressed in, and a genetic model may be specified for, general population data. Most problems quickly become too complex for such a simple approach. In particular, the causal spectrum is presented here as if only one locus is involved, but often many loci contribute to a trait. Realistic genetic models are often too complicated to test in random samples from the population, and special samples are collected that will reveal patterns of phenotypic variation that are consistent with evolutionary principles and are distinguishable from other kinds of risk process.

By far the most important samples in use in human genetics today are *family* data. Mendelian segregation and independent assortment of alleles generate strong structure in the relationship between parents and their offspring and, by extension, other relatives. The analysis of family data to establish a genetic model is the subject of Chapters 5–8.

APPENDIX 4.1: **The likelihood method and some principles of genetic inference**

Many books and papers discuss methods for estimating the parameters of genetic models and determining whether the models are acceptable by the usual statistical criteria {Elandt-Johnson, 1971; Hartl and Clark, 1989; Vogel and Motulsky, 1986; Weir, 1990}. One general approach, the *likelihood method*, is so widely used that it is important to sketch its basic principles here, for readers who may not be familiar with it {Cavalli-Sforza and Bodmer, 1971 (Appendix); Ott, 1991; Weir, 1990}. The examples here are rather trivial and chosen to make the points.

Estimating parameter values

A statistical model usually has one or more parameters that describe the population from which the data were sampled; these may include such things as the allele frequency. Denote the set of parameters collectively by Θ. If we knew the true value(s) of Θ we could specify the probability of observing the data. For example, for a fair coin with a probability of being 'heads' $p = 0.5$, the probability of two successive heads is $(\frac{1}{2})(\frac{1}{2}) = \frac{1}{4}$. In fact, we usually do not know the value(s) of Θ, which we wish to estimate. The *likelihood* of a given model – value(s) of Θ – *given the data*, $\mathscr{L}(\Theta|\text{data})$, is related to the probability of the data, given value(s) of the parameter(s):

$$\mathscr{L}(\Theta|\text{data}) \propto \Pr(\text{data}|\text{model with parameter values } \Theta). \quad [A4.1]$$

The likelihood of a model (e.g., parameter value) is the probability of the data, assuming that model to be true. If observations are independent, their individual likelihoods can be multiplied to form the likelihood of the set of observations, i.e., $\mathscr{L} = \Pi\mathscr{L}_i$. For example, the likelihood of two heads if we did not know the coin was fair is $\mathscr{L} = p^2$, which can be evaluated for all possible values of p from 0 to 1.

Generally, for categorical data, if we collect n_i observations of type $i = 1, 2, \ldots, m$ and the probability that a given observation will be of this type under our model is $\Pr(i)$, the likelihood has the form

$$\mathscr{L} \propto \prod_{i=1}^{m} \Pr(i)^{n_i}, \quad [A4.2]$$

where the product is over all possible observation types. The likelihood contains one term for each observation, but if n_i observations are of the same type, their product can be grouped as $\Pr(i)^{n_i}$. The actual likelihood contains a combinatorial term reflecting the number of possible permutations of the observed outcomes, i.e., orders in which they could arise; however, this term is omitted when, as often happens, it does not affect the use of the likelihood, but its existence is flagged by the notation $\propto$ (is proportional to).

Actually, it is often easier to work with the log-likelihoods, which generate the same results as the likelihoods themselves

$$\ln \mathscr{L} = \Sigma \ln \mathscr{L}_i = \Sigma\, n_i \ln \Pr(i). \quad [A4.3]$$

Here 'ln' refers to the 'natural' logarithm to the base e.

Expressions [A4.1] to [A4.3] can be used to find the best estimate of the parameters consistent with the data we have actually collected. Under broadly applicable conditions, the estimate we want is the value of a parameter, among all its possible values, that maximizes $\mathscr{L}$ for these data. Such an estimate is known as the *maximum likelihood estimate* (*MLE*), denoted Θ_{MLE}.

Example: estimating allele frequencies from phenotype frequencies
Consider the problem of estimating the allele frequency p from a set of sampled phenotypes, under a dominant diallelic model, with complete penetrance, in HWE. With dominance, we cannot estimate p by gene counting as given earlier in [4.1] because we cannot tell the AAs from the Aas. We could estimate q from the frequency of recessive homozygotes, as $\hat{q} = \surd(n_{aa}/N)$, but can illustrate likelihoods by constructing one to estimate q, as follows.

Under this model the probability of the dominant phenotype is $P_{AA+Aa} = p^2 + 2pq = 1 - q^2$. For a sample that includes $n_{AA,Aa}$ dominant and n_{aa} recessive phenotypes, the likelihood of the data is

$$\mathcal{L} \propto [P_{AA+Aa}]^{n_{AA+Aa}}[P_{aa}]^{n_{aa}} = (1 - q^2)^{n_{AA+Aa}}q^{2n_{aa}}. \qquad \text{[A4.4]}$$

For those familiar with the calculus, the value of q_{MLE} can be found by taking the derivative of $\mathcal{L}$ with respect to q, equating the derivative to zero, and solving the resulting equation for q. In this simple case the solution is $q_{\text{MLE}} = \surd(n_{aa}/N)$, identical with the method given in the previous paragraph. For non-mathematicians, [A4.4] could be plotted, e.g., in a spreadsheet program, for q at narrow intervals across its range from 0 to 1, and the maximizing value found visually, where the curve for $\mathcal{L}$ (or ln $\mathcal{L}$) has a peak.

If there are several parameters, things are more complex, although the idea is the same; the various partial derivatives of $\mathcal{L}$ are found and equated to zero, yielding a set of equations to be solved. For complex likelihoods with many parameters, computer methods are usually used to find the MLEs numerically.

Hypothesis testing
For a genetic model to be convincing it must be shown in some way to be consistent with the data. The likelihood method cannot prove the truth of a given hypothesis (an impossible task for statistical methods), but is well suited to *compare* the relative support the data give to various hypotheses.

Example: co-dominant diallelic locus
The simplest test of the hypothesis that the genotypes at a given locus are in HWE is a standard *goodness of fit* test. For example, for the pair of allele frequencies, we can compute the *expected* numbers of each genotype in a sample of size N as $E(AA) = N\hat{p}^2$, $E(Aa) = 2N\hat{p}\hat{q}$, and $E(aa) = N\hat{q}^2$, and evaluate the standard chi-squared statistic (if p is unknown, $\hat{p}$ estimated from the data is used):

$$\begin{aligned}\chi^2_{\text{d.f.}} &= \Sigma_g (O_g - E_g)^2/E_g \\ &= \Sigma_g (n_g - NP_g)^2/NP_g,\end{aligned} \qquad \text{[A4.5]}$$

which has d.f. equal to the number of classes minus 1, minus the number of parameters estimated from the data (here, three genotypes and one estimated allele frequency, so d.f. = 1). If the value of [A4.5] exceeds the chosen cutoff level (e.g., 3.84 for 5% significance), the test is significant, meaning the data are not in the expected HWE distribution.

Another way to test the hypothesis that the population is in HWE with allele frequency p_{MLE} is to compare the likelihood of the observed genotypes under that hypothesis with the simple possibility that the observed genotype fequencies in the sample are the true ones in the population, i.e. $\hat{P}_{AA} = n_{AA}/N$ rather than p^2_{MLE}, etc. The maximum likelihoods of the two hypotheses are

$$\left.\begin{aligned} &\text{H}_1 : \mathcal{L}_1 \propto [p^2]^{n_{AA}}[2p(1-p)]^{n_{Aa}}[(1-p)^2]^{n_{aa}} \\ &\text{H}_0 : \mathcal{L}_0 \propto \hat{P}^{n_{AA}}_{AA}\,\hat{P}^{n_{Aa}}_{Aa}\,\hat{P}^{n_{aa}}_{aa}, \end{aligned}\right\} \qquad \text{[A4.6]}$$

(writing p for p_{MLE} for simplicity). Now we compute the *likelihood ratio*, $\lambda = \mathcal{L}_0/\mathcal{L}_1$, where $\mathcal{L}_0$ is a less restricted hypothesis in that it has more parameters that are free to vary rather than being constrained by the model. Under some assumptions that are generally acceptable, if two hypotheses are *nested*, i.e., one can be derived from the other by fixing one or more of its parameters, then a general result from statistics is that $2 \ln (\lambda)$, has a chi-squared distribution, with d.f. equal to the difference in the number parameters (variables to be estimated) by which the two hypotheses differ. In our example, the likelihood ratio has 1 d.f. If the chi-squared value is large, the test fails, and we infer that the numerator hypothesis is significantly more likely than the denominator (see e.g., Sokal and Rohlf, 1981).

Example: estimation of linkage disequilibrium by likelihoods

A sample of gamete frequencies is unlikely to be exactly in HWE, but is the observed linkage disequilibrium significantly different from zero or is it just sampling error? The first hypothesis is that the true amount of disequilibrium in the population is δ_{MLE} as estimated by the likelihood

$$\mathcal{L}(\delta | p_D, p_M) = \prod_{i=1}^{4} \gamma_i^{n_i} \qquad \text{[A4.7]}$$

where the γs have the form given in [4.2]. The MLE allele frequencies must be determined as well as δ_{MLE}. The second hypothesis is that δ is zero and the observed deviation in the data is due simply to sampling variation. The likelihood of that value, given the data, is expression [A4.6], with $\delta = 0$ used in the expressions for the gamete frequencies, e.g., $\gamma_{ij} = p_i p_j + 0$; these are just the HWE two-locus gamete frequencies, which would obtain if there were no disequilibrium. These must be estimated separately under each hypothesis to compute the maximum likelihood. The hypothesis test is based on the ratio of the two maximized likelihoods.

Part II *Introduction to genetic epidemiology: inference from observational data*

5 *Segregation analysis: discrete traits in families*

And so do his sisters, and his cousins and his aunts!
W. S. Gilbert, *H.M.S. Pinafore* (1878)

We can understand the basic principles of genetic epidemiology by studying the behavior of alleles at a single locus in nuclear families. We take advantage of evolution-based constraints on the distribution of genetic variation in families. The analysis of trait distributions in families is known as *segregation analysis* after Gregor Mendel's Law of Segregation of individual alleles at a locus. The idea is to see if the pattern of phenotypes in families is consistent with a genetic model.

Families are ascertained via one or more index individuals, or *probands* (*propositi*), who may be either randomly identified, or chosen because of their disease or other phenotype status. The nature of the sampling must be built into the genetic model. This chapter illustrates the principles of segregation analysis by the study of genotypes that are strongly associated with categorical phenotypes {Cavalli-Sforza and Bodmer, 1971; Elandt-Johnson, 1971; Levitan, 1988; Li, 1961; Morton, 1982; Thompson and Thompson, 1986; Vogel and Motulsky, 1986}.

Appendix 5.1 provides some basic probability theorems for those who need them.

Nuclear families and sibships

The distribution of traits in families

A diploid, sexually reproducing organism has two sets of genes, one inherited from each parent. Each time that individual produces his/her own gamete (sperm or egg), one of his/her inherited alleles, at each locus, will be randomly chosen and transmitted in the gamete. There is thus a probability of $\frac{1}{2}$ that an offspring will inherit a specific parental allele. This *probabilistic aspect of inheritance* is not a matter of ignorance on our part, but is a fundamental aspect of our biology.

Transmission probabilities

Consider the pattern of allele transmission from parent to offspring for a single diallelic locus, with alleles A and a. We define *transmission probabilities*, $\tau(\alpha|g)$, as the probability that a parent of genotype g produces a gamete with allele α. These are *conditional* probabilities because they depend on the genotypic state of the parent. For autosomal loci, they are $\tau(A|AA) = 1$, $\tau(A|Aa) = \frac{1}{2}$, and $\tau(A|aa) = 0$; the transmission probabilities for the a allele are the complement of these. With more than two alleles, a separate set of transmission probabilities is needed for each allele. For an X-linked locus, the transmission probabilities for the A allele are 1, $\frac{1}{2}$, and 0 for the three maternal genotypes, and 1, 0 for the two paternal (y) genotypes (Ay, ay). Other mechanisms can produce non-mendelian transmission probabilities (Appendix 5.1 and Chapter 14).

Mating types

The probability that an individual has a given genotype is determined by the genotypes, or *mating types*, of its parents. A nuclear family is a set of repeated selections of offspring genotypes from the mating type, M_{ij}, of parents with genotypes i and j. In a population (or sample), the probability of an $i \times j$ mating, $\Pr(M_{ij})$ can take any value between 0 and 1, so long as the total, summed over all genotypes, $\Sigma_i \Sigma_j \Pr(M_{ij}) = 1$. These mating frequencies can be estimated empirically from a sample of N matings, $\hat{M}_{ij} = n_{ij}/N$.

If there is random mating relative to the locus in question, the mating type frequencies are determined by the genotype frequencies, just as the latter are determined by the allele frequencies; that is, $\Pr(M_{ij}) = P_i P_j$. If we ignore which parent has a given genotype, then if $i \neq j$ $\Pr(M_{ij}) = P_j P_i + P_i P_j = 2P_i P_j$. Table 5.1A provides this basic 'mating table'. Note that the chance you would encounter a given mating type in the population or sample depends fundamentally on the allele frequency.

Transition probabilities

The core of family data consists of parent–offspring triads. The conditional probabilities of genotypes in offspring given those in the father and mother, written $P_{g_o | g_f, g_m}$, are called the *transition* probabilities. These are built from the transmission probabilities. For a diallelic locus,

Table 5.1A. *Genotypic mating table for an autosomal diallelic locus*

Mating type	Empiric mating frequency	Under random mating	Offspring genotype probabilities: Conditional AA	Conditional Aa	Conditional aa	Unconditional AA	Unconditional Aa	Unconditional aa
$AA \times AA$	M_{11}	$p^2p^2 = p^4$	1	0	0	p^4	0	0
$AA \times Aa$	M_{12}	$2(p^2)(2pq) = 4p^3q$	$\frac{1}{2}$	$\frac{1}{2}$	0	$2p^3q$	$2p^3q$	0
$AA \times aa$	M_{13}	$2p^2q^2$	0	1	0	0	$2p^2q^2$	0
$Aa \times Aa$	M_{22}	$(2pq)(2pq) = 4p^2q^2$	$\frac{1}{4}$	$\frac{1}{2}$	$\frac{1}{4}$	p^2q^2	$2p^2q^2$	p^2q^2
$Aa \times aa$	M_{23}	$2(2pq)q^2 = 4pq^3$	0	$\frac{1}{2}$	$\frac{1}{2}$	0	$2pq^3$	$2pq^3$
$aa \times aa$	M_{33}	$q^2q^2 = q^4$	0	0	1	0	0	q^4
Total:	1	1				p^2	$2pq$	q^2

Note: Here, $p_A = p$, $p_a = q = 1 - p$. It is left as an exercise to show that the last line of the table is correct. The conditional probabilities are those with the mating type known; unconditional probabilities are those in the general random-mating population, weighting the conditional probabilities by the mating-type probabilities.

Table 5.1B. *Parent to offspring transition probabilities for a diallelic locus*

		Father's genotype AA	Father's genotype Aa	Father's genotype aa
Mother's genotype	AA	$\{1 \ \ 0 \ \ 0\}$	$\{\frac{1}{2} \ \ \frac{1}{2} \ \ 0\}$	$\{0 \ \ 1 \ \ 0\}$
	Aa	$\{\frac{1}{2} \ \ \frac{1}{2} \ \ 0\}$	$\{\frac{1}{4} \ \ \frac{1}{2} \ \ \frac{1}{4}\}$	$\{0 \ \ \frac{1}{2} \ \ \frac{1}{2}\}$
	aa	$\{0 \ \ 1 \ \ 0\}$	$\{0 \ \ \frac{1}{2} \ \ \frac{1}{2}\}$	$\{0 \ \ 0 \ \ 1\}$

Note: Each entry {a b c} represents the probability that an offspring of the specified parents will produce an AA, Aa, or aa offspring, respectively.
Source: Elston and Stewart, 1971.

there are three possible offspring (AA, Aa, aa) for which the transition probabilities are

$$[\tau(A|f)\tau(A|m),\ \tau(A|f)(1 - \tau(A|m)) + \tau(A|m)(1 - \tau(A|f)),\ (1 - \tau(A|f))(1 - \tau(A|m))]. \quad [5.1]$$

These values are given for a diallelic locus in Table 5.1B. Similar values can be derived for multiple alleles or loci {Elston, 1981; Elston and

Table 5.2. *Phenotypic mating table for an autosomal diallelic locus*

Mating type	Random mating frequency	Offspring segregation proportions (ω)			
		Conditional		Unconditional	
		D	R	D	R
Dominant by dominant matings					
$AA \times AA$	p^4	1	0	p^4	0
$AA \times Aa$	$4p^3q$	1	0	$4p^3q$	0
$Aa \times Aa$	$4p^2q^2$	$\frac{3}{4}$	$\frac{1}{4}$	$3p^2q^2$	p^2q^2
All D × D	$(1 - q^2)^2$	$(1 + 2q)/(1 + q)^2$	$q^2/(1 + q)^2$	$p^2(1 + 2q)$	p^2q^2
Dominant by recessive matings					
$AA \times aa$	$2p^2q^2$	1	0	$2p^2q^2$	0
$Aa \times aa$	$4pq^3$	$\frac{1}{2}$	$\frac{1}{2}$	$2pq^3$	$2pq^3$
All D × R	$2q^2(1 - q^2)$	$1/(1 + q)$	$q/(1 + q)$	$2pq^2$	$2pq^3$
Recessive by recessive mating					
$aa \times aa$	q^4	0	1	0	q^4
All R × R	q^4	0	1	0	q^4

Note: D, dominant; R, recessive. Lines labeled 'All {mating phenotype}' give values for all matings of specified phenotype, appropriately weighted by their population frequencies (Snyder's ratios) {see Li, 1961}.

Stewart, 1971; Morton, 1982}. In practice, existing segregation analysis computer packages do this automatically.

All essential genetic relationships in families follow from the mating table as given in Table 5.1A for an autosomal locus. The table provides the basis for quantifying the probability of various family structures in a population and, by extension, in a whole genealogy, in terms of (1) the allele frequencies, (2) the mating type frequencies, and (3) the transmission probabilities.

Segregation analysis: segregating phenotypes in families

We have been discussing genotypes, but the problem we want to solve is how to infer the action of unseen genotypes from the pattern of phenotypes in families. The general principles can be illustrated by considering a diallelic locus, with complete dominance for allele A. Table 5.2 recasts the mating table in terms of phenotypes. Each line provides the *segregation proportion* (or segregation ratio), ω, the prevalence of specific offspring phenotypes in sibships produced by each mating genotype (also interpretable as the probability that a random offspring will be affected).

From these, the segregation proportions with respect to each mating *phenotype* (called *Snyder's ratios*) can be computed. For example, on the basis of the rule for conditional probability (Appendix 5.1), the probability that a dominant × recessive mating produces a recessive offspring is:

$$\begin{aligned}
&\Pr(\text{R child}|\text{D} \times \text{R mating}; p, \Omega\text{s}, \text{HWE}) \\
&\quad = \Pr(aa \text{ child and } AA \times aa \text{ or } Aa \times aa \text{ mating})/ \\
&\qquad \Pr(AA \times aa \text{ or } Aa \times aa \text{ mating}) \\
&\quad = [M_{13}(0) + M_{23}(\tfrac{1}{2})]/(M_{13} + M_{23}) \\
&\quad = 2p(1-p)^3/(2p^2(1-p)^2 + 4p(1-p)^3) \\
&\quad = 2p(1-p)^2(1-p)/2p(1-p)^2(p + 2(1-p)) \\
&\quad = (1-p)/(2-p) = q/(1+q). \qquad [5.2]
\end{aligned}$$

The first line of this expression indicates that the conditional probability also depends on the (often unstated) genetic model, which includes the assumption of random mating (HWE). If penetrance is incomplete, each term in the conditional proportion section of Table 5.2 (as well as the mating genotype frequency) is multiplied by the appropriate penetrance (Ω_{AA}, Ω_{Aa}, Ω_{aa}), which will alter the segregation proportions.

These are population ratios that give the family types at disease-related loci as a function of the allele frequency. Historically, these ratios were important for documenting a genetic cause for various rare childhood diseases, for which the allele frequency would be at most a small percentage. Often, rather than population data, we work with a subset of family types ascertained by way of affected individuals. The segregation proportion in such a non-random subset of the population can affect how we think of genetic disease – and can be misleading if we are not careful.

Consider a recessive trait. Only mating types M_{22}, M_{23}, and M_{33}, with frequency in the population $4p^2q^2 + 4pq^3 + q^4$, can produce affected offspring. The fraction of these families that are of the type M_{22}, i.e., Nor × Nor, is

$$4p^2q^2/(4p^2q^2 + 4pq^3 + q^4) = 4p^2/(p^2 + 2p + 1). \qquad [5.3]$$

If the disease-producing allele is rare (i.e., $p \approx 1$) [5.3] approaches 1; that is, most of the matings that produce any affected offspring will be of the $Aa \times Aa$ type. Their segregation proportion is $\frac{1}{4}$. This is the source of the common belief that for a recessive trait a quarter of all siblings are affected. Similarly, the idea that $\frac{1}{2}$ of all siblings are affected for dominant traits derives from the rare-allele assumption.

The rare-allele assumption traditionally has been acceptable in the study of diseases because most diseases are, fortunately, rare. For accurate genetic counseling purposes (e.g., to estimate risk to a new offspring of parents who have already produced an affected child), however, precision is important {Murphy and Chase, 1975; Thompson and Thompson, 1986; Vogel and Motulsky, 1986}. Methods for significance and hypothesis testing for segregation proportions were given by Elandt-Johnson (1971).

Not all traits, or even all diseases, are rare. For traits that are not rare, the frequency of affected sibships depends on the allele-frequency-dependent mating type frequencies, and can be difficult to intuit. Fortunately there are better, more general, methods that do not rely on characterizing a disease by a single proportion. Before turning to these methods, however, we must consider an important aspect of family analysis, namely how to account for the way in which the data were ascertained.

Ascertainment bias and its correction: sibship data

The way in which families are ascertained can have major effect on the interpretation we make of the data. If we do not adjust for this, *ascertainment bias* may result; that is, no matter how large the sample, the estimate of ω may not approach its true value. False inference can result.

This can be illustrated by a simple nuclear family situation. Suppose we ascertain all affected children through the school system, and collect data on all siblings of these affected children. Consider a given sibship size, s, in families that can produce affecteds with segregation proportion ω. The probability that such a family produces r affected children is given by the *binomial distribution* of statistics, and is

$$\Pr(r|s,\omega) = C(r;s)\omega^r(1-\omega)^{(s-r)}, \qquad [5.4]$$

where $C(r;s) = s!/r!(s-r)!$, a combinatorial term giving the number of ways r things can be chosen from s 'trials'. From [5.4], the probability that such a family will be fortunate enough to produce s normal children is $(1-\omega)^s$. These families can never be identified if we ascertain sibships through affected school children.

Suppose that we are trying to use these family data to see whether the segregation proportion that we estimate from the data is consistent with our causal model, which specifies a value ω. Clearly our estimate will be biased, since many normal children from these families will never be identified by our sample design. We must correct for this. The simplest way to do this is to recognize that our sample can contain all families except those with no affecteds; that is, represents a fraction $1 - (1 - \omega)^s$ of the total population of sibships of the given type that we are considering. The corrected probabilities of r affecteds of a total s are found by adjusting [5.4] to include only the ascertained families:

$$\Pr(r|s;\omega) = C(r;s)\omega^r(1 - \omega)^{(s-r)}/[1 - (1 - \omega)^s]. \quad [5.5]$$

If, in our sampling, the probability that a given affected child is actually ascertained is *small*, the corrected distribution equals the probability of $r - 1$ affected among $s - 1$ children, so if we ignore the affected probands and just look at the remaining siblings our estimate of ω will be unbiased (for a derivation of this result, see Elandt-Johnson, 1971).

Actually, this oversimplifies the situation greatly. Because families with many affecteds may have a higher chance of being ascertained by a given sampling scheme, there will be several different potentially informative mating types, etc. Corrections for some simple sampling situations have long been known in medical genetics {Cavalli-Sforza and Bodmer, 1971; Elandt-Johnson, 1971; Li, 1961; Morton, 1982}, but methods for complex situations are still inexact.

Segregation analysis in pedigrees using likelihood methods

Modification of the genetic model

The models just presented show how the basic principles of mendelian inheritance enable us to predict a single value in a simple family structure, the segregation proportion among siblings that can be used to infer the presence of genetic risk factors. These models have been fundamental for most of this century in identifying classic, single-locus traits (often referred to as 'mendelian' traits) such as many severe pediatric diseases. However, sets of siblings represent only a fraction of the data we might collect, and for more complicated problems we would like to include all the information that is contained in extended family structures (known as *genealogies* or *pedigrees*) with persons from multiple generations, however large or variable those family structures that are available to us.

There is no simple expected 'proportion' of cases for an arbitrary, especially complex, family structure. Even if there were, each family differs in structure, age and sex composition etc., and would generate its own family-specific segregation proportion. What is the expected proportion of affecteds in a family whose sampled members include two grandparents, three of their offspring and the parents of two of their spouses, and five cousins? How different does the observed distribution of cases among these family members have to be to be inconsistent with an hypothesized genetic model? The idea of a simple goodness-of-fit test (e.g., an observed proportion against its expectation) does not really apply in such situations.

Fortunately, there is a better, more powerful, unified approach to inferring the presence of genetic factors in the segregation of risk in families of arbitrary size or structure and for any genetic model. This is known as the *general likelihood method* (Elston, 1981, 1986; Elston and Stewart, 1971; Morton, 1982). The method is conceptually straightforward, although in practice the mathematics become complex and computer programs are needed, and available, to do the job.

The idea of this approach is to compute the likelihood for the observed occurrence of phenotypes in a given sampled family structure, according to a proposed genetic model. To compute this likelihood requires that we specify the three basic components of the genetic model, in terms of the pedigree structure. Those components are: (1) the unconditional genotype probability for individuals randomly sampled from the population, which specifies the mating structure in the population among other things; (2) the inheritance pattern, i.e., the conditional probability of a given genotype in an offspring given the genotype of his/her parents; and (3) the penetrance function for the model, i.e., the probability that someone of a given genotype will have his/her actual observed phenotype. We must also account for how the data were ascertained.

We specify these as follows:

(1) *Unconditional genotype*[1] *probabilities*, P_g, for each *founder* individual. Founders are probands (members of the family at the top of the ascertained pedigree, or at the bottom if the analysis is from the present generation backwards), and incoming spouses unrelated to other members of the pedigree. For each such person the probability of a given genotype is simply that such a genotype occurs in the population; usually, we assume random mating and use HWE genotype probabilities, but non-random mating or sampling can in principle be included.

(2) *Conditional genotype probabilities* for the other members of the pedigree; that is, the relatives of the founders. The conditional genotype probabilities are just the transition probabilities given earlier, for genotypes of offspring given genotypes of their parents, $P_{o|f,m}$. This is the core of the genetic model, because it requires a specification of how many loci are involved, how many alleles, etc.

(3) *Penetrance probabilities* for each genotype, $\Omega_g(\phi)$, which specify the genotype-to-phenotype relationship.

(4) *Ascertainment correction* to account for how the family data were collected.

The likelihood on a nuclear family

Since the definition of the likelihood is $\mathcal{L}(\text{model}|\text{data}) = \Pr(\text{data}|\text{model})$ (Appendix 4.1), our problem is to express the probability of the observed set of family phenotypes in terms of the genetic model. We then need to find the values of the parameters of the model that maximize this likelihood.

The algebra is gory, but worth following to understand the underlying ideas. Consider two parents as founders, with phenotypes ϕ_{f} and ϕ_{m}, respectively. From [4.9], defining a genetic model, the probability of phenotype ϕ in a randomly sampled individual is

$$\Pr(\phi) = \sum_{g \in G} P_g \Omega_g(\phi). \qquad [5.6]$$

(Since the individual's actual genotype is unknown, we must account for all genetic ways that he/she might end up with his/her observed phenotype, ϕ.) Assuming random mating, the probabilities of observing the two parental phenotypes are independent, so the probability of observing the mating is $\Pr(\phi_{\mathrm{f}}\phi_{\mathrm{m}}) = \Pr(\phi_{\mathrm{f}})\Pr(\phi_{\mathrm{m}})$, each term having the form of [5.6]; non-random mating can in principle be accounted for (e.g., Bonney, 1986; Elston, 1981).

Now consider an offspring of these two parents. The probability of observing the offspring phenotype, ϕ_{o}, is conditional on the parents' genotypes and the penetrance, $P_{go|\mathrm{f,m}}\Omega_{go}$, using go to denote the offspring genotype. The likelihood of the model (H), given a set of two parents and one child is

$$\mathcal{L}(H|\text{parents, child}) \propto \Pr(\phi_{\mathrm{f}})\Pr(\phi_{\mathrm{m}})\Pr(\phi_{go}|\phi_{\mathrm{f}}, \phi_{\mathrm{m}})$$
$$= \sum_{\mathrm{f}} P_{\mathrm{f}}\Omega_{\mathrm{f}} \sum_{\mathrm{m}} P_{\mathrm{m}}\Omega_{\mathrm{m}} \sum_{go} P_{go|\mathrm{f,m}}\Omega_{go}. \qquad [5.7]$$

Here the sums are over all possible genotypes for each individual.

Unless there is selective abortion, parental stopping rules and the like (non-trivial issues that generate non-mendelian transitions and should be, but rarely are, addressed in using these models), the phenotypes of each child of these parents are independent, and the likelihood for the whole nuclear family with n offspring becomes proportional to

$$\sum_{\mathrm{f}} P_{\mathrm{f}} \Omega_{\mathrm{f}} \sum_{\mathrm{m}} P_{\mathrm{m}} \Omega_{\mathrm{m}} \prod_{o=1}^{n} \sum_{go} P_{go|\mathrm{f,m}} \Omega_{go}. \qquad [5.8]$$

The product on the right side is over all offspring, indexed by o, and the rightmost summation is done separately for all possible genotypes for each child.

Let us extend this to another generation by letting one of the children have a spouse; a summation term for that spouse is added adjacent to the child's term on the right of [5.8], and a product and sum term for the grandchildren (gc). The likelihood becomes

$$\sum_{\mathrm{f}} P_{\mathrm{f}} \Omega_{\mathrm{f}} \sum_{\mathrm{m}} P_{\mathrm{m}} \Omega_{\mathrm{m}} \prod_{o=1}^{n} \sum_{go} P_{go|\mathrm{f,m}} \Omega_{go} \sum_{s} P_{s} \Omega_{s} \prod_{gc}^{ngc} \sum_{ggc} P_{ggc|k,s} \Omega_{ggc}. \qquad [5.9]$$

In this way, a likelihood for an entire pedigree, no matter how complex, can be written as nested sums of products. In schematic form, the likelihood of a pedigree is

$$\mathscr{L} \propto \Sigma \ldots \Sigma_{\text{genotypes}} [\Pi_{\text{all founders}} P_i \times \Pi_{\text{all non-founders}} P_{j|\mathrm{m}_j,\mathrm{f}_j} \times \Pi_{\text{all individuals}} \Omega_k]. \qquad [5.10]$$

The first set of sums in [5.10] is over all possible genotype combinations for the entire set of individuals in the pedigree; i indexes founders, j non-founders, and k all individuals. This method was described in detail by Elston (1981, 1986), and Thompson (1986) also considered sampling.

Obviously, computers have to be used for this! Fortunately, convenient programs exist (see Notes, p. 116) that can handle complexities such as inbreeding loops, and complications such as age-, sex- and other effects can be built into the penetrance functions. These packages require only that the user understand and properly specify the model, and provide straightforward input files (e.g., phenotypes and relationships among sampled individuals).

Evaluating this likelihood can be daunting even using modern computers, since [5.10] includes all k^n genotype arrays that are possible in a pedigree with n family members and a model with k genotypes. Various computation-reducing algorithms have been built into existing programs [Cannings *et al.*, 1978; Lange and Elston, 1975; Morton, 1982; Thompson, 1986].

Correction for ascertainment bias

When families are ascertained via some characteristic that is relevant to the study (e.g., having some disease), a correction must be made in the likelihood, as discussed earlier, to avoid biased inferences. Since, for events A and B, Pr(A|B) = Pr(AB)/Pr(B), the general idea is that

$$\Pr(\text{the data}|\text{ascertained part}) = \frac{\Pr(\text{data and ascertained part})}{\Pr(\text{ascertained part})}. \qquad [5.11]$$

If the probability that an individual with a given phenotype will be ascertained is small, a corrected likelihood is the entire [5.10] divided by a likelihood of the same form as [5.10] but including only terms for the individuals who were part of the ascertainment of the data (see e.g., Cannings and Thompson, 1977). Other authors have suggested methods that involve specifying (and estimating) a parameter representing the probability that an affected individual or family will be ascertained (e.g., Elston and Sobel, 1979). Ewens and Shute (1986; Shute and Ewens, 1988a,b) suggested removing from consideration all information on members of the ascertained family who *could have* been involved in the ascertainment.

The subject of ascertainment correction is intricate, tricky, and not well worked out. It is beyond the scope of this book (and this author). In practice there is a problem beyond correcting for the way the data were ascertained: it is knowing *how* that ascertainment occurred. Except in the case of pure, random ascertainment of probands from the population, it is often difficult to specify why a family came to your attention. Is a family more likely to appear in a specialty clinic if several of its members have been affected? If two affected cousins are ascertained, how do we correct for the demographic structure of a family that produced them? Often, it is simply unrealistic to expect an investigator to ignore a deliciously informative family, regardless of how it became known to him/her.

Often such complications are handled by 'hand waving' – acknowledging the problem but hoping it does not matter much. Perhaps this is often true; even if there is unaccounted ascertainment bias, the bias may affect only a small fraction of all the information contained in the set of family data, and may lead only to a small probability of false inference or slightly erroneous parameter estimates. I know of no major instance in which *subtle* ascertainment biases (as opposed to gross ones) have been shown to lead to incorrect inference, but the literature is replete with reported findings that subsequent studies could not confirm; some of these may be due to ascertainment bias.

Ascertainment issues are most important for rare traits or extreme phenotypes, because then it is expensive to sample the general population; otherwise, not enough cases are ascertained per unit of investment. For some phenotypes, however, population sampling is possible and is preferred. The best way to approach the ascertainment problem is: (1) to know clearly how a given data set was ascertained; (2) to design studies based on population or single-proband ascertainment; (3) to follow up existing studies by collecting data on new individuals from informative parts of the same pedigrees (this can be done without bias so long as no new individual is ascertained because of information known about his/her phenotype); (4) to collect new families to confirm results from other studies (but beware of etiological heterogeneity!).

Parameter estimation and hypothesis testing

How is this general family likelihood used? If, as is usual, data from separate families are independent, the likelihoods [5.10] of each family can be multiplied together to get the likelihood of the total sample:

$$\mathcal{L} = \Pi\, \mathcal{L}_i \quad \text{or} \quad \ln \mathcal{L} = \Sigma \ln \mathcal{L}_i. \qquad [5.12]$$

This expression can then be analyzed: (1) to estimate the MLEs of the parameters of the model, and (2) to assess how well various models fit the data. The parameters to be estimated include the allele frequencies (ps), penetrances (Ωs), and perhaps other aspects of the model. For example, some authors estimate the values of the transmission probabilities (τs) rather than specifying them based on mendelian rules; this allows some aspects of 'noise' in the data to be accommodated in the parent–offspring relationships, and an important test would be whether the estimated values are consistent with mendelian transmission (see e.g., Demenais and Elston, 1981). Other forms of inheritance do exist (Chapter 14), and shared environments can induce non-mendelian correlations among family members, as could parental reproduction 'stopping' rules, and other factors.

Computer programs use efficient algorithms to search the possible values of all the parameters to find that set which maximizes the value of [5.12]. The likelihood evaluated with these MLE parameter values represents the best fit of the model to the data. But is this a good fit? Unfortunately, there is usually no goodness-of-fit test for the kinds of likelihood typically required in genetic inference. Family data are usually too complex for us to specify how extreme the deviation of the observed phenotype distribution is from some 'expectation'[2]. Instead, the significance of family data is tested by comparing the support those data provide for each of a set of plausible genetic hypotheses.

To do this, the likelihood ratio method is used; likelihoods for two competing hypotheses, each evaluated for its respective parameters' MLEs, are determined, and their ratio computed:

$$\lambda = \mathcal{L}_1/\mathcal{L}_2 = \mathcal{L}(\mathrm{H}_1|\mathrm{data})/\mathcal{L}(\mathrm{H}_2|\mathrm{data}). \qquad [5.13]$$

Here $\mathcal{L}_1$ is the likelihood in [5.12] for hypothesis 1; $\mathcal{L}_1$ is the product of the individual family likelihoods (or sum up the log-likelihoods), each computed under hypothesis 1. If the two models are *nested*, then as stated in Appendix 4.1, $2 \ln \lambda$ has an approximate chi-squared distribution with d.f. equal to the number of parameters by which the two tested models differ (see e.g., Ott, 1991).

For example, suppose one wants to test whether the data are more likely to have been sampled from a single-locus mode of control than a totally sporadic one. In the genetic model, we estimate the allele frequency p; the sporadic model can be specified by setting $p_A = 1$ (or, equivalently, specifying that all genotypes have the same penetrance). $2 \ln \mathcal{L}(p_{\mathrm{MLE}})/\mathcal{L}(p = 1)$ is a χ^2 with 1 d.f. When the models are not nested in this way, the distribution of the likelihood ratio can be approximated by simulation (Schork and Schork, 1989).

The likelihood ratio method compares the relative support for models that the author chooses to test, but of course can say nothing about models that are not tested but might have even higher support. Fortunately, experience has shown that a core repertoire of basic models does a good job of identifying strong genetic mechanisms in properly sampled data, even if some situations currently elude the method. In any case, a positive segregation result must be confirmed, especially given the level of heterogeneity that exists (as may be seen later).

Likelihood computation for hypothetical family

The basic idea can be illustrated from Table 5.3, which shows the terms in the likelihood (ignoring any ascertainment term in this simplified example), for a hypothetical diallelic dominant locus and a triad consisting of an affected father, unaffected mother, and affected child. Even in this simplest case there are $3^3 = 27$ possible genotypes. HWE genotype probabilities are used for founders and mendelian probabilities for the offspring, and the penetrances are applicable to each person's possible genotypes. These terms show how complex a full likelihood for a family,

Table 5.3. *Illustration of likelihood method for parent–offspring triad*

Family phenotypes: father affected, mother unaffected, daughter affected. Model: single diallelic locus, 'affected' dominant.

Possible genotypes			Genotype probabilities			Penetrance probabilities		
FA	MO	CH	FA	MO	CH	FA	MO	CH
AA	*AA*	*AA*	p^2	p^2	1	1	0	1
		Aa			0	1	0	1
		aa			0	1	0	0
	Aa	*AA*		$2pq$	$\frac{1}{2}$	1	0	1
		Aa			$\frac{1}{2}$	1	0	1
		aa			0	1	0	0
	aa	*AA*		q^2	0	1	1	1
		Aa			1	1	1	1
		aa			0	1	1	0
Aa	*AA*	*AA*	$2pq$	p^2	$\frac{1}{2}$	1	0	1
		Aa			$\frac{1}{2}$	1	0	1
		aa			0	1	0	0
	Aa	*AA*		$2pq$	$\frac{1}{4}$	1	0	1
		Aa			$\frac{1}{2}$	1	0	1
		aa			$\frac{1}{4}$	1	0	0
	aa	*AA*		q^2	0	1	1	1
		Aa			$\frac{1}{2}$	1	1	1
		aa			$\frac{1}{2}$	1	1	0
aa	*AA*	*AA*	q^2	p^2	0	0	0	1
		Aa			1	0	0	1
		aa			0	0	0	0
	Aa	*AA*		$2pq$	0	0	0	1
		Aa			$\frac{1}{2}$	0	0	1
		aa			$\frac{1}{2}$	0	0	0
	aa	*AA*		q^2	0	0	1	1
		Aa			0	0	1	1
		aa			1	0	1	0

Note: FA, father; MO, mother; CH, child.

or a more complex penetrance situation, will be and why computers must be used. (For another example, see Thompson (1986).)

Example: segregation analysis for breast cancer

Breast cancer is a serious health problem in most industrialized populations. The relative risk of breast cancer among relatives of a case is greater than 1, but evidence for any simple kind of genetic inheritance has been elusive; penetrance may be too low, at-risk individuals may not manifest the disease by chance or because they die of something else first, and so on. However, there may be a *precursor* trait with earlier onset that

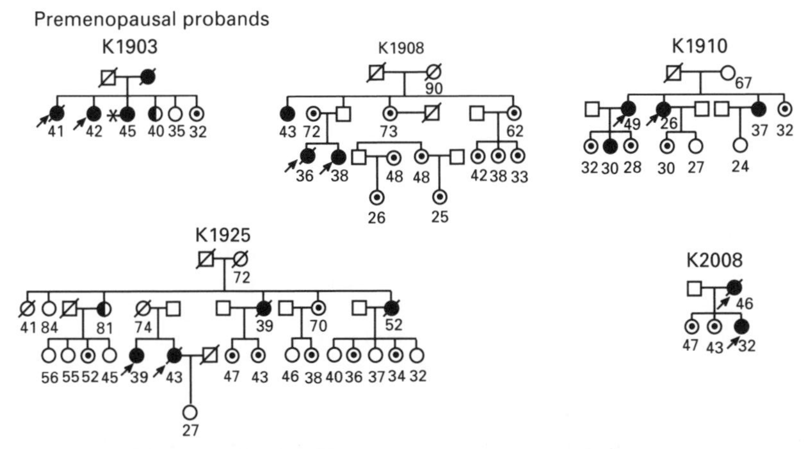

Figure 5.1. Sample of pedigrees in proliferative breast disease study. For explanation of symbols, see Appendix 5.1, p. 90. (From Skolnick *et al.*, 1990, *Science* **250**: 1715–1720. Copyright 1990 by the AAAS.)

may be informative of risk of later breast cancer. Proliferative breast disease (PBD), the benign excess growth of some epithelial cells in the breast duct tissue, may be such a precursor. But is PBD a heritable trait? To test this, a study was conducted of 19 pedigrees identified through a clinic in Utah (Skolnick *et al.*, 1990), a few of which are shown in Figure 5.1.

Segregation analysis was done with the likelihood method. Random mating was assumed to be relative to a single diallelic locus, with allele frequency p, for the susceptibility allele, and mendelian transmission probabilities. Male susceptibility status was always taken to be 'unknown' (penetrance set to 1 for all male phenotypes regardless of genotype), a way to ignore male phenotypes (but not genotypes) in the pedigree, because breast disease in males is so rare. Penetrance was specified as a function of age using the definitions of [3.3], with survival probabilities and hazard rates genotype specific, using a proportional hazard model (Chapter 3).

The likelihood analysis is given in Table 5.4, omitting some details from the original study. The table shows the estimated parameter values for a single-locus (1) dominant, (2) recessive, (3) purely sporadic, and (4) a general model in which all parameters were estimated. In the first two models, the penetrances for the two dominant genotypes were fixed to be equal to each other, but the other genotype was allowed to have non-zero penetrance. The sporadic model is specified by assigning the same penetrance to all 'genotypes'. The first three models are subsets of the

Table 5.4. *Segregation analysis for proliferative breast disease*

Model	Allele frequency	Parameter estimates (MLEs) penetrance			χ^2	d.f.
		Ω_{AA}	Ω_{Aa}	Ω_{aa}		
General	0.14	1.00	0.47	0.00		
Dominant	0.12	**0.59**	**0.59**	0.07	1.6	2
Recessive	0.47	**0.00**	**0.00**	0.60	2.1	2
Sporadic	1.00	**0.13**	**0.13**	**0.13**	64.3	6

Note: Bold type identifies parameters constrained by the model to be equal.
Source: Simplified from Skolnick *et al.*, 1990.

general model, in that the same parameters are included but the values for some of them are specified, and hence not estimated.

Because it involves more free (estimable) parameters, the likelihood of the general model will exceed that of the other models, and the test compares those models to the general one. The dominant and recessive models differ by two parameters from the general model, and generate approximately the same chi-squared log-likelihood ratio test relative to that model. Because this chi-squared is small, the general model does not fit significantly better than either of these simpler genetic models, despite having extra parameters to work with. By the principle of parsimony, an agreement among scientists to accept simpler models whenever possible, the single-locus model is preferred over the more elaborate model that does not fit appreciably better. The sporadic model, i.e., that there is no excess familial risk of either PBD or breast cancer, has a much lower likelihood than do the other models ($\chi^2 = 64.3$, 6 d.f.) and can be rejected relative to those models.

In this study, neither the dominant nor the recessive model receives stronger support than the other; a rare dominant with penetrance of about 0.6, and zero penetrance for the recessive genotype, generates almost the same likelihood as a common recessive with penetrance of 0.6 and zero for the dominant genotype. Thus, while the study suggests genetic effects on PBD, it does not help to resolve their mode of expression.

In the same study (not shown in the table) Skolnick *et al.* (1990) rejected single-locus susceptibility for breast cancer itself, but not for PBD, and concluded that they have found evidence for a locus leading to PBD and affecting breast cancer risk as a precursor. Having the susceptibility allele raises the breast cancer hazard by a factor of about 10. Of course one study does not a proof make, and a potentially important finding requires strong confirmation.

The general likelihood approach is the method of choice in most genetic epidemiological studies, although the theory and its application are far too complex to develop further in this book {Elston, 1981, 1986; Morton, 1982; Ott, 1991; Thompson, 1986}; current methods and details are in the human genetics journal literature and in the documentation of the computer packages for segregation analysis.

Families or genealogies?

How much data are needed for segregation and other family analysis? Should a few large, extended genealogies be collected, or a larger number of smaller ones? The answer is that each situation is different. The cost of ascertaining a family must be balanced against the added information multi-generation families can provide. Large families may have many uninformative branches (e.g., no one inheriting a risk allele), and the larger the family the more likely that various errors, such as non-paternity, may creep in.

If a disease is rare, large pedigrees are likely to be relatively homogeneous with respect to genetic risk, but if the causation is more complex or common, small pedigrees may be more likely only to be segregating a single high-risk allele. In large pedigrees, individuals in deep ancestral generations have died before diagnostic criteria were good; their phenotypes may be unknown or unreliable.

There is no simple answer. One possible strategy is *sequential sampling*: collect small, say three-generation pedigrees, and only extend these by collecting data in potentially informative branches of the family (see e.g., Boehnke *et al.*, 1988). Such extensions must be made only on the basis of data already collected; an individual is never ascertained because of his/her phenotype itself, but relatives of affecteds may be more informative as to risk than relatives of unaffecteds. For example, a family is ascertained from an affected proband. That person's father is unaffected, but his mother is affected. On a dominant single-locus hypothesis, it is most likely that only the mother carries the risk allele; her branch of the family would be followed up. This approach was used in the PBD study.

Epidemiological approaches: risk in relatives of affected individuals

The model-free measures of risk used commonly in epidemiology (Chapter 3) can be expressed in terms of genetic models, resulting in model-

Table 5.5. *Probability relative (Y) of proband (X) is affected as function of gene frequency and proband affection status*

	Probability relative is affected	
State of relative	Autosomal dominant	Autosomal recessive
Parent X–offspring Y		
Pr(Y Aff|X Aff)	$(1 + p - p^2)/(2 - p)$	q
Pr(Y Aff|X Nor)	p	$q^2/(1 + q)$
Sibling X–sibling Y		
Pr(Y Aff|X Aff)	$(4 + 5p - 6p^2 + p^3)/4(2 - p)$	$(1 + q)^2/4$
Pr(Y Aff|X Nor)	$p(4 - p)/4$	$q^2(3 + q)/(4(1 + q))$
Avuncles		
Pr(Y Aff|X Aff)	$\frac{1}{2}((1 - q^2) + (p/(1 - q^2)) + (pq/(1 + q)))$	$q(1 + q)/2$
Pr(Y Aff|X Nor)	$p(2 + q)/2$	$q^2p/(1 - q^2)$
First cousin X–first cousin Y		
Pr(Y Aff|X Aff)	$(p/4(1 - q^2))(4p^3 + q(15p^2 + 14pq + 1))$	$q(p + 4q)/4$
Pr (Y Aff|X Nor)	$p(4 + 3q)/4$	$(q^2p/(1 - q^2))(1 + 3q/4)$

Note: Based on the ITO-matrix method. p is the frequency of the dominant gene (so $q = 1 - p$ is frequency of the pathogenic allele for recessive trait).
Source: Modified from Weiss *et al.*, 1982.

based *RR*s and *SMR*s for relatives of affected probands. This can be useful since many non-genetic epidemiological studies often report phenotype data on first-degree relatives without collecting other family data; *RR*s often provide a first look at a problem.

Table 5.5 gives affection probabilities for various types of relative (parent–offspring, siblings, avuncles[3], first cousins) conditional on the affection status of a proband, under single-locus recessive and dominant autosomal genetic models. The values depend on allele frequencies, and are related to Snyder's ratios. The table was computed using the ITO method, an easy way to compute such probabilities (see e.g., Li, 1976; Majumder *et al.*, 1983; Weiss *et al.*, 1982).

The resulting *RR*s and *SMR*s in relatives exposed to a case's genes are given in Table 5.6 (Chakraborty *et al.*, 1982; Easton and Peto, 1990; Weiss *et al.*, 1982), and illustrated in Figure 5.2. For example, the *RR* for a dominant trait is $(1 + p - p^2)/p(2 - p)$ in offspring of an affected person; if the disease is rare ($p \approx 0$), the *RR* can be very high, but for a common trait, *RR* approaches only 1. Thus, even small *RR* values can be compatible with strong genetic control. The reason is that, for a fairly

Table 5.6. *Expected relative risks of disease in relatives of proband by proband affection status*

	Autosomal	
Relative types	Dominant	Recessive
Offspring (O) of probands (P)		
(O Aff|P Aff)/(Pop risk)	$(1 + p - p^2)/p(2 - p)^2$	$1/(1 - p)$
(O Aff|P Nor)/(Pop risk)	$1/(2 - p)$	$1/(2 - p)$
(O Aff|P Aff)/(O Aff|P Nor)	$(1 + p - p^2)/p(2 - p)$	$(2 - p)/(1 - p)$
Siblings (S) of probands (P)		
(S Aff|P Aff)/(Pop risk)	$(4 + 5p - 6p^2 + p^3)/4p(2 - p)^2$	$(2 - p)^2/4(1 - p)^2$
(S Aff|P Nor)/(Pop risk)	$(4 - p)/4(2 - p)$	$(4 - p)/4(2 - p)$
(S Aff|P Aff)/(S Aff|P Nor)	$(4 + 5p - 6p^2 + p^3)/p(4 - p)(2 - p)$	$(2 - p)^3/(1 - p)^2(4 - p)$

Note: p is the frequency of the dominant allele.
Source: Adapted from Chakraborty *et al.*, 1982.

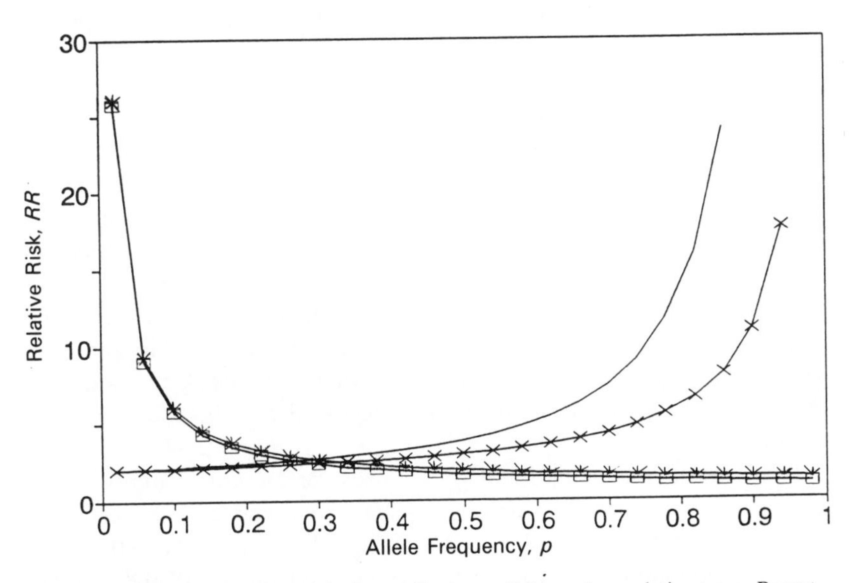

Figure 5.2. Relative risks for qualitative trait for various relative types: Parent–offspring dominant (□); sibling–sibling dominant (∗); parent–offspring recessive (×); sibling–sibling recessive (—).

common trait, knowing that a proband is affected does not provide much more information than starting from a random proband, relative to the risk in a relative – with large p even a random proband is likely to be a gene-bearer.

It can often be difficult to distinguish between different models of inheritance. This is especially true if all genotypes have some risk of the trait, there are substantial environmental effects, and/or there is incomplete penetrance (Edwards, 1969; Risch, 1990a; Vogel and Motulsky, 1986).

Epidemiologists should be aware of the risk of incorrect inferences in studies of elevated relative risks of disease in relatives that fail to consider both environmental and genetic factors (Khoury *et al.*, 1986). Indeed, for some diseases, the *RR*s in relatives are too high to be explained solely by the usual kinds of environmental risk factor, even if exposure to such factors is correlated among close family members (Khoury *et al.*, 1988). Many epidemiological methods usually applied model-free by epidemiologists can be adapted to increase our understanding of genetic traits (see e.g., Khoury *et al.*, 1986; Susser and Susser, 1989; Weiss *et al.*, 1982).

Caution: discrete traits need not have single-locus control!

Not all traits that seem to be specific and discrete need have single-locus control, and some traits that seem to be sporadic may in fact be purely genetic.

Emergenetic traits

Some traits caused by multiple locus genotypes can present difficult family patterns. A particular example is multiple-recessive traits. If the alleles are relatively rare at both loci, neither the parents nor the offspring of an affected person are likely to be affected. The traits will appear to 'emerge' and just as suddenly disappear (fail to recur), and for that reason have been called *emergenetic* (Li, 1987). Examples of multiple-recessive traits are prelingual deafness (Majumder *et al.*, 1989) and a skin condition known as vitiligo, which produces color blotches on the skin (Majumder *et al.*, 1988). A study of the latter suggested a four-locus recessive etiology.

If alleles at one or more of these recessive loci are common – for example, in some population isolates – the trait will be less 'emergenetic'. A two-locus recessive could 'become' a single-locus recessive in this way; that is, a population fixed for the recessive allele at one of the loci will appear to segregate risk only at the single remaining locus. There is evidence that IDDM may be due to variation at a locus in the insulin gene region as well as in the HLA DR4 locus, but will show clearer segregation patterns only when one of the alleles is fixed (see e.g., Julier *et al.*, 1991).

Conclusion

Segregation analysis has a long history of successes in recognizing simple single-locus disorders, sex-linked traits, and other 'easy' cases for which only a single major parameter – the segregation proportion – needs to be estimated. Subtle situations are often more problematic to solve by segregation analysis alone, and must be followed by linkage analysis.

This chapter has concentrated on models of discrete traits in nuclear families or pedigrees. However, it may sometimes be advantageous to use some other kind of sample. Each has its own literature too extensive to summarize here. One of the most often used types of sample is *twins* {Thompson and Thompson, 1986; Vogel and Motulsky, 1986}. Monozygous (MZ) twins are genetically identical and their phenotypic differences can be attributed to environmental or developmental factors, whereas the phenotypic differences between full siblings and differences between dizygous (DZ) twins can be attributed to both genetic and environmental effects.

Another common sampling strategy is to study small isolates, such as rural religious sects, or inbred populations. These groups sometimes have elevated frequencies of rare diseases due to chance events in their founding, and often have an elevated frequency of some recessive diseases due to the effects of inbreeding. Indeed, Garrod's (1908) pioneering studies of recessive metabolic defects, or 'inborn errors of metabolism', which launched formal genetic epidemiology, were based on their increased frequency in inbred individuals.

Notes

1. The resulting phenotype might depend on underlying categorical risk factors other than genotypes. Cannings *et al.* (1978) introduced the term *ousiotype* to refer to such categories; in most genetic usages, the investigator suspects that they are highly correlated with genotypes.
2. One might try simulating the distribution. Using the sample family structures, one can simulate the genotypes and phenotypes in the families by computer, assuming that the model in question is true. The parameter MLEs are estimated from this simulated data. The simulation is repeated hundreds of times, resulting in hundreds of different parameter estimates. If the MLEs derived from the *observed* data are very unusual relative to the simulated values, the data can be said to be unlikely relative to the model. Such simulation, however, is usually an impractical amount of work.

3. *Avuncle* is an apt term coined by Crow (1986) to denote uncle/aunt–nephew/niece relationships.

APPENDIX 5.1: Some miscellany

Some basic rules of probability

A few basic rules of probability are given in [A5.1] to [A5.3]. The first shows that for two exclusive events A and B, the probability that *either* will occur is the sum (or union) of the probabilities of the individual events [A5.1]. Similarly, the probability that *both* events occur, their *joint* probability, is the product (intersection) of their separate probabilities [A5.2]. The third rule specifies the *conditional probability*, that even A will occur given that event B has occurred [A5.3].

$$\mathrm{Pr(A\ or\ B) = Pr(A \cup B) = Pr(A) + Pr(B)} \qquad [A5.1]$$

$$\mathrm{Pr(A\ and\ B) = Pr(A \cap B) = Pr(AB) = Pr(A)Pr(B)} \qquad [A5.2]$$

$$\mathrm{Pr(A|B) = Pr(A\ and\ B)/Pr(B) = Pr(AB)/Pr(B).} \qquad [A5.3]$$

The genotype probabilities under HWE given in Chapter 4 were joint probabilities; for example, $p_{AA} = p_A p_A$. The probability for a heterozygote also uses unions: $p_{Aa} = p_{Aa} + p_{aA} = 2p_A p_a$. Equation [5.5] used a conditional probability to account for ascertainment bias.

Making and reading pedigree diagrams

For readers unfamiliar with pedigree diagrams, a few symbols are conventionally used. Figure 5.1 exemplifies some of these rules. Squares represent males, circles females. Horizontal lines between symbols represent a mating, and a vertical line from the mating to a 'comb' over several symbols represents the sibship of children descending from the mating. A diamond denotes multiple offspring with one symbol. An inverted V connects twins.

Probands are identified by arrows. Affection status is marked on the symbol; open symbols are unaffected, darkened symbols affected. Various partly darkened symbols can be used for intermediate phenotypes. Deceased individuals have diagonal lines through their symbol. Other information, such as age, can be listed under each individual.

For convenience, generations may be labeled with Roman numerals, individuals numbers in arabic, left to right within generation. Thus, individual 7 from generation II is denoted II-7. Siblings are arrayed in birth order from left to right.

Basic characteristics of mendelian (single locus) traits

The basic characteristics of traits caused by rare alleles at a single locus are summarized here, for readers who may not be aware of them.

Autosomal dominant: With complete penetrance all affected individuals have affected parents, except for rare cases caused by new mutations. Transmission does not skip generations. For rare traits, most affected children are produced by $Aa \times aa$ matings, with a risk of $\frac{1}{2}$ to each offspring. Both sexes are affected with equal probability.

Autosomal recessive: If rare, most matings that produce affecteds are between two normal *carriers*; that is, $Aa \times Aa$, with risk $\frac{1}{4}$ to each of their children. Consanguineous (inbred) matings are more likely to produce cases that are usually rare in other collateral relatives. Sexes are affected equally.

Sex-linked dominant: Affected mothers transmit equally to sons and daughters; if the trait is rare, most are heterozygotes, and the risk is $\frac{1}{2}$ to offspring. Affected (*hemizygous*) males transmit the trait to all daughters but never to sons. Except for new mutations, all affecteds have affected parents. If the trait is rare (q small, p near 1), prevalence in females will be approximately twice that in males: $(1 - p^2)/q = (1 - p)(1 + p)/(1 - p) = 1 + p \approx 2$.

Sex-linked recessive: If rare, relatives except maternal uncles and other male ancestors in the female ancestral line are unaffected. Affected males do not transmit the trait to any offspring, but the daughters are all carriers. Heterozygous carrier females transmit the trait to half their sons but, if the allele frequency is low, daughters are rarely affected (though $\frac{1}{2}$ are carriers). Affected females come mainly from matings of carrier females and affected males ($Aa \times ay$). Their frequency is about the square of that in males. Most affected males come from carrier females. The trait appears to skip generations.

Y-linked inheritance: Except for new mutations, all males are sons of affected fathers, and all their sons will be affected. The trait is never found in females.

Mitochondrial inheritance: Mitochondrial DNA (mtDNA) is inherited only through the maternal line. Fathers never transmit the trait. Affected mothers transmit the trait to all their children, of either sex.

It is a worthwhile exercise to draw a few sample pedigrees, label the parents with the genotypes appropriate to the preceding discussion, and trace inheritance by informally 'sampling' parental genes to produce the offspring. At least three generations should be included.

6 *Segregation analysis: quantitative traits in families*

One side will make you grow taller, and the other side will make you grow shorter. One side of *what*? The other side of *what*? thought Alice to herself.

L. Carroll, *Alice in Wonderland* (1865)

The previous chapter dealt with the genetic basis of categorical phenotypes such as the presence or absence of a disease. But how do we assess the genetics of *quantitative* phenotypes that cannot be broken down into distinct categories? Such traits may be affected by a large number of loci acting together, as well as by environmental factors. Important disease-related traits such as blood pressure, obesity measures, or cholesterol and triglyceride levels are examples. For such traits, what we need to understand is the *effect* of the genotypes, and the environment, on the phenotype.

This chapter introduces many important concepts and models for quantitative traits {Crow and Kimura, 1970; Falconer, 1989; Hartl and Clark, 1989; Hedrick, 1985}. The material is rather dense and relentless, but the concepts are important to understand, as they enable us to look further at the genetic causal spectrum, and to decompose it into its constituent parts.

Einstein's trains: assigning phenotypic effects to genotypes

Allelic and genotypic values

Albert Einstein made himself famous by looking out of a train window and speculating as to why he could not tell whether his train was moving north or the train on the next track was moving south. This led to his theory of relativity, which in essence stated that one motion can only be judged relative to a specified platform or frame of reference. The effects of genotypes on quantitative phenotypes are relative in a similar way. Does genotype AA increase the phenotype, or does aa decrease it?

The simplest measure of genetic effect is the *genotypic value*, the mean phenotype observed among individuals with a given genotype in the population of reference:

$$\mu_g = \sum_i \phi_i \Omega_g(\phi_i) \qquad [6.1]$$

(for continuous phenotypes, the major subject of this chapter, the summation would be replaced by an integral, $\int \phi \Omega_g(\phi)\, \mathrm{d}\phi$, over the range of possible phenotypes). Genotypic values in the population are estimated by their values in a random sample from the population.

The mean phenotype in the population is the weighted average of these values (summed over all genotypes),

$$\mu = \sum_g P_g \mu_g, \qquad [6.2]$$

which for a diallelic locus is

$$\mu = \Sigma\, P_g \mu_g = p^2 \mu_{AA} + 2pq\, \mu_{Aa} + q^2 \mu_{aa}.$$

Genetic variation in quantitative traits

The *genotypic variance* is defined as the variance among the genotypic values in the population:

$$\sigma_g^2 = \Sigma\, P_g(\mu_g - \mu)^2 = \Sigma\, P_g \mu_g^2 - \mu^2. \qquad [6.3]$$

The rightmost version is just a computationally easier expression.

Because it is the relative rather than absolute genotypic value that is important in any given population (and to evolution), it is often convenient to express genotypic values as *deviations* from the population mean, which we can denote by $\nu_g = \mu_g - \mu$.

In the simplest situation, which reasonably approximates many loci, the effects of the individual alleles are *additive*, and the genotypic value is the sum of the effects of the two alleles in the genotype, a direct dose–response relationship to the phenotype, much as Alice was able to regulate her height by the proper number of nibbles from the two sides of the caterpillar's mushroom.

In this case each allele contributes its own *allelic value*, α_i, to the genotype, which can be expressed as the average deviation of a phenotype containing the i-th allele from the population mean. Since allele A is paired with another A a fraction of p of the time, and with a for q of the time,

$$\left.\begin{aligned}\alpha_A &= p\nu_{AA} + q\nu_{Aa}\\ \alpha_a &= p\nu_{Aa} + q\nu_{aa}.\end{aligned}\right\} \qquad [6.4]$$

An important characteristic of effects expressed as deviations is that their average over all items in the population (e.g., genotypes) must be zero; here, this means $p\alpha_A + q\alpha_a = 0$. When the allelic effects are additive, the *breeding value*, or average deviation, of genotype ij is $\alpha_i + \alpha_j$ (the term derives from agricultural genetics). The sum of the squares of these values, weighted by the genotype frequencies, is the *additive genotypic variance*, σ_A^2

$$\sigma_A^2 = p^2(2\alpha_A)^2 + 2pq(\alpha_A + \alpha_a)^2 + q^2(2\alpha_a)^2 = 2(p\alpha_A^2 + q\alpha_a^2). \qquad [6.5]$$

The position of the heterozygote relative to the two homozygotes is known as the *dominance displacement*, $d = (\mu_{Aa} - \mu_{aa})/(\mu_{AA} - \mu_{aa})$; if the effects are purely additive, the heterozygote genotypic value will be exactly midway between those of the homozygote and $d = \frac{1}{2}$[1]. The *dominance variance*, i.e., the variance due to *dominance deviations* from additivity, equals

$$\sigma_D^2 = p^2(\nu_{AA} - 2\alpha_A)^2 + 2pq(\nu_{Aa} - \alpha_A - \alpha_a)^2 + q^2(\nu_{aa} - 2\alpha_a)^2. \qquad [6.6]$$

The mean number of 'doses' of a given allele, say A, in genotypes in a population is $\mu_g = 2p^2 + 2pq(1) + q^2(0) = 2p$, and its variance is $\sigma_g^2 = 2pq$, the heterozygosity at the locus.

Environmental effects on quantitative phenotypes

The genotypic values are only the *means* of the phenotypes found among individuals with a given genotype. As implied in Chapter 4, there is usually a distribution of phenotypes among those individuals (the penetrance function). What is responsible for this *within genotype* variance? One factor is certainly the variable exposure the individuals have to environmental factors such as diet, climate etc.

In Chapter 12 we see how to account for those environmental factors that we can identify and measure. Here, we deal only with the simplest way to account for environmental variance, namely as an aggregate of unmeasured *effects* on the phenotype. There is no formal theory for the distribution of exposures to such an aggregate of environmental effects, but it has proven satisfactory to assume that they have a normal distribution. Scaling this in terms of *deviations* from the mean environmental exposure, the probability density of environmental exposures, E, is $f(E|0,\sigma_E^2)$, here as elsewhere using the non-rigorous shorthand of employing the same notation (e.g., E) for a variable and its specific values. The variance (σ_E^2) is estimated from data.

We can now express the determination of the phenotype as

$$\phi = A + D + E, \qquad [6.7a]$$

with variance

$$\sigma_\phi^2 = \sigma_A^2 + \sigma_D^2 + \sigma_E^2. \qquad [6.7b]$$

The environmental effects can be additive, i.e., act similarly on each genotype, or there can be a *genotype by environment* ($G \times E$) *interaction* if the same environmental exposure affects different genotypes differently. If there is $G \times E$ interaction, an additional term, σ_{GE}^2, for the variance produced by that interaction is added to [6.7b].

The standardized effect of a locus

The effects of alleles at a diallelic locus on the phenotypic variance can be quantified in a way that is useful for comparing loci or even traits. The *standardized effect of a locus*, ξ, is defined as

$$\xi = (\mu_{AA} - \mu_{aa})/\sigma_{PG+E}; \qquad [6.8]$$

that is, the difference between the two homozygote genotypic values, expressed in phenotypic standard deviation units. In [6.8] σ includes only the variation *within* the genotypes, i.e., among individuals with the same genotype at the locus in question; this variation is presumably due to environmental variation, E, and to other unspecified genes, PG, (discussed below) that affect phenotypic variation. The standardized effect measures the strength of the 'signal' due to the locus in question compared to the 'noise' of the other factors. Some authors do this slightly differently, and use [6.8] with σ_ϕ, the entire phenotypic standard deviation, including variance due to the test locus itself, in the denominator (e.g., Falconer, 1989).

Table 6.1. *Genetic characteristics of apo-AI levels in Minnesota families*

Allele frequencies: $p_A = 0.806$, $q_a = 0.194$

Genotypic values: $\mu_{AA} = 127.97$
$\mu_{Aa} = 143.55$
$\mu_{aa} = 177.27$

Dominance displacement: $d = (143.55 - 127.97)/(177.27 - 127.97) = 0.316$

Allelic values: $\alpha_A = -3.71$, $\alpha_a = 15.39$

Genotypic mean = 134.7

Total phenotypic variance, $\sigma_\phi^2 = 402.35$

Total genotypic variance = 122.12
$\sigma_A^2 = 114.07$; $\sigma_D^2 = 8.05$

Source: Moll *et al.*, 1989 (based on best-fitting parameter estimates for model of single co-dominant locus, plus environment, no polygenes).

Example: apolipoprotein concentrations

Table 6.1 provides values for some of the above parameters for plasma levels of apo-AI, the major protein in HDL particles. These estimates were derived from segregation analysis of a sample of 1880 individuals from families in Rochester, Minnesota (Moll *et al.*, 1989). More on this study will be given below (Table 6.6, and Chapter 12).

Mendelian transmission and the quantitative resemblance between relatives

Just as mendelian transmission determines the probabilities of genotype sharing between parents and offspring, and by extension among all relatives, mendelian transmission also induces a quantitative relationship among relatives. The structure that this introduces is used to infer gene action for quantitative traits.

Kinship and inbreeding coefficients: probabilities of shared genes

Several quantities are used to measure the genetic relationship between two individuals. The *coefficient of kinship* (*co-ancestry*, or *consanguinity*), F_{AB}, between individuals A and B, is the probability that two alleles at the same locus, one chosen randomly from each individual, are *identical by descent* (*i.b.d.*) from some common ancestor (or from one relative to the other if they are related by lineal descent) {Crow and Kimura, 1970; Hartl and Clark, 1989; Hedrick, 1985; Jacquard, 1974; Li, 1976}. For example, a parent and child must share one allele by descent, so we can denote their genotypes Ax and Ay. If we randomly pick one of

Table 6.2. *Genetic relationships among various types of relative*

Relative type	'Degree of relationship'[a]	Coefficient of Kinship (F)[b]	Coefficient of Relationship (r)
MZ twins	—	—	1
Parent–offspring	1st	$\frac{1}{4}$	$\frac{1}{2}$
Full siblings (and DZ twins)	1st	$\frac{1}{4}$	$\frac{1}{2}$
Half-siblings	2nd	$\frac{1}{8}$	$\frac{1}{4}$
Avuncles[c]	2nd	$\frac{1}{8}$	$\frac{1}{4}$
Half-avuncles	3rd	$\frac{1}{16}$	$\frac{1}{8}$
First cousins	3rd	$\frac{1}{16}$	$\frac{1}{8}$
Double first cousins	2nd	$\frac{1}{8}$	$\frac{1}{4}$
Half first cousins	4th	$\frac{1}{32}$	$\frac{1}{16}$
First cousins once removed	4th	$\frac{1}{32}$	$\frac{1}{16}$
Second cousins	5th	$\frac{1}{64}$	$\frac{1}{32}$

Notes: [a] Degree of relationship is the common-use terminology for the relationship.
[b] Coefficient of kinship for two individuals equals the inbreeding coefficient of any children they might have.
[c] *Avuncle* refers to uncle/aunt–nephew/niece combinations (Crow, 1986).

the two alleles from each individual, the chance that we will get the A in both cases is $F_{PO} = \frac{1}{2} \times \frac{1}{2} = \frac{1}{4}$.

Related to this is the *inbreeding coefficient*, F, of an individual, the probability that his/her two alleles at a locus are i.b.d. Since an offspring is formed by conducting the 'kinship experiment' in its parents (i.e., randomly drawing one allele from each parent), the inbreeding coefficient of an individual equals the kinship coefficient of its parents. The *coefficient of relationship*, $r = 2F_{AB}$, is the fraction of genes shared i.b.d. by two individuals.

The inbreeding coefficient of someone in a known pedigree includes terms for all possible paths from the individual up through a common ancestor and down again. Table 6.2 gives kinship F coefficients for various important kinds of relative pair. However, two people may share many pathways not seen in a small pedigree (e.g., at any locus we are all descended from some common gene at some point in history) so in human genetics the kinship coefficient usually refers to close pedigree relationships.

Genotypic correlation between relatives

Genetic relationships can be used to express the phenotypic relationships among relatives once a model of allelic effects is assumed. The question is: what does the mendelian relationship between two individuals tell us

Table 6.3. *Parent–offspring relationships*

Parent			Offspring				
			Genotype probabilities				
Genotype	Probability	Value	*AA* 2	*Aa* 1	*aa* 0	Total	Mean value
AA	p^2	2	p	$1-p$	0	1.0	$p+1$
Aa	$2p(1-p)$	1	$p/2$	$\frac{1}{2}$	$(1-p)/2$	1.0	$p+\frac{1}{2}$
aa	$(1-p)^2$	0	0	p	$1-p$	1.0	p

about the relationship of their phenotypes, if the latter are genetically controlled?

To illustrate this we look first only at additive effects at a diallelic locus. First we define two general statistical measures. The *covariance* between two variables x and y is the average of the products of the deviation of x and y from their respective means, over all pairs x,y in the population:

$$\left.\begin{aligned}\mathrm{Cov}(x,y) &= \Sigma_i \Pr(x_i, y_i)(x_i - \mu_x)(y_i - \mu_y) \\ &= \Sigma \Pr(x_i, y_i)(x_i y_i) - \mu_x \mu_y.\end{aligned}\right\} \qquad [6.9]$$

The second version is an easier computational form (cf. [6.3]). The *correlation* between the variables is defined as

$$\rho(x,y) = \mathrm{Cov}(x,y)/\sigma_x \sigma_y, \qquad [6.10]$$

which removes the effects of measurement scale from the covariance. In a population in HWE, the genotype distributions in each generation (e.g., parents and offspring) are identical, and $\rho(\mathrm{P,O}) = \mathrm{Cov(P,O)}/\sigma_g^2$.

Now consider the genotypic values of parents and offspring, for an additive diallelic locus. For a locus with three genotypes, there are nine possible parent–offspring genotype pairs; Table 6.3 gives these probabilities.

For example, the probabilities for an *AA* parent and an *AA*, *Aa*, or *aa* child are p, $(1 - p)$, 0, respectively (first row in table): all offspring receive an *A* from the father, and they get an *A* from the mother (making them *AA*) with probability p; an *a* from their mother with probability $1 - p$; since they must receive an *A* from their father they cannot have the *aa* genotype.

Using genotypic dose values of 2, 1, and 0, and [6.9], the genotypic covariance between parent (P) and offspring (O) is

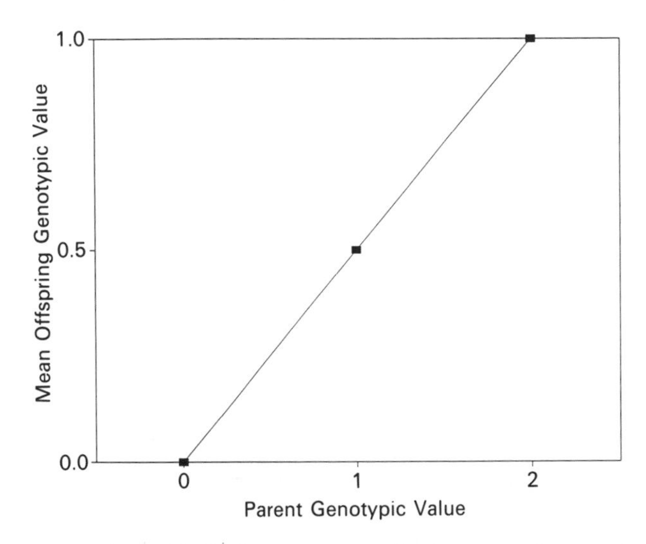

Figure 6.1. Single-locus parent–offspring genotypic regression.

$$\begin{aligned}\mathrm{Cov}_g(\mathrm{P,O}) &= [p^3(2)(2) + p^2(1-p)(2)(1) + p^3(0)(2)(0) \\ &\quad + 2p(1-p)(p/2)(1)(2) + 2p(1-p)(1/2)(1)(1) \\ &\quad + 2p(1-p)^2/2(1)(0) + (1-p)^2(0)(0)(2) \\ &\quad + (1-p)^2p(0)(1) + (1-p)^3(0)(0)] - (2p)(2p) \\ &= p(1-p) = (1/2)\sigma_g^2, \end{aligned} \qquad [6.11]$$

recalling that $\sigma_g^2 = 2p(1-p)$, from page 94.

Generally, the additive genetic covariance between two types of relative is defined by the kinship coefficient between them as $\mathrm{Cov}_g(A,B) = 2F_{\mathrm{AB}}\sigma_g^2$, and the genotypic correlation between two individuals is their coefficient of relationship, r. From [6.10] and [6.11], the genotypic parent–offspring correlation is $\rho_g(P,O) = \frac{1}{2}$. This is intuitively understandable, since parents and offspring share half of their genes. The deviation of an offspring genotypic value from its generation's mean is half that of the parent from his/her mean.

Figure 6.1 shows the parent–offspring relationship in a way that illustrates another important concept related to quantitative phenotypes. On the X-axis are the three phenotypes (0, 1, and 2) for the parent, and on the Y-axis the mean phenotypes for their offspring. These were given in the rightmost column of Table 6.3 in terms of an unspecified value of p;

Table 6.4. *Components of genetic covariance for various types of relative*

	Coefficient of	
Relative type	σ_A^2	σ_D^2
MZ twins	1	1
Full siblings (and DZ twins)	$\frac{1}{2}$	$\frac{1}{4}$
Parent–offspring	$\frac{1}{2}$	
Mid-parent–offspring	$\frac{1}{2}$	
Half-siblings	$\frac{1}{4}$	
Avuncles[a]	$\frac{1}{4}$	
Double first cousins	$\frac{1}{4}$	$\frac{1}{16}$
First cousins	$\frac{1}{8}$	
General	r[b]	u[b]

Notes: [a] '*Avuncles*' refers to uncle/aunt–nephew/niece combinations (Crow, 1986).
[b] r is the coefficient of relationship and u is defined in [6.13]; for details, see text.

the values on the Y-axis can be plotted relative to p, whatever that is. These points in Figure 6.1 lie on a straight line. This is the parent–offspring genotypic *regression*. The slope of the regression line, the *regression coefficient*, is defined by

$$b = \mathrm{Cov}_g(x, y)/\sigma_x^2. \qquad [6.12]$$

When the variance is the same in both generations,

$$b = \rho_g(\mathrm{P,O})\sqrt{\sigma_g \sigma_g}/\sigma_g^2 = \tfrac{1}{2};$$

that is, the slope of the regression line is equal to the correlation between parent and offspring.

The covariances between any pair of relatives, P and Q, can be expressed as a weighted combination of additive and dominance effects. Let the parents of P be denoted A and B, and of Q, C and D. Then,

$$\mathrm{Cov(P,Q)} = r_{\mathrm{PQ}}\,\sigma_{\mathrm{A}}^2 + u_{\mathrm{PQ}}\,\sigma_{\mathrm{D}}^2 \qquad [6.13]$$

where $u_{\mathrm{PQ}} = F_{\mathrm{AC}}F_{\mathrm{BD}} + F_{\mathrm{AD}}F_{\mathrm{BC}}$. Values for these coefficients are given in Table 6.4.

The additive or breeding value of a genotype was defined earlier as the average amount by which a gene from that individual deviates from the mean genotypic value. This is what an offspring inherits from a parent, so that the covariance between parent and offspring involves only the additive variance. This is true for other lineal relatives. However collateral relatives, such as siblings, share genotypes combining the breeding values of more than one person (e.g., two parents). The covariance among such relatives thus involves both additive and dominance variation. In such cases, the value of at least one term in u_{PQ} is non-zero.

Extension to multiple loci: polygenic traits

Biology was deeply split around the turn of the century over whether evolution proceeded through mutations for the kinds of discrete phenotype that Gregor Mendel had studied, or on quantitative traits such as height. A famous paper by R. A. Fisher, one of the founders of both modern statistics and population genetics (Fisher, 1918) showed that the single-locus genetic relationships among relatives were preserved for multiple additive loci, and that quantitative phenotypes could be reconciled with mendelian inheritance in this way. This unified mendelian and quantitative genetics and marked the beginning of the modern era in evolutionary biology and genetics.

The essence of the argument is as follows. If the allelic effects at a locus are additive, the genotypic values correspond to the sum of the doses provided by its various alleles. At a single locus, there are three genotypes and three genotypic 'dose' values (0, 1, and 2). If the phenotype is the sum of the genotypic values at two such loci, there are nine genotypes (e.g., *aabb*, *aabB*, . . . *AABB*), and five different genotypic values (0, 1, 2, 3, 4). In general, for n such loci there are 3^n genotypes and $2n + 1$ genotypic values. As the number of loci becomes even modestly large, the distribution of additive genotypic values more and more closely resembles the continuous distribution of a quantitative trait. Figure 6.2(a) illustrates this distribution for three loci, and (b) shows the corresponding normal distribution. In practice, the distribution of summed additive effects can be approximated by a normal distribution, and Alice could nibble her way to essentially *any* height.

Here we have introduced an important new assumption, namely *allelic equivalence*. The phenotype is equated to dose, but there is no longer a one-to-one relationship, or 'mapping', between specific genotypes and phenotypes. The same phenotype can arise in many ways (e.g., *Aabbcc*, *aaBbcc*, or *aabbCc* have phenotype 1). We can express only the *total*

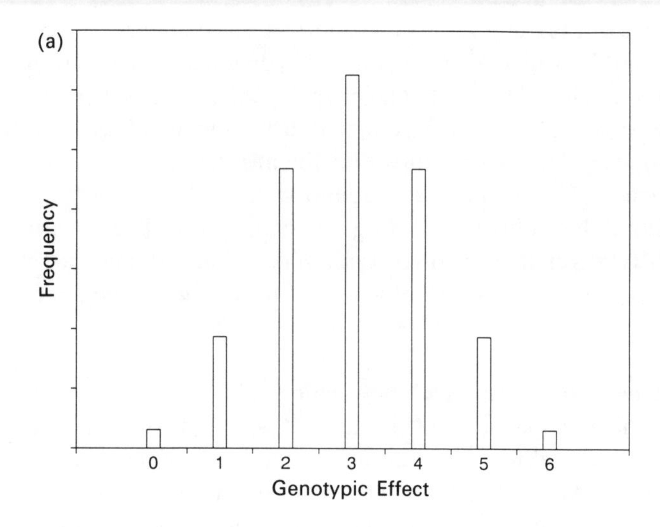

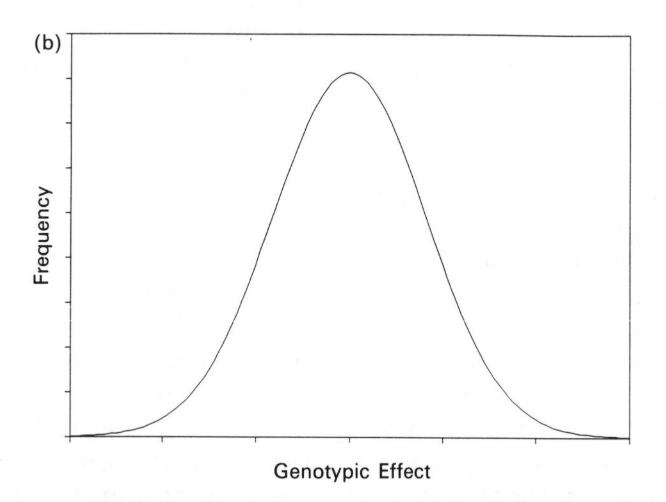

Figure 6.2. (a) Phenotype type distribution for three additive diallelic loci, compared to (b) normal distribution with same mean and variance.

probability of such a phenotype in the population as the sum of the probabilities of the individual multilocus genotypes that can generate it. Using the normal distribution as an approximation, the probability density of a given phenotype becomes $f(PG|\mu,\sigma^2)$, where the mean and variance are determined empirically from the population. Traits modeled in this way as the contribution of the effects at many loci treated in aggregate are called *polygenic* traits (hence, *PG* for polygenic effects).

Polygenic correlation between relatives

The genotypic correlations between relatives discussed above also hold for multiple additive loci. Intuitively, if a parent has a genotypic value of, say n allele-doses, his/her average offspring inherits half, or $n/2$ of these. With random mating, the average spouse has a genotypic value of μ doses, and provides $\mu/2$ to the offspring, who thus has genotypic value $(n + \mu)/2$. The parent's genotypic value differs from the population mean by $n - \mu$ and the offspring's by $(n + \mu)/2 - \mu = (n - \mu)/2$; that is, the offspring deviates by half as much as the parent.

The polygenic parent–offspring genotypic regression is $\frac{1}{2}$, as is their genotypic correlation. The mean offspring genotypic value is the *mid-parent* value, $(PG_m + PG_f)/2$.

Epistasis

Dominance refers to non-additive (*interaction*) effects between alleles at the same locus. *Epistasis* is the term applied to interactions among alleles at *different* loci, which, when present, add an additional term to the expression for the determination of the phenotype:

$$\phi = PG + E = A + D + I + E, \quad [6.14a]$$

with variance

$$\sigma_\phi^2 = \sigma_{PG}^2 + \sigma_E^2 = \sigma_A^2 + \sigma_D^2 + \sigma_I^2 + \sigma_E^2 \quad [6.14b]$$

Epistatic effects must be biologically fundamental. For example, structural or enzyme-coding gene expression is controlled by other, regulatory gene(s), and there is no reason why variation at these loci needs to act additively to produce the structural or enzymatic product. However, we have little *statistical* evidence on epistatic effects in humans. Probably, only loci with very marked effects on a trait will typically also have marked epistatic effects.

The covariances, correlations, and regressions discussed above are estimated from samples of data, usually by equating the unknown true values to the values of these parameters in random samples from the population. The machinery of linear statistics models (standard biostatistics) has been developed to make such estimations from various kinds of data (see e.g., Sokal and Rohlf, 1981).

Demonstrating a role for genes in a quantitative trait

When polygenic effects are treated as a normally distributed aggregate (here expressed as deviations from the mean), and environmental deviations are also assumed to have a normal distribution, the phenotype distribution in the population has the form

$$\begin{aligned}\Pr(\phi) &= \int_{PG} P_{PG}\,\Omega_{PG}(\phi)\,\mathrm{d}PG \\ &= \int_{PG} f(PG|0,\sigma^2_{PG})f(\phi - PG|0,\sigma^2_E)\,\mathrm{d}PG \\ &= f(\phi|0,\sigma^2_{PG+E}).\end{aligned} \qquad [6.15]$$

Here, we are treating the genotypic distribution as continuous. Under that assumption, $f(PG)$ refers only to an exact PG value, for which there is no 'probability', and is thus referred to as a probability 'density'; the integral (with its accompanying dPG) is used to denote summation of a continuous variable over an interval, which must be done to relate the density to our usual sense of 'probability' (see e.g., Sokal and Rohlf, 1981). Recall also that the integration limits ($-\infty$ to ∞) are a reasonable approximation to reality. The penetrance term in the middle line means that, given the polygenotype, phenotype ϕ results if the environment $E = \phi - PG$.

It would seem difficult to discern much about the nature of genetic control of a trait if we have to treat genetic effects as an aggregate. However, polygenes must still follow mendelian rules, and generate predictions about phenotypic relationships among relatives for traits that are genetically controlled.

The concept of heritability

For a purely additive polygenic trait ($\sigma^2_D = \sigma^2_I = 0$), we can rewrite [6.14], to express the *fraction* of the phenotypic variance that is due to genetic and environmental effects:

$$1 = \sigma^2_{PG}/\sigma^2_\phi + \sigma^2_E/\sigma^2_\phi. \qquad [6.16]$$

The first term is known as the *heritability*, $h^2 = \sigma^2_{PG}/\sigma^2_\phi$. Dominance and other complications can be accounted for (Falconer, 1989).

The heritability is strictly relative and population specific; if one of the components in [6.16] changes, the heritability changes. For example, if the distribution of exposure to an environmental factor changes, the heritability will change even in the same population, with the same set of

Table 6.5. *Heritability estimates for systolic blood pressure*

Population	h^2	Reference
Montreal	0.34	Annest *et al.*, 1979
Northeastern Brazil	0.41	Krieger *et al.*, 1980
Tecumseh, Michigan	0.42	Longini *et al.*, 1984
Detroit		
Whites	0.32	Moll *et al.*, 1983
Blacks	0.13	Moll *et al.*, 1983
Japanese-Americans	0.24	Morton *et al.*, 1980
Tokelau Islands		
Migrants	0.34	Ward *et al.*, 1980
Non-migrants	0.22	Ward *et al.*, 1980

genotypes. It is thus interesting that for many traits, such as blood pressure, the heritability is similar across populations that live in presumably different environmental circumstances (Table 6.5). The reason may involve a balance among evolutionary forces (Chapter 9).

Heritability can be estimated by two kinds of data, as follows.

Heritability: response to selection

In experimental and agricultural genetics, heritability can be estimated directly from the extent to which a trait responds to selection under controlled conditions. Individuals from a selected part of the phenotype distribution in one generation are selected as the parents of the next generation, as occurs when a breeder chooses, say, the woolliest sheep. The distribution of the progeny phenotypes will be shifted from that of the whole parental generation by an amount that is related to the intensity of the selection exercised in choosing the parents, and the genetic component of the phenotypic variance.

The *intensity of selection* is the difference between the population mean phenotype and the mean among the chosen parents, $S = \mu_S - \mu_\phi$. The *response to selection* is the difference between the original population mean and that in offspring of the selected parents, $M = \mu_o - \mu_\phi$. These two parameters of a selection experiment are related to the heritability by

$$M = Sh^2. \qquad [6.17]$$

Artificial breeding might seem an irrelevant subject for a book about human beings, because we do not usually manipulate marriages with

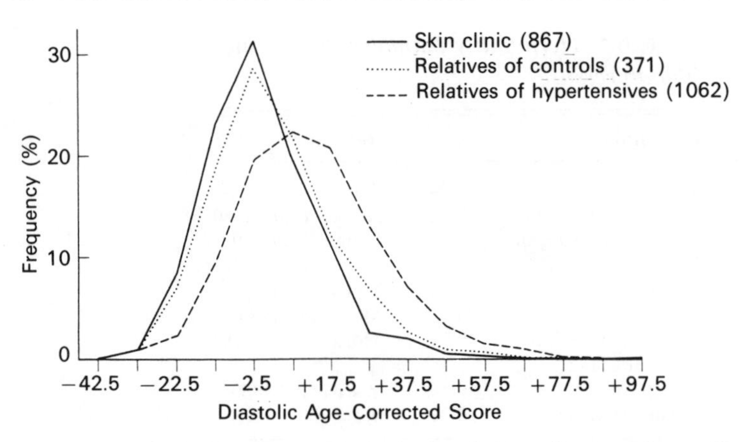

Figure 6.3. Blood pressure distribution in relatives of hypertensives and in the normal population. (From Hamilton *et al.*, 1954.)

specific genetic objectives in mind. However, routinely we do an 'artificial breeding' experiment just by sampling. If individuals are ascertained because they are in a restricted part of the phenotype distribution in one generation (e.g., hypertensives), their offspring simulate the offspring of a selective breeding experiment. This has been a common approach in genetic epidemiology; an example of the associated shift in the phenotype distribution is shown in Figure 6.3.

Heritability: similarities between relatives

Heritability can also be estimated in humans by taking advantage of the correlation between relatives in naturally occurring families. In general, ignoring selection, dominance variation, and other complications, the heritability is defined as the ratio of the *observed phenotypic correlation* to the *theoretical genotypic correlation*.

$$\rho(A,B) = 2F_{AB}h^2 \quad \text{or} \quad h^2 = \rho(A,B)/2F_{AB}. \qquad [6.18]$$

The parent–offspring regression is

$$b_{PO} = \mathrm{Cov}(P,\mathrm{O})/\sigma_\phi^2 = (1/2)\sigma_G^2/\sigma_\phi^2 = (1/2)h^2. \qquad [6.19]$$

Figure 6.1 shows the parent–offspring regression for a 'perfect' system, in which every offspring inherited the 'average' genotye from its parent. In actual data, there are environmental effects and so on, and it is from the scatter of the actual sampled points (p_i, o_i), each consisting of a parent and an offspring phenotype, that we estimate the regression between the relatives, and, on the basis of their known mendelian relationship, the heritability.

Other methods for polygenic inference

Heritability concepts can be tested in other kinds of data. For example, the genetic variance between MZ (identical) *twins* is zero, but DZ twins are as variable as full siblings. The heritability of a quantitative trait is $h^2 = (\sigma^2_{DZ} - \sigma^2_{MZ})/\sigma^2_{DZ}$ (see e.g., Plomin *et al.*, 1990). Twins share environments to a greater extent than siblings sometimes do, but this can be controlled by studying the natural experiment in which twins are reared apart (e.g., in adopted homes). The twins' phenotypes are compared to those of their natural siblings, natural parents, adopting parents, and 'adopting' siblings. Siblings reared apart can also provide a control on common sibship environmental effects. The children of twins are genetic half-siblings reared in separate environments, and can provide an informative natural experiment.

Some qualitative traits seem to be due to the action of underlying polygenic mechanisms. Such traits as cleft lip/palate, pyloric stenosis, and some chronic diseases such as stroke, are familial but do not segregate in a neat mendelian way. These can sometimes be modeled as *threshold traits*. A normally distributed polygenically determined risk, known as the *liability*, is assumed to exist, and the trait appears in persons whose liability exceeds some threshold value, T. *Meristic* traits, or counts, such as the number of vertebrae in fish or of flower petals also may be produced in this way.

In this model, the prevalence of the trait is the fraction of individuals whose liability exceeds the threshold; that is, the area of the liability distribution above T. Liability is not measured directly, but the prevalence in offspring of a series of affected probands mimics the design of a selection experiment (described above), from which heritability can be estimated {Elston, 1981; Falconer, 1989}.

Understanding more of the causal spectrum

Polygenic models as described above treat genetic factors as an aggregate of unidentified locus-specific effects modified by the environment. Obviously, it is important to try to identify loci that are responsible for phenotypic variation, or at least whether they make a substantial contribution to that variation. That is what working out the causal spectrum is all about.

Mutations vary in the strength of their effects. Most of these seem to be quite small (Chapter 9), but we would like at least to be able to identify the existence of those mutations that have large effects. Loci segregating alleles with substantially different effects on a trait are known as *quantitative trait loci* (*QTLs*). Chapter 8 discusses how to find QTLs once their existence is known. Here, I describe segregation analytic methods that infer, from phenotypic data in families, that such loci exist. These methods work well if the standardized effect of the locus (defined above) is substantial (a rule of thumb has been $\xi > 2.5$); that is, if there is little overlap between the penetrance functions of at least two genotypes (usually, the homozygotes). Such loci are known in human genetics as *major genes*. Although the allele associated with disease usually has a low frequency in the population, its frequency may be high in families that are ascertained through probands with extreme phenotypes (e.g., hypertensives, persons with elevated cholesterol levels).

Segregation analysis and the mixed model

We assume that in a sample of families only one major locus (or at most some small specified set of specific genotypes) has major effects on the trait, and all the remaining genetic effects are assumed to be *individually* so small by comparison that they can be treated as a polygenic aggregate. The model, which also allows for environmental effects, is

$$\phi = G + PG + E, \qquad [6.20]$$

where G refers to the set of major locus (or loci) *oligogenotypes*. This is known as a *mixed model*, because it includes identified major genotypes as well as aggregate polygenotypes and environmental effects {Elston, 1981, 1986; Elston and Stewart, 1971; Lalouel *et al.*, 1983; Morton, 1982; Morton and Maclean, 1974}.

To test whether the mixed model is a reasonable explanation of the distribution of quantitative phenotypes family data, I extend the general likelihood method given in Chapter 5 for qualitative traits. The polygenic and environmental effects are usually expressed as normally distributed deviations from their mean effects in the population, i.e., $\mathrm{Nor}(0,\sigma^2_{PG})$ and $\mathrm{Nor}(0,\sigma^2_E)$. These are assumed to be uncorrelated with major-locus genotypes; that is, the distribution of PG and E effects is assumed to be the same for all major genotypes. Epistasis, dominance, and familial environmental correlations can be added but will not be discussed here.

For each individual in the family, we use the same likelihood ideas developed in Chapter 5, but now we have to specify probabilities for both the major locus oligogenotypes and the polygenotypes. Under the usual assumptions, the terms are:

(1) *Unconditional genotype probabilities for founders, probands, and spouses*: The oligogenotype G and polygenotype PG are assumed to be unlinked and independent. The joint probability density, for genotypes in founders of the pedigree is

$$P_G f(PG|0,\sigma^2_{PG}). \qquad [6.21]$$

The first term is just the familiar genotype probability for the major locus (see e.g., Chapter 5). The second is the probability density for polygenotype deviation of the exact amount PG.

(2) *Conditional genotype specifications for other family members*: Transition terms must be included to relate major locus and polygenic genotypes between parents and offspring. For the major locus, the transition probabilities given in Chapter 5 are used (sometimes, investigators choose to estimate these rather specifying mendelian probabilities). The offspring's polygenotype distribution, given the PGs of his/her parents, is a normal distribution, whose mean as we saw earlier equals the mid-parent; that is, $PG_o = (1/2)PG_f + (1/2)PG_m$. In each parent, the polygenic variance is σ^2_{PG}. However, a general result in statistics is that if variable $z = ax + by$, then $\sigma^2_z = a^2\sigma^2_x + b^2\sigma^2_y$. Thus, the variance for the offspring is $\sigma^2_o = \frac{1}{4}\sigma^2_{PG,f} + \frac{1}{4}\sigma^2_{PG,m} = \sigma^2_{PG}/2$. Thus, we have the conditional offspring genotype distribution

$$\begin{aligned}&\Pr[(o|f,m) \quad \text{and} \quad (PG_o|PG_f,PG_m)]\\ &\quad = P_{o|f,m} f(PG_o|(PG_f + PG_m)/2,\sigma^2_{PG}/2).\end{aligned} \qquad [6.22]$$

The constraints placed on this by the polygenic assumptions can be relaxed, for example by the regressive models discussed below.

(3) *Penetrance probabilities*: An individual with genotypic values G and PG can be observed to have phenotype ϕ if his/her environmental exposure has an effect $E = \phi - (\mu_G + PG)$. Since we assume that the environmental exposures are normally distributed, the penetrance probability density is

$$\Omega_{G,PG}(\phi) = f(\phi - (\mu_G + PG)|0,\sigma^2_E). \qquad [6.23]$$

Given all this, for a diallelic locus the likelihood term for a founder is

$$\begin{aligned}\Pr(\phi) \propto & \int_{PG} p^2 f(PG|0,\sigma^2_{PG}) f(\phi - (\mu_{AA} + PG)|0,\sigma^2_E)\,\mathrm{d}PG \\ & + \int_{PG} 2pqf(PG|0,\sigma^2_{PG}) f[\phi - (\mu_{Aa} + PG)|0,\sigma^2_E]\,\mathrm{d}PG \\ & + \int_{PG} q^2 f(PG|0,\sigma^2_{PG}) f[\phi - (\mu_{aa} + PG)|0,\sigma^2_E]\,\mathrm{d}PG \\ = & \, p^2 f(\phi - \mu_{AA}|\sigma^2) + 2pqf(\phi - \mu_{Aa}|\sigma^2) + q^2 f(\phi - \mu_{aa}|\sigma^2), \quad [6.24]\end{aligned}$$

where $\sigma^2 = \sigma^2_{PG+E}$. Here, in addition to the sums of [5.10] for the major genotypes, there is also an integral that ranges over all polygenotypes.

The final likelihood for a family is

$$\begin{aligned}\mathcal{L} \propto & \sum_{G_1} \cdots \sum_{G_n} \int_{PG_1} \cdots \int_{PG_n} [\Pi_{\text{all founders}} P_G \Pr_i(PG_i) \\ & \times \Pi_{\text{all non-founders}} \Pr(G_j|G_{mj},G_{fj}) \Pr(PG_j|PG_{fj},PG_{mj}) \\ & \times \Pi_{\text{all individuals}} \Omega(\phi_k|G_k,PG_k,\boldsymbol{x}_k)]\,\mathrm{d}PG_1 \ldots \mathrm{d}PG_n. \quad [6.25]\end{aligned}$$

This has one sum and one integral for each individual, corresponding to his/her possible oligo- and polygenotypes. There is one penetrance term for each individual. I included $\boldsymbol{x}$ in the penetrance to suggest that other risk factors, measured on individuals, might be considered.

The parameters we have to estimate in this model are the major-locus allele frequencies, σ^2_{PG}, σ^2_E, and μ_Gs and τs if they are not specified with mendelian values. Different authors and program packages use numerically equivalent variations of this model[2,3].

(4) *Ascertainment bias correction*: The ascertainment bias correction is as described in Chapter 5, a likelihood such as [6.25] in form but restricted to those individuals that were responsible for the ascertainment of the pedigree.

Statistical inference with the mixed model

The mixed model contains many other plausible models as subsets. A systematic approach to parameter estimation and hypothesis testing that takes advantage of this is known as *complex segregation analysis* {Elston, 1981, 1986; Lalouel *et al.*, 1983}. The likelihood is first computed for the full model and then for a set of sub-models and likelihood ratio criteria, based on the chi-squared approximation (see Appendix 4.1) are used to evaluate the relative support the data offer to the different models. For example, if the allele frequencies at the major locus are set to 1 and 0, or all the oligogenotype means are set equal, there is no major gene effect, and the likelihood in [6.25] reduces to the polygenic model (no discernible major-locus effects). The latter could be removed from the model by setting $\sigma^2_{PG} = 0$. Removing both leaves only environmental effects.

Most program packages are designed specifically to evaluate sets of nested models, for which the convenient chi-squared ($2 \ln (\mathscr{L}_1/\mathscr{L}_2)$, where $\mathscr{L}_1$ is the less restricted of the two hypotheses, i.e. has fewer variables with constrained ranges) provides an approximate significance test for hypothesis testing. Unnested models can be compared by simulation methods (see e.g., Schork and Schork, 1989). This is a powerful tool in contemporary statistics that is finding increased use in human genetics for problems with complex models. Essentially, in this case, the models to be compared are *simulated* in families with the same structure as those observed, many times, and the likelihood ratio of these simulated results is compared to the one observed in the real data. If the observed λ is larger than some chosen $100\,\alpha\%$ of the simulated λs, it is taken as significant support for one hypothesis over the other.

The penetrance functions in the likelihood can be modified to account for threshold phenotypes, age- and sex-effects, and other complications, readily possible with available programs. However, each study may require its own tailoring.

Example of the use of mixed modeling for segregation analysis

Complex segregation analysis was done on the apo-A1 levels in members of randomly ascertained Minnesota families (Moll *et al.*, 1989). Table 6.6 shows the parameter estimates and likelihood results for several of the models tested by Moll *et al.* in this study (in the full paper, the authors pursued details of the data not given here). The authors ask what model best fits the distribution of plasma apo-A1 levels in families ascertained through random children in the population. The table compares the likelihoods for several models, each evaluated at the MLEs for their respective parameters.

Parameter values fixed by the investigators to specify a particular model are in parentheses. The general model allows all parameters to 'float' (be estimated rather than being assigned user-specified values). The 'major environmental model' stipulates a non-transmitted factor (NTF); that is, an environmental determinant with major effects to which every individual is exposed with the same probability, regardless of position in the pedigree. A single-locus model with polygenes stipulates

Table 6.6. *Analytic results from the mixed model: apolipoprotein-AI levels*

Parameter	General	Major environmental	Mixed Dominant	Mixed Co-dominant	Polygene [sporadic]
μ_{AA}	129.65	134.66	133.39	131.06	134.6
μ_{Aa}	143.55	129.90	133.39	140.59	134.6
μ_{aa}	182.27	175.47	178.05	184.48	134.6
p	0.784	0.818	0.829	0.851	1.0
h^2	0.345	0.418	0.369	0.292	[0.367]
σ	17.46	18.31	18.59	18.16	19.96
$\tau_{A/AA}$	0.922	(p)	(1.0)	(1.0)	—
$\tau_{A/Aa}$	0.734	(p)	(0.5)	(0.5)	—
$\tau_{A/aa}$	0.262	(p)	(0.0)	(0.0)	—
log ($\mathscr{L}$)	−8230.49	−8232.91	−8234.7	−8233.18	−8253.15 [−8295.54]
χ^2 from general		4.85	3.08	5.38	45.3*** [130.1***]
d.f. from general		3	2	3	6 [7]

Note: *** significant at $p = 0.001$.
Source: Moll *et al.*, 1989.

mendelian transmission probabilities. The simplest models are the last two, which generate a simple normal phenotype distribution in the population, the difference being that there is a pattern of inheritance in the polygenic model, but not in the sporadic model.

Comparison of the log-likelihoods of these various models gives the relative support that they receive from the data. Recall that the significance of a difference in log-likelihoods is related to the number of d.f. by which the tested models differ. The general model is taken as the standard, because it invokes more 'free' parameters than any of the restricted models to explain the same data.

Models fitting significantly worse than the general model are identified by stars. Single-locus, polygenic, solely environmental and various combination models were tested (several other models tested by the authors are not shown here). Of the models shown, only the full mixed models, with mendelian-constrained transmission parameters, fit as well as the general model. Models that involve fewer parameters are usually preferred. However, it was not possible to choose between a dominant and co-dominant model based on these data.

Figure 6.4 plots the actual 1880 apo-AI levels (data supplied courtesy of P. P. Moll) and the *population* predictions of the general, mixed dominant, and purely polygenic or sporadic models. This shows that all of these models generate similar-looking distributions, which is why family data are needed. Also, note that, although the frequency of the rarer allele is only about 0.15, so that the population distribution (or histogram of actual data) does not appear to be multimodal, the separations of the genotypic values (e.g., $\mu_{AA} - \mu_{Aa} = 0.5\sigma$, $\mu_{AA} - \mu_{aa} = 2.9\sigma$) are quite substantial and would be easy to detect in enriched samples, for example families ascertained through probands with high apo-AI levels. The full report by Moll *et al.* (1989) provides a much more thorough and subtle analysis than does the summary presented here.

Tractability, flexibility, and 'regressive' models

The mixed model, which has led to great advances in human genetics, especially when combined with linkage analysis, was made possible by advances in the speed and capacity of computers and by software developed to make the required calculations reasonably efficiently. However, for large pedigrees or many pedigrees, the computational problems can still be a problem. One reason is the restrictive assumptions we make about non-oligogenic factors, especially polygenes, and their constrained behavior in families.

Easier and more flexible models have been developed. These are known as *regressive* models {Bonney, 1988; Bonney *et al.*, 1988, 1989}. Rather than make explicit polygenic assumptions, the penetrance function in these newer models is changed to a regression of the form $\Pr(\phi|G,Y_A,X)$, where G is the genotype of the individual, Y_A are the observed phenotypes of a set of ancestors of the individual, and X is a set of other measured risk factor variables.

Bonney has developed several types of model in which, for example, the set A of relatives includes only the spouse, the spouse and parents, those plus older siblings, etc. The algebraic form of this function can be likened to the regressions given in Chapter 3, and can vary depending on which relatives' phenotypes and which risk factors (e.g., sex) are included. Logistic regression is used for categorical phenotypes. More than one outcome phenotype can be included, and the regressive model allows complex family relationships, and spouse correlations, to be modeled (see e.g., Bonney *et al.*, 1988).

For example, with no assortative mating (no phenotypic correlation between spouses) conditional on the j-th individual's major genotype G, we might have:

$$\phi_j = \mu_G + \beta_f\phi_{fj} + \beta_m\phi_{mj} + \Sigma\beta_i E_{ij}, \qquad [6.26]$$

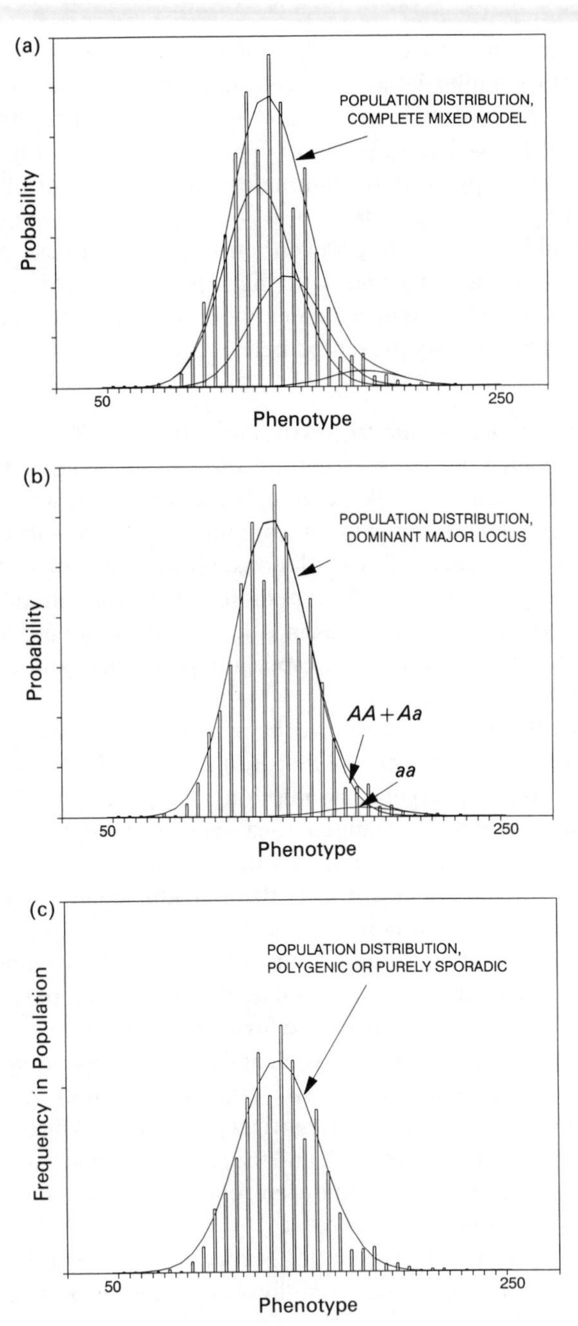

Figure 6.4. (a) to (c) Plot of expected population distribution of apo-AI levels under various fitted models. (From Moll *et al.*, 1989.)

with one regression coefficient to estimate for the effect of father's phenotype on the offspring, one for mother's, one for each measured environmental variable, plus a residual into which fall other unmeasured effects. In the simplest model all variables are measured (directly observed) except the major genotype. The regression parameters must be estimated. A detailed comparison of the standard polygenic models and regression models was given by Demenais and Bonney (1989), who showed the formal equivalence of the two approaches under some conditions, and that the simpler and more flexible regressive models will correctly identify the existence of a major gene.

A comment

In recent years, methods such as the general likelihood for pedigree data have enabled geneticists to identify many genes that have statistically significant effects on quantitative phenotypes such as lipid levels. However, progress in understanding the genetics of such phenotypes has been slow, because their causation is complex and heterogeneous.

Real traits are caused by a much more mixed process than the simple oligogenic + small polygenic model given above, although in some families only one major locus may be at work. Heterogeneity is likely, with different major genes segregating in different families or populations. Such situations place a premium on our methods of sampling and analysis, the topic of several later chapters in this book. In these chapters there is more on the importance of the relativistic nature of our models for quantitative effects.

Notes

1. Some authors parameterize dominance effects by assigning relative values of z, $z + td$, and $z + t$, to the three genotypes at a locus, where $z = \mu_{aa}$, $t = \mu_{AA} - \mu_{aa}$, and d is the dominance displacement.
2. *Segregation methods*: For the polygenic and environmental component of the mixed model, some programs estimate the variance, σ^2_{PG+E}, and heritability, $h^2 = \sigma^2_{PG}/\sigma^2_{PG+E}$, within each major genotype; some authors alternatively use the term heritability to refer to our $\sigma^2_{PG}/\sigma^2_{G+PG+E}$ (e.g., Lalouel *et al.*, 1983, 1985).

   Finding the parameter MLEs in the family likelihood in [6.25] involves one summation and one integration for each member of the pedigree, for every trial set of parameter values tested. For k oligogenotypes and n pedigree members, this means k^n major-locus genotypes and n integrals for the pedigree members in the data set (including all families). This is done numerically. However, the numerical demands are very great and many implementations use an approximation for the integrals due to Hasstedt (1982).

One simplification that is sometimes useful is to rewrite the likelihood of the purely polygenic model as a multivariate normal density, a computationally easier, but numerically equivalent form that does not involve multiple integration (see e.g., Lange *et al.*, 1976; Ott, 1991):

$$\{(2\pi)^n|\Sigma|\}^{-1/2} \exp[-\{(\boldsymbol{\phi} - \boldsymbol{\mu}\}' \Sigma^{-1} (\boldsymbol{\phi} - \boldsymbol{\mu})\}/2], \qquad [6.27]$$

where $\boldsymbol{\phi} - \boldsymbol{\mu}$ is an n-dimensional vector of deviations of the phenotype of each of the n family members from the mean. The covariance matrix, Σ, is an $n \times n$ matrix whose elements are $2F_{ij}\sigma^2_{PG} + \sigma^2_E\text{I}$ where F is the kinship coefficient between family members i and j, and I is the identity matrix (1s on the diagonal, 0s elsewhere). The notation $(\boldsymbol{\phi} - \boldsymbol{\mu})'$ indicates the transpose of the vector of deviations. Lange *et al.* (1976; see also Boerwinkle *et al.*, 1986a) provide an algorithm for determining these elements for a particular family structure.

3. All parameters specific to major genotypes must be estimated separately for each such genotype. For models more complex than [6.25], like the $G \times E$ models of [12.3] and [12.10] in Chapter 12, one must estimate full *sets* of β_Gs for each genotype. Appealingly simple-looking models can present daunting estimation problems, even for Bonney's regressive models.

Software: Several widely used packages for segregation or linkage analysis are available at little or no cost {see Ott, 1991, for description of some of these}. These include: Pedigree Analysis Package (PAP) from S. Hasstedt, Department of Human Genetics, University of Utah (used in the apo-A1 example and several others in this book); S.A.G.E. from R. Elston, Department of Biostatistics, Louisiana State University, New Orleans (includes regressive models); LINKAGE from J.-M. Lalouel, Department of Human Genetics, University of Utah, Salt Lake City (for genetic linkage and gene-mapping analysis); MENDEL, FISHER, and GENE, from K. Lange, Department of Biostatistics, UCLA, Los Angeles (does various tasks, especially estimating variance components in genetic models); LIPED from Jurg Ott, Sloan-Kettering Foundation, New York (for genetic linkage analysis); MAP-MAKER, E. Lander, Whitehead Institute, Harvard University, Cambridge, Massachusetts (special-purpose QTL analysis).

7 *Linkage analysis: finding and mapping genes for qualitative traits*

Segregation analysis provides evidence that a disease has a genetic causation, but this is only the beginning, since our real goal is to find the genes involved. An important objective is to *map* the genes to their chromosomal location.

Modern molecular biology provides a rapidly increasing repertoire of methods for the *physical mapping* of genes. Genes can be isolated and cloned using many different techniques (e.g. see Appendix 1.2), and variants can be related to function and physiology both in experimental systems and in natural settings. Physical mapping will probably become the overwhelming method of choice for identifying human genes.

That day is not yet here, however, and we currently still rely on statistical epidemiological methods. In some cases simple association studies will suffice, but usually we rely on *linkage mapping* methods. These take advantage of the fact that evolution has produced a genome consisting of genes concatenated on chromosomes. Probably we owe this arrangement to evolution by gene duplication and to the need to coordinate regulation {De Duve, 1991}. In any case, linkage mapping methods search for association relationships that are due to linkage, between genes that have already been mapped and genetically controlled phenotypes whose genes we have not yet identified. The bible on linkage mapping methods is the text by Ott {1991} {Conneally and Rivas, 1980; Lathrop and Lalouel, 1984; White and Lalouel, 1987}. These methods are so powerful that scarcely a week passes without their being used to map some new gene[1].

Association between loci

Chapter 3 provided examples of association between disease and specific genotypes, such as *Gm* types and Amerindian diabetes, and diseases associated with HLA and ABO types. Association studies look for significant differences in disease risk with exposure to different genotypes. While such studies usually test candidate genes, genetic markers can also be used (defined in Chapter 3).

One statistical association never implies causation; positive results must be followed by confirmatory studies in other populations. Confirmation can take two important forms. If the same genotypes are found associated with disease risk in different populations, those genotypes may be causally associated (in the physiological sense) with the trait. Alternatively, if an association involves the same locus, but different genotypes, in different populations, the locus may not itself be involved, but may be closely *linked* to a locus that is, i.e. the alleles at the tested locus are in *linkage disequilibrium* with the disease locus.

Linkage disequilibrium between two loci tells us little about how close the two loci may be. In part, this is because demographic events such as fluctuations in population size, admixture, or population subdivision can induce disequilibrium even among physically unlinked loci. Thus, to find the location of the gene of interest, we switch from population to family data, in which recombination naturally controls the relationships among alleles at the loci in ways that allow us to map the disease locus relative to the marker. Although linkage studies are more powerful than association studies, the price is that the former require that we specify a genetic model, which usually means knowing how the phenotypes appear in families.

Linkage mapping in families with marker loci

General principles

Recombination typically occurs at least once per meiosis, on both sides of the centromere of each chromosome. The probability of a recombination between two loci is denoted θ, the *recombination distance* between them, expressed in units called *morgans*, after T. H. Morgan, who pioneered linkage analysis. A morgan is the length of chromosome between the ends of which recombination occurs essentially at every meiosis; a *centimorgan* (cM) is the corresponding 1% recombination distance. In humans, 1 cM is on average about 1000 kb, i.e., about 1 mb long, and the total human genome is about 33 morgans in length.

The recombination distance is only approximately linearly related to the *genetic*, or *physical*, map distance in kilobase-pairs between two loci, because recombination rates differ near the centromere, between the sexes, and owing to chemical factors. One factor affecting recombination is *interference*; recombination at one site may inhibit a second crossing-over at nearby sites in the same meiosis. The relationship between physical and recombination distance is specified by the *mapping function*. Linkage analysis is usually done with computer packages written for the

Table 7.1. *Offspring genotypes in double heterozygote × double homozygote mating with recombination probability θ*

		Mother ($MmDd$) Gametes			
	Prob, if phase: MD/md	MD	Md	mD	md
Father ($mmdd$)	md	$(1-\theta)/2$	$\theta/2$	$\theta/2$	$(1-\theta)/2$
	Md/mD				
	md	$\theta/2$	$(1-\theta)/2$	$(1-\theta)/2$	$\theta/2$

Note: Entries are probability of genotype specified by first parent gamete types (given phase) plus *md* gamete from second parent, with θ recombinations between M (marker) and D (disease) per gamete.

purpose, which can implement a variety of mapping functions; we are not concerned with these in this book, but they are important {Ott, 1991}.

Several standard terms that originated in experimental and agricultural genetics are frequently used in human genetics. The term *cross* refers to a mating between two inbred strains homozygous for different alleles; in human genetics, the term refers to matings such as $AA \times aa$, for one locus, $AABB \times aabb$ for two, etc. The parental generation is denoted P, their offspring the F_1 generation, and subsequent generations F_2, etc. A *backcross* is a mating between the F_1 offspring of a cross, and an individual with one of the parental genotypes (e.g., $Aa \times AA$). A *double backcross* is a backcross at two loci, $AaBb \times AABB$. An *intercross* is a mating between two members of the F_1 generation produced by a cross, $AaBb \times AaBb$.

Linkage mapping is based on the offspring haplotypes produced by the various parental mating types via recombination. We can illustrate the principles by the example of a double backcross mating, for example the mating between one parent who is heterozygous for both a disease locus and a marker locus, and the other parent homozygous at both loci. Table 7.1 lists the parental gametes and possible offspring genotypes. Recombination between marker m,M and disease d,D loci is assumed to occur with a probability θ per meiosis (note that recombination can be ignored relative to the *mmdd* parent).

The rows in Table 7.1 give the probabilities of the offspring genotypes for each possible *phase*, or chromosomal haplotype, in the heterozygous parent. The first row refers to MD/md mothers, the second to Md/mD. An offspring who inherits one of the parental haplotypes is said to be in *coupling* phase, and, if inheriting a recombinant chromosome, in *repulsion* phase (a common convention is to refer 'capital letter' parental haplotypes such as *MD* as in coupling, and 'mixed' haplotypes such as *Md* as in repulsion).

In population data (and for founders and spouses in pedigrees) two-locus gamete probabilities depend on the allele frequencies at the two loci. In family data in which the diploid genotype, but not the haplotype phase, of the parent is known, it is usually assumed that the two possible phases are equally likely; that is, coupling and repulsion phases each have 'prior' probability of $\frac{1}{2}$. However, for offspring of founders, the conditional genotype probabilities of the possible offspring haplotypes given the parental haplotypes depend on only the mendelian transmission probabilities and θ, as in Table 7.1. These are the conditional probabilities used in likelihoods in families. For unlinked genes, regardless of whether an individual provides a given gamete with a copy of his/her paternal or maternal allele at one locus, he/she will provide that gamete with the paternal or maternal copies of alleles at an unlinked locus with the mendelian probability of $\frac{1}{2}$.

The basic purpose of a linkage study is to determine whether a marker and locus responsible for a genetic disease segregate as if they are physically linked, i.e., if the probability of recombination between them is $\theta < \frac{1}{2}$, or whether, on the contrary, they segregate independently with $\theta = \frac{1}{2}$.

In our double backcross example, the likelihood of observing N children of which n_i, $i = 1, 2, 3, 4$, have genotype i (*MmDd*, *Mmdd*, *mmDd*, and *mmdd*) is

$$\begin{aligned}
&\mathscr{L}(\theta, \text{single locus dominant}|\text{data}) \\
&= (1/2)[(1-\theta)/2]^{n_1}[\theta/2]^{n_2}[\theta/2]^{n_3}[(1-\theta)/2]^{n_4} \\
&\quad + (1/2)[\theta/2]^{n_1}[(1-\theta)/2]^{n_2}[(1-\theta)/2]^{n_3}[\theta/2]^{n_4} \\
&= (1/2)^{N+1}[(1-\theta)^{(n_1+n_4)}\theta^{(n_2+n_3)} \\
&\quad + \theta^{(n_1+n_4)}(1-\theta)^{(n_2+n_3)}]. \qquad [7.1]
\end{aligned}$$

The term on the left of the plus sign in the first version of this equation is the probability of observing the data with parental phase *MD*/*md* and that to the right with phase *Md*/*mD*, each line having a coefficient of $\frac{1}{2}$ to reflect our assumption of equal phase probability, assuming unknown parental phase. If phase were known, for example, from grandparental data, only the relevant half of [7.1] would be needed, without a coefficient of $\frac{1}{2}$.

The value θ_{MLE} can be estimated in various ways. For simple situations such as [7.1] we can simply try values of θ at small intervals across the permissible range (0 to 0.5) into [7.1] and determine which value approximately maximizes the likelihood; θ_{MLE} is the best estimate of the map distance between the two loci these data can provide. Approximate confidence limits for θ are sometimes taken as those levels above and below θ_{MLE} at which the likelihood is 1.0 less than its value at θ_{MLE}; the value of [7.1] can be accurately maximized by numerical or calculus-based methods {Ott, 1991}.

The usual next step is to compare this likelihood to that with $\theta = \frac{1}{2}$ in [7.1], representing the null hypothesis that the two loci are not linked. Inspection of [7.1] shows that $\mathcal{L}(\frac{1}{2}) = (\frac{1}{4})^N = (\frac{1}{2})^{2N}$. The base-10 log-likelihood ratio is known as the logarithm of the odds, or *LOD score*, which is

$$\begin{aligned}\mathrm{LOD}(\theta, \tfrac{1}{2}) &= \log_{10}[\mathcal{L}(\theta_{\mathrm{MLE}})/\mathcal{L}(\tfrac{1}{2})] \\ &= \log_{10}2^{N-1}[(1-\theta)^{(n_1+n_4)}\theta^{(n_2+n_3)} + \theta^{(n_1+n_4)}(1-\theta)^{(n_2+n_3)}].\end{aligned} \quad [7.2]$$

The chi-squared test for $2 \ln \lambda$ is not accurate here, so the convention is to accept linkage at θ_{MLE} v. no linkage if LOD ≥ 3.0; that is, if the hypothesis $\theta = \theta_{\mathrm{MLE}}$ is more than $10^3 = 1000$ times more likely than no linkage. This approximate significance level is based roughly on the fact that, with 22 autosomes (after adjusting for the varying length of the actual pairs), a random marker has a chance of about 5% of being on the same chromosome as a random disease locus. For details and explanations see Ott {1991}.

The strength of the evidence for linkage can be seen graphically by plotting the LOD score from 0 to 0.5; a sharply peaked likelihood 'surface' (curve) around θ_{MLE} reflects strong support for that value. If the curve is flat, other recombination values are about as likely.

Likelihood for linkage analysis in pedigrees

The above illustrative examples involve simple offspring sets, but the general family likelihood can be used to do linkage studies in complex pedigrees, with only a slight modification of the form given in Chapter 5. The founder genotype probabilities are based on HWE and multilocus rather than single-locus frequencies (e.g., of marker and disease locus). However, since the founders' genotypes have been determined at the marker locus, only the term applicable to that genotype need be used; for example, for an *MM* founder, we sum over only the unknown disease-

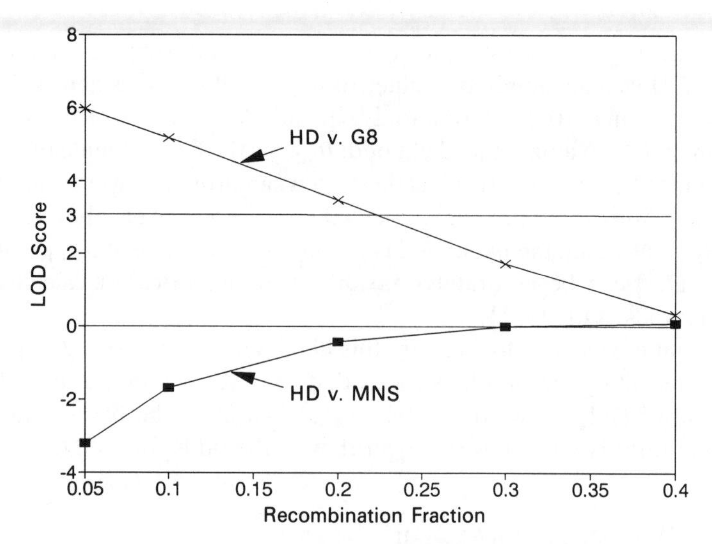

Figure 7.1. LOD score plot for Huntington's disease (HD) against two candidate marker polymorphisms (G8, MNS). (Data from Gusella *et al.*, 1983.)

locus genotypes: *MMDD*, *MMDd*, and *MMdd*, assigning r^2 as the component of the unconditional genotype probability applicable to the marker locus. For the conditional genotype probabilities, the offspring marker genotypes are known; the transmission probabilities for the disease locus genotypes depend on each possible parental trait genotype and the recombination probability θ between those and the known parental marker genotypes. This was elucidated in Table 7.1. Penetrance functions refer to the disease locus as before. An example was given by Conneally and Rivas (1980). Segregation analysis programs usually also do linkage analysis {Ott, 1991}.

Figure 7.1 shows an example of this kind of linkage analysis, for Huntington's disease (HD), from a classic study of two markers, *G8*, a random DNA RFLP, and the MNS blood group system (Gusella *et al.*, 1983). This likelihood was computed for a large Venezuelan pedigree in which the HD locus was first mapped – near to *G8* on chromosome 4.

Multigene families and candidate genes

It is usually necessary to test many markers before a linked locus is found. The odds of a successful result can be small unless this number is quite large (see e.g., Appendix 7.1). Increasingly, it is possible to do such tests at reasonable cost, but the job can still be challenging. It may be important therefore to try to shorten the odds. An obvious way to do this is to choose candidate genes as markers.

If something is known about the phenotype, candidate genes, i.e. genes whose physiology is at least generally related to that of the trait in question, should have at least better than random odds of being related to the trait. Actually, such a statement is true even if the candidate gene itself turns out *not* to be related. The reason has to do with evolution by gene duplication and the organization of the genome. As reviewed earlier in this book, it is common to find genes that are coordinately expressed or regulated, or of related function, to be located in contiguous gene blocks; indeed, the human and mouse genome have many conserved gene arrangements, after about 100 million years of separation (Searle *et al.*, 1989).

For example, the lipid-system genes discussed in Appendix 2.1 are good candidate loci for cardiovascular disease (see e.g., Lusis, 1988). This approach has been used in many studies. For example, hypobeta-lipoproteinemia (HBLP) is a disease in which little or no apo-B is produced. In HBLP heterozygotes, serum lipids are reduced to less than 50% of normal values, LDL and triglyceride concentrations are low; this leads to fat malabsorption, neuropathies and other consequences. Because apo-B is a major constituent of LDL and the ligand for LDL receptors it was tested as a candidate marker. HBLP cases were shown to co-segregate with an apo-B RFLP haplotype in a large affected family; no recombinant genotypes were observed (all affecteds inherited the same apo-B haplotype), suggesting that the apo-B gene is itself involved in HBLP (Leppert *et al.*, 1988).

Linkage mapping using multiple markers and other strategies

The concept of a linkage map

A mighty effort is under way to develop a complete human *linkage map*; that is, highly polymorphic RFLPs evenly spaced at short recombination intervals (ideally, 10 cM or less), in a known order, along the entire human genome. Given the frenzied competition and massive funding infusion of the Human Genome Project, such a map is imminent, and will include most functional genes, as well as hundreds of non-coding RFLPs including dinucleotide repeats and other VNTRs {White and Caskey, 1988; White and Lalouel, 1988}. With this resource almost any well-defined genetic trait for which adequate family data can be collected will be mappable. (For current status, see note 1.)

Currently this work is being undertaken largely using a series of about 40 *reference families*, known as 'CEPH' families (for *Centre d'Etude du Polymorphisme Humain*). These families consist of two parents, about eight offspring, and many grandparents (Dausset *et al.*, 1990). From each family member a small blood sample was drawn, and the white cells were infected with a virus that 'immortalizes' them; that is, allows them to keep dividing just as leukemia cells do. Cell lines have been established in this way to supply an endless supply of CEPH family DNA. These are being typed to establish the location of functional genes as well as markers. Currently, over 95% of the genome is closely linked to at least one marker locus, though not yet uniformly so. Unmapped genes for which a DNA probe is available can often be mapped quickly by multipoint linkage methods in these families. Other methods for mapping are also being developed, including YAC libraries, *in situ* hybridization of karyotype preparations with fluorescently labeled markers, and so on.

Our current ability quickly to map new genes rests largely on the huge Human Genome effort to establish a polymorphic marker map. However, this effort has to date been concentrated largely on the CEPH set of Caucasian families. Although the map location of genes are the same in all human populations, as may be seen in Chapters 9 and 10, the evolutionary process generates tremendous differences in sequence polymorphism among populations. Thus, it is important to obtain and type samples of the genome from the many populations that make up the human species as a whole. There is currently an initiative under way to do this.

Multipoint mapping: general principles

The power of linkage tests is greatly enhanced if several markers are tested simultaneously in the same family data, especially when these markers are themselves linked. We can obtain a significant LOD score for linkage, and a more precise linkage location, from the same number of meioses than can be achieved by testing those meioses one marker at a time.

As stated earlier, the linkage map (recombination) distance is not additive for a set of markers along a chromosome. For example, for three loci with gene order *M-D-N*, even if all recombinations are independent (no interference), the probability of observing recombination between M and N is

$$\begin{aligned}\theta_{MN} &= \theta_{MD}(1-\theta_{DN}) + \theta_{DN}(1-\theta_{MD})\\ &= \theta_{MD} + \theta_{DN} - 2\theta_{MD}\theta_{DN}\\ &\leq \theta_{MD} + \theta_{DN}.\end{aligned}$$

Table 7.2. *Basic multipoint recombination probabilities*

	Loci D and N		
Loci M and D	Recombination	No recombination	Total
Recombination	g_{11}	g_{10}	θ_{MD}
No recombination	g_{01}	g_{00}	$1-\theta_{MD}$
Total	θ_{DN}	$1-\theta_{DN}$	1

(this is like the 'triangle inequality' of distance in geometry). Because of statistical variation in a finite sample, the observed pairwise recombination frequencies may not even generate the inequality; e.g., the largest physical distance may not manifest a correspondingly large number of recombinations.

Methods for joint (simultaneous) analysis of recombination among more than two loci are generally known as *multipoint mapping*, and play a central role in the development of the human gene map. We can illustrate multipoint mapping by looking at our three loci, markers M and N, and trait locus D (for more loci, indeed even for three, computer packages are usually used). Here, assuming no interference for simplicity, and making no assumptions about gene order, four different gamete haplotypes can be produced by an informative parent, depending on the recombination(s) that do or do not occur between M-D and D-N. Letting g_{ij} be the probability of the ij-th event, the haplotype probabilities are given in Table 7.2, from which, solving for the gs from the θs yields

$$\left.\begin{aligned} g_{11} &= (\theta_{MD}+\theta_{DN}-\theta_{MN})/2 \\ g_{10} &= (\theta_{MD}-\theta_{DN}+\theta_{MN})/2 \\ g_{01} &= (-\theta_{MD}+\theta_{DN}+\theta_{MN})/2 \\ g_{00} &= 1-(\theta_{MD}+\theta_{DN}+\theta_{MN}). \end{aligned}\right\} \qquad [7.3]$$

From the likelihood of a set of n offspring haplotypes the values of the recombination probabilities can be estimated {see e.g., Ott, 1991}. The largest of these estimated values is generally taken to represent that between the flanking pair of loci (i.e., the third locus in between). Likelihood ratios are used to decide whether this is significantly better than other possible gene orders [Lathrop *et al.*, 1984, 1985; Lathrop and Lalouel, 1984, 1988; Ott, 1991; White and Lalouel, 1988].

Multipoint linkage relative to a known marker map

These days, when multiple markers are tested against a putative disease, the relative positions of the markers, i.e. the *marker map*, are often already known. Under these circumstances, the likelihood has only one parameter, namely the location of the putative locus relative to the map itself (i.e., the other recombinations are already known). If there are three markers, for example, A, B, and C, in that order, the likelihood for *A-D-B-C specifies* θ_{AD} as a parameter to be estimated, but θ_{BC} is fixed, since it is already known (given the tested value of θ_{AD}, θ_{DB} follows from the known value θ_{AB}).

The log-likelihood ratio comparing the likelihood for the location of the new gene along a known marker map, relative to the likelihood that the new locus is unlinked to the entire set of markers, is called the *location score* for the new gene (Lathrop *et al.*, 1984). For any tested point x on the map the two likelihoods are a ratio $\mathcal{L}(x)/\mathcal{L}(\frac{1}{2})$. Location scores are often plotted in published reports of linkage studies because they show graphically the strength of the evidence (the scale, usually labeled w, is adjusted to physical distance by applying a mapping function [Ott, 1991]).

An example of the early successful use of location scores for multipoint mapping is the search for the gene responsible for *cystic fibrosis* (*CF*) (Chapter 10), long a challenge to biomedical research because, despite enormous effort, its physiology was poorly understood. Figure 7.2 shows the location scores for CF, on a map of two marker loci, the enzyme paraoxinase (PON) and the random DNA probe DOCRI-917 (Tsui *et al.*, 1985). CF is shown to be about 5 cM from *PON* and 15 from *DOCRI-917*. The gene has subsequently been cloned (see below).

The higher the point on the curve the more likely is the location; the horizontal line at LOD = 3 marks the conventional cutoff level for statistical significance. Location score curves are often multimodal (when the gene is somewhere on the tested map). There will be an 'antimode' at the exact location of each marker for which there is any evidence for recombination with the tested disease locus that cannot be attributed to incomplete penetrance; in such cases, the likelihood that D is exactly at the marker has to be 0 (log-likelihood negative infinity).

Interval mapping

A current objective of many human geneticists is to develop a complete dense map with fully polymorphic markers (i.e., so each founder in a family will be a unique heterozygote) at closely spaced intervals (10–20 cM or even less) across the genome. Any new locus would have to lie in

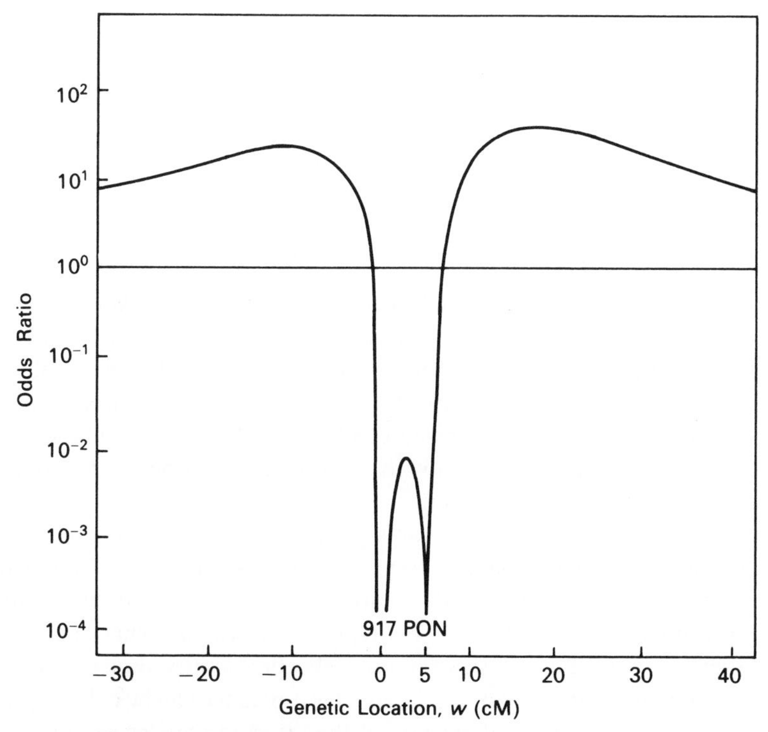

Figure 7.2. Early map location score plot for cystic fibrosis. PON, paraoxinase. (From Tsui *et al.*, 1985, *Science* **230**: 1054–1057. Copyright 1985 by the AAAS.)

one – and only one – of the intervals constituting such a map. Linkage mapping with a full marker map is known as *interval mapping*, and can be statistically very efficient {Lander, 1988; Lander and Botstein, 1986a,b, 1989}. Essentially, location scores are computed all along the genome, to search for the one interval in which a test locus lies. A complete map does not yet exist for humans, but the widely dispersed hypervariable regions that are being found (dinucleotide repeats, VNTRs, etc.) move us rapidly towards such a map. Meanwhile, dense marker maps already exist for many areas of the human genome, in which important disease and other loci are known to lie. It may be possible at least to determine convincingly if a given test locus does *not* lie in such a region; that is, to do *exclusion mapping*.

Some special linkage mapping strategies

Incomplete penetrance can cause serious problems for studies attempting to map a gene by comparing phenotypes and marker genotypes in families {see e.g., Ott, 1991}. For example, for a locus that may be responsible for a late-onset disease, the phenotypes of children will be essentially uninformative (all will be unaffected) regardless of their genotype at the locus we want to map. Several sampling strategies are available to circumvent some of these problems.

Affected pairs

One model-free way to search for linkage to one or more marker loci is to build on the fact that if two genetic traits are linked, pairs of relatives will be 'concordant' for both traits more often than if the traits are unlinked {Haseman and Elston, 1972; Li, 1961; Penrose, 1935; Suarez *et al.*, 1978}. Such association should be detectable even if the genetic model for the trait is not understood, although the *amount* of concordance does depend on the genetic mechanism.

In particular, we can considerably reduce problems of incomplete penetrance by restricting a study to pairs of *affected* relatives – siblings, for example. If a marker is linked to a disease locus, affected sibling-pairs who (presumably) share disease genotypes i.b.d. will also share marker haplotypes more often than by chance. The probability that two such siblings share no, one, or two, marker alleles by chance can be calculated, and a chi-squared test can determine if the affected pairs share marker alleles more often than predicted.

To illustrate this idea, suppose two parents are of marker genotypes *ij* and *kl* for a locus so polymorphic that it can be assumed that each allele is distinct. These parents can produce four kinds of offspring marker type: *ik*, *il*, *jk*, and *jl*. Choose an offspring at random, and then a second. With probability $\frac{1}{4}$, the two genotypes will be identical. With probability $\frac{1}{4}$ the two will share no alleles, and with probability $\frac{1}{2}$ the two will share one marker allele. If another locus, say a disease locus, is segregating independently of the marker locus, then the fractions of times that two *affected* siblings share no, one, or two marker alleles will be as just given. But if the loci are linked, siblings who share disease phenotypes will be more likely to share marker genotypes.

For example, for a dominant disease locus, in a $Di/dj \times dk/dl$ mating, a fraction $1 - \theta$ of the time no recombination will occur and affected offspring will either be Di/dk or Di/dl, each with probability $\frac{1}{2}$. The remaining θ of the time, affected offspring will carry either of the recombinant genotypes, Dj/dk and Dj/dl, again with equal probability. For example, suppose $\theta = 0$ (marker *is* the disease locus). All affected offspring will be of marker type *ik* and *il*, and among pairs of affected siblings $\frac{1}{2}$ will share two marker alleles, $\frac{1}{2}$ will share one allele, and no pair will share no alleles. Sibling-pair data can be used in likelihoods to estimate θ and test for linkage.

The affected pair method has been modified to consider a variety of types of relative pairs and of genetic models [Amos and Elston, 1989; Amos *et al.*, 1990; Chakravarti *et al.*, 1987; Ewens and Clarke, 1984; Haseman and Elston, 1972; Hodge, 1981, 1984; Risch, 1990a,b,c; Suarez and Hodge, 1979; Suarez and Van Eerdewegh, 1984; Suarez *et al.*, 1978; Weeks and Lange, 1988]. Although not as statistically efficient, the sibling-pair method can be less costly, computationally easier, and more effective for finding linkage than pedigrees. The model depends on etiologic homogeneity in the pedigrees, and often on the rare-allele assumption that the disease allele can come only from one parent (or that both are heterozygotes, for a recessive trait). Cases due to mechanisms other than a putative causal allele can seriously undermine the power of the method.

The sibling-pair method has been used to search for linkage between the HLA system and a number of important diseases. The HLA system is very polymorphic so that, even if the parents' HLA types are not known, shared HLA types among offspring are likely to be shared i.b.d. from the same parent. This makes the system ideal for sibling-pair testing, and the method has been successful in showing linkage to multiple sclerosis (see e.g., Suarez *et al.*, 1982) and insulin-dependent diabetes mellitus (see e.g., Suarez and Hodge, 1979), as well as other diseases.

HLA is more polymorphic than most marker loci currently available, although the mapping of hypervariable loci is rapidly making this statement less true. With less variable loci, it may not be possible to distinguish marker types i.b.d. from those simply identical by state (i.b.s.), i.e., separate copies, *not* descended from a common ancestor in the pedigree. The affected-pair method can be adapted for this situation (Bishop and Williamson, 1990; Lange, 1986a,b; Weeks and Lange, 1988); i.b.s. mapping can be effective in some circumstances, even when the mode of inheritance cannot be assumed, but the choice of the best kind of sample to collect is not easy because the relevant variables interact in a complex way (Bishop and Williamson, 1990). Obviously, the closer the markers the better the method will work, because with closely spaced markers recombination is rare and i.b.s. is nearly identical with i.b.d. in the pedigree.

Affected-pair tests are similar in spirit to epidemiological association studies, discussed in Chapters 3 and 5 (Khoury *et al.* 1986, 1990; Majumder *et al.*, 1983; Weiss *et al.*, 1982). Khoury *et al.* (1990) provide

RR values for recurrence in close relatives for single-locus diseases as a function of haplotype sharing for various levels of disease prevalence. The relative risk of disease in relatives of an affected person, compared to that in the general population (an *SMR*-like measure), is a determinant of the potential effectiveness of a linkage study (Risch, 1990a). The ability of affected-pair data to reveal linkage locations in practice depends on the types of relative used, number of marker loci, recombination distance, and polymorphism level of the markers (Risch 1990b,c).

In a recent example of the successful use of the affected pair method, a full segregation analysis has failed to find linkage for Alzheimer's disease (AD) in a set of families. However, the affected-pair method found strong evidence for linkage to a locus on chromosome 19 in the same data (Pericak-Vance *et al.*, 1991). The reasons appear to involve problems that etiologic heterogeneity and incomplete penetrance cause for segregation analysis in pedigrees because AD has been thought to be typically more closely associated with the *APP* locus on chromosome 21.

Homozygosity mapping: affected inbred individuals

A long-standing source of material for the study of rare recessive disease has been inbred families, because inbreeding raises the probability of homozygosity i.b.d., and hence the frequency of recessive traits. Affected individuals in inbred families are likely to be homozygous i.b.d. for the disease-causing alleles, especially if the trait is rare (low allele frequency in the general population so that spouses are unlikely to bring in another copy of the allele). The same inbred individual will also be homozygous i.b.d. for many regions, in fact, the probability of homozygosity is the same for every locus, and is determined by the inbreeding coefficient, F. The probability that any n independent loci are all homozygous i.b.d. is F^n, which quickly becomes small as n increases.

This is also the probability that n *consecutive* markers will be homozygous i.b.d. if the markers are segregating independently. However, at a region – say, around the disease locus in an affected inbred individual – for which two chromosomes are i.b.d., the probability that consecutive markers will also be homozygous i.b.d. becomes quite substantial, and depends on the number of recombinations that have occurred from the common ancester to the affected individual. This suggests that we should screen a map of polymorphic marker loci for regions of contiguous marker homozygosity in affected inbred individuals, such as offspring of a cousin-to-cousin mating. A stretch of consecutive marker homozygosity strongly suggests a location for the disease-causing gene. This statistically appealing approach can detect linkage in only a few tens of affected inbred children (Lander and Botstein, 1986a,b). For example, one affected offspring of first cousins is roughly as informative as a nuclear family with three affected children.

The idea can be extended to more distant relatives, even members of a population in which it can be assumed that all copies of a disease are *clonal*; that is, are i.b.d. from some original mutant (this is not as unrealistic as it may sound, as may be seen in Chapter 10). Except as broken up by recombination, homozygosity should be retained among affecteds. In sets of affected individuals, for a series of markers near a disease locus, the pattern of shared haplotypes may indicate where a disease gene lies, and can be useful in fine-mapping a trait (A. Chakravarti *et al.*, unpublished results).

For example, if a new mutation quickly became fixed 1000 generations ago, there would still be about a 2–5 cM stretch of largely unrecombined DNA. This might be relevant to the problem of diabetes in Amerindians, for example; it could be that most affected individuals, of whatever tribal affiliation, share such a region (A. Connor, unpublished result).

The step from genetic linkage to physical mapping

Once the general location of a gene has been established by genetic linkage mapping, the next step we would hope to take is to identify the gene itself, and to show its physiological role by sequencing it, studying its coded protein, expression and the like. Usually, genetic linkage methods – tracing recombination in families – can do no better than locate the gene to within several megabase-pairs. Obtaining a more precise location often requires fine-structure or *physical* mapping.

Physical mapping is approached with a variety of molecular biological techniques. All one has to do is spend a short time in an active genetics laboratory to sense the energy and innovation of efforts to make mapping quicker and more effective. Here I can only give a very cursory taste of the current options[1,2].

Various methods use overlapping cloned pieces of DNA from a chromosome region of interest to map a gene. Libraries of mapped clones exist; that is, clones of known chromosomal origin and relationship to each other. A DNA probe containing sequences from the general target region will hybridize to cloned fragments that contain its sequence. By working along overlapping fragments with probes, one can 'walk' or 'jump' along the chromosome until a small fragment is found that is always associated with disease (e.g., in family members).

Another recently developed technique is radiation hybrid mapping. Mouse cell lines containing only randomly generated pieces of single human chromosomes, generated by exposing hybrid cell lines to X-ray irradiation, can be typed to see how frequently the radiation breaks a chromosome between two markers known to be on the same chromosome (see e.g., Boehnke *et al.*, 1991).

Synthesized oligonucleotides tagged with fluorescent molecules, representing a gene of interest can be hybridized to whole-cell karyotypes, along with similar fluorescent probes from a known marker map. The hybridization pattern will reveal the relative map locations of marker and target DNA.

After the gene for CF was mapped to chromosome 7 by linkage methods, intensive chromosome walking eventually identified the locus itself, which was called the cystic fibrosis transmembrane conductance regulator (CFTR) (Kerem *et al.*, 1989; Riordan *et al.*, 1989; Rommens *et al.*, 1989). *CFTR* shares sequence homology and function with other genes involved in transporting ions across cell membranes, and mutational analysis shows this is consistent with the dysfunction of mucus membranes in the respiratory tract in CF.

Duchenne muscular dystrophy (DMD) was known from segregation patterns to be X-linked (Whorton and Thompson, 1988), and deletions or translocations near the band Xp21 have been observed, the location supported by linkage mapping. A few families were known in which DMD patients also were affected with a white blood cell gene deficiency known as chronic granulomatous disease (CGD). Messenger RNA from healthy white cells was compared with mRNA from a patient with these conditions by *subtraction hybridization*, to isolate mRNA found in normal but not in CGD cells. This revealed a gene from a small deleted region of the X-chromosome (Kunkel *et al.*, 1985; Monaco *et al.*, 1985; Whorton and Thompson, 1988). The DMD gene, now known as the dystrophin gene, was then found, an antibody to part of the gene product was synthesized on the basis of the DNA sequence, and the dystrophin gene is now known to code for a cytoskeletal protein present on normal muscle cell surfaces that is missing in DMD patients (Whorton and Thompson, 1988).

Down's syndrome (DS) has long been known from chromosomal studies to involve trisomy (three copies) of chromosome 21. It has also been known that DS patients can experience premature aging and dementia, similar to that in AD, and even that DS-free relatives of DS patients have an elevated risk of AD. This phenotypic correlation suggests that AD may be linked to the DS region on 21. A small region has in fact been identified that is responsible for DS (triplication of this part alone is causal), and another gene, that for amyloid precursor protein (APP), the precursor of the β-amyloid protein that is concentrated in the neurofibrillary tangles in the brain of AD patients, is near this region on chromosome 21. In at least some early-onset families, mutant APP proteins lacking proper receptor-binding function appear to cause AD (Goate *et al.*, 1991; recent history reviewed by Wright *et al.*, 1991).

This is an appropriate place to note that one important factor in gene mapping is simply to be aware of the medical literature. On many occasions a single rare family or even one patient with a disease of interest, perhaps associated with a more complex syndrome, or with some rare features, proves to have a chromosomal deletion or rearrangement that reveals the gene location. One example is *retinoblastoma*, a rare cancer of the retina. Deletions on chromosome 13 shown to be linked to the trait revealed the *RB* locus, which has subsequently been of fundamental importance in our understanding of cancer genetics (Chapter 15).

Various ways to examine normal and aberrant tissue-specific gene expression exist, for example comparing cDNA from affected and normal individuals using protein-specific antibodies to search for normal and abnormal tissue patterns of expression. There are also many precedents for the use of animal models, based on genetic similarities between humans and laboratory mammals such as rodents. For example, a specific mutation in the HLA system has been shown to be involved in insulin-dependent diabetes in rodents and humans (Todd *et al.*, 1988b).

Conclusion: methods triumphant

This chapter discusses three general strategies for identifying genes. When enough is known of the biology of the trait to suggest candidate genes that might be involved, those are the first genes tested for a relationship to the trait. This strategy, that begins with genes and works towards the phenotype, has been called a *bottom up*, or **G** $\rightarrow$ **P**, approach (Sing *et al.*, 1988; Sing and Moll, 1989). Its simplest form is an association study, the rationale being that a positive finding for a physiologically suspect gene is strong evidence for a causal role. But linkage or physical mapping would have to follow to demonstrate that the association was not simply due to chance linkage disequilibrium, recalling that related genes are often closely linked.

For most of the history of medical genetics too little was known about diseases to take the candidate gene approach. Instead, classic segregation and linkage was done, using genetic markers. Fortuitous chromosomal anomalies or other indicator events discovered by chance were also exploited. This has been called *reverse genetics* because it is a *top down* strategy that works from the phenotype down to the unknown genotype, i.e. **G** ← **P**, reversing our usual notion of physiological causation. Most successful linkage mapping has used this seemingly backward approach. Association studies, especially based on random genes, are often not very convincing. However, the large series of markers that can be screened in families, and the power of linkage and physical mapping methods, provide a strong rationale for mapping the human genome and for this approach.

Despite their success, both methods rely essentially on a rather simple, **G** → **P** model of *physiological* causation. However, we are beginning to confront the fact that real genetic causation is complex and heterogeneous, involves complex feedback loops, gene regulation and even autoregulation, and interaction among loci. Nonetheless, despite complexities, even our oversimplified approach makes it possible to be able to identify genes with at least modestly strong effects on essentially *any* human trait. Complications will arise, but the success of the methods for mapping genes of major effect constitutes a dramatic change that has transformed biology.

Notes

1. Details on the human gene map are available electronically through data bases GBD and OMIM, run by the Welch Medical Library at Johns Hopkins University. *Genomics* and *Human Genetics and Cytogenetics* regularly publish details on the human gene map, also *Science* **258** (5079), 1992.
2. The current status of many physical mapping methods is given by a series of papers in *Current Opinion in Genetics and Development* (1992), vol. 1, no. 1.

APPENDIX 7.1: Some statistical and sampling issues in linkage studies

Planning a linkage study requires consideration of many important points. In particular these include sample size and molecular typing aspects of the study design. These subjects were covered with elegant detail and clarity by Ott (1991), who provided an ample bibliography. Here, I want merely to make the reader aware of what the issues are and to provide some very crude rules of thumb.

How much variation is needed for practical studies of linkage?

A marker locus is informative only if disease status and marker types in the offspring, compared to those in the parents, provide information about linkage. The standard measure of the informativeness of a marker locus is the *PIC* (Chapter 4), essentially, the probability that an affected person will be a heterozygote and his/her mate will have a different genotype. This is the minimal requisite for mapping a rare dominant disease. Often, grandparent marker types are important in establishing phase. *PIC* depends on allele frequency and is thus population dependent (see e.g., Ott, 1991; Roychoudhury and Nei, 1988). Heterozygosity is another measure of informativeness.

Hypervariable regions have maximal *PIC*. The informativeness of a marker also depends on the recombination distance between marker and trait locus, and the prevalence of the trait as expressed, for example by the risk in the type of relative to be studied compared to that in the general population {Khoury *et al.*, 1990; Risch, 1990a; Weiss *et al.*, 1982}.

How many people?

The statistical power of a linkage study, or its ability to detect a true linkage $\theta < \frac{1}{2}$, depends on the sample size, and the true value of θ. If a representative family structure can be assumed, one way to estimate how many such structures are needed is to compute the *expected LOD* (*ELOD*) for each structure; the ELOD is the average LOD score over all possible marker disease–locus combinations, weighting each type by the probability of its occurrence:

$$\mathrm{ELOD} = \Sigma_i \Pr(i) \log [\lambda_i(\theta, \tfrac{1}{2})], \qquad \text{[A7.1]}$$

where $\lambda_i(\theta, \frac{1}{2})$ is the likelihood ratio comparing linkage at recombination θ to $\theta = \frac{1}{2}$, for the *i*-th combination. While one never knows what outcomes *will* occur in actual data, this provides an idea of how much evidence each structure might provide if θ is true. To reach a significance level T one needs at least n of these structures, where $n = T/\mathrm{ELOD}$.

For example, an offspring of an informative, phase-known double backcross has two outcomes: (1) recombination, whose probability equals, in this example, the average 'number' of occurrences per meiosis; and (2) no recombination, with probability and mean frequency of occurrence $1 - \theta$. The resulting ELOD comparing θ with $\frac{1}{2}$ is $\log [\theta^{\theta}(1 - \theta)^{(1-\theta)}/(\frac{1}{2})]$. For T to exceed the standard value of 3, to detect θ ranging from 0.1 to 0.2, a sample of 19 to 36 such offspring is needed. Methods for estimating the power of a family structure to produce significant linkage tests have been worked out for various conditions {Bishop and Williamson, 1990; Elandt-Johnson, 1971; Ott, 1991; Ploughman and Boehnke, 1989; Risch, 1990b,c; Weeks and Lange, 1988}.

Until a tight marker map is readily available, it will remain difficult to detect values of θ greater than about 0.15. With large θ there are too many recombinants to provide evidence that $\theta < 0.5$ in modest family structures. Many linkage studies have reported $\theta_{\mathrm{MLE}} = 0$, i.e., no recombinants, between marker and a disease trait, which seems unlikely to be literally true. Probably, when there are even a few recombinants the evidence that $\theta < \frac{1}{2}$ becomes non-significant, even if linkage is true.

How many markers?

Beside the question of *detecting* a linkage, how many marker loci must be used to achieve a given probability that there is linkage to detect; that is, that one of the markers will be close to the trait locus? With 22 pairs of variable length autosomes (sex-linked traits can be identified from their segregation pattern), it has been estimated that the prior probability of linkage between a random locus and a random marker is about 0.05, and that the two are θ units apart, approximately $0.06\,\theta$ {Ott, 1991}. Roughly, an estimate of the number, n, of marker loci required to have a prior probability of at least P that the nearest marker is no farther than θ away is

$$n = \ln(1 - P)/\ln(1 - 0.06\,\theta) \approx -[\ln(1 - P)]/(0.06\,\theta) \qquad [A.7.2]$$

(Ott, 1991; see also Boehnke, 1986; Conneally and Rivas, 1980; Elston and Lange, 1975). Thus, even to have an 80% chance that the nearest marker is ≤ 0.10 units away requires about 270 markers, a substantial number even for a large laboratory.

The situation is improved with marker (interval) maps. With a complete 20 cM marker map, far fewer than 100 markers may suffice if families are fully informative for a single-locus trait, in which case only about half as many families would be needed to detect the linkage as with single-marker studies (Lander, 1988; Lander and Botstein, 1986a,b).

Rarely will any family be informative for all typed markers, and there is no easy current escape from the need for a large number of markers if a whole-genome search must be done. Usually, we hope to find other evidence, such as candidate genes, to help us to identify the most likely regions; such partial guesswork has been successful in a number of cases.

The effects of multiple testing

If the aim is not to be misled very often, one would want to be conservative in setting significance levels for linkage tests, and this means adjusting significance levels when multiple markers are tested. This is because the more tests that are done the more likely chance positives are to arise. In linkage there are countervailing effects. More tests mean a higher chance of false positive, but more (randomly located) markers also increase the prior probability that at least one is on the same chromosome as the disease, which decreases the chance of false linkage claims. The generally accepted criterion of LOD ≥ 3 seems to be acceptable for multiple markers for testing single genetic models. For discussions of this issue see the text by Ott (1991), the article by Risch (1991), and relative to the philosophy of searching for causes, that by Rothman (1990). A linkage assignment never rests on a single study, and should always be confirmed (or ruled out) by further studies in new families or extensions of the existing sample to include other relatives.

Joint segregation and linkage analysis

Linkage testing makes no sense if a disease is not genetic, and segregation analysis to show that a trait is genetic is quicker and easier than linkage analysis because it does not require any actual genotyping. However, these days genotypes may be easily available (e.g., by PCR), and there are sometimes statistical reasons for doing a joint analysis that fits the parameters of a genetic model and estimates marker trait recombination values in family data at the same time. It usually seems to make little statistical difference which approach is taken, but an initial segregation analysis may help an investigator to 'purify' the family data so that only those families most likely to be segregating a major locus are used in linkage analysis (Boehnke and Moll, 1989).

Ott {1991} provides an excellent discussion of how to deal with some of the problems that affect both segregation and linkage, such as ascertainment bias, incomplete penetrance, and so on.

8 *Linkage analysis: finding and mapping genes for quantitative traits*

The genes, or QTLs, that contribute to variation in quantitative traits can be found by methods analogous to those used for qualitative traits. Association and linkage studies are both possible. Both 'top down' and 'bottom up' approaches can be also taken (see Chapter 7). The methods rest essentially on an extension of the mixed model presented in Chapter 6. This chapter considers how individual major genes are found and mapped; later, in Chapter 11, we see how to map multiple loci relative to the same trait.

Association studies for quantitative traits

The bottom up (G → P) *approach: candidate genes in populations*

Association studies for qualitative traits test for different trait *frequencies* among candidate locus genotypes. The analog for quantitative traits is to search for differences in the *distribution* (penetrance function) of the trait among candidate genotypes. The only differences usually tested for are in the genotypic values (means) at the locus. Chapter 6 showed how to use the mixed model to *infer* the effects of a major gene acting in the context of other polygenic and environmental effects treated as aggregates, in segregation analysis. Here, we examine the effects of a candidate major locus that has *already* been identified and for which genotypes are determined on sampled individuals.

The basic model is still that $\phi = G + PG + E$, except that here G is a candidate locus known (or guessed at on physiological grounds), and a sample of individuals has been genotyped at the candidate locus. Polygenic and environmental effects are modeled as normally distributed deviations about the major locus genotype values. The likelihood of observed phenotype ϕ_i for individual i is $\mathcal{L}_i = f(\phi_i - \mu_G | 0, \sigma^2_{PG+E})$. The estimated genotypic means (μ_{MLE}) can be estimated; the hypothesis that

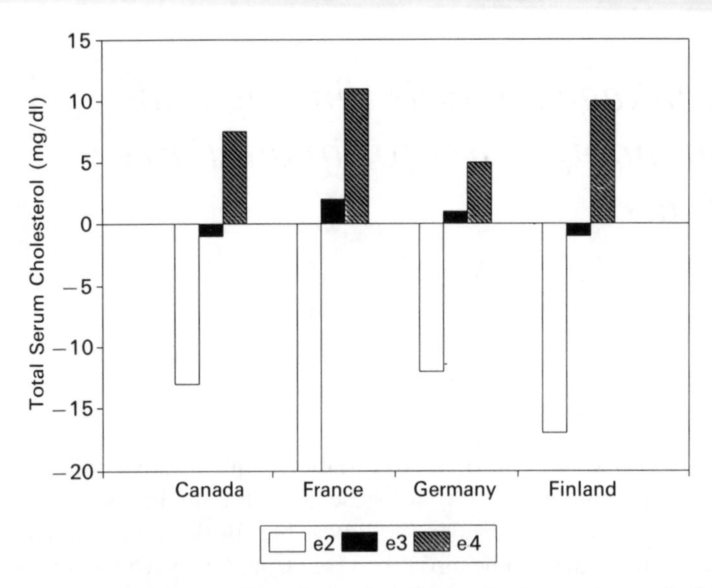

Figure 8.1. Average effects of substituting the three major apo-E alleles in several populations, showing consistent direction but different magnitude of effects for different alleles. (From Sing and Boerwinkle, 1987.)

these differ among genotypes can be tested against the hypothesis that they are equal, by computing the likelihood ratio. An alternative method is to do a one-way analysis of variance, with the observations classified by genotype, in which the significance of differences is tested by the standard *F* ratio test {Armitage and Berry, 1987; Sokal and Rohlf, 1981}.

Unlike segregation analysis where we are trying to infer an unidentified, and hence unmeasured locus, here our individuals are directly genotyped at that locus; for that reason this has been termed a *measured genotype* approach (Boerwinkle and Sing, 1987; Boerwinkle *et al.*, 1986a, 1987).

Examples

Participants in a population-based preventive health program in Nancy, France, were typed for electrophoretically detectable alleles at the apo-E locus, in a study of the possible effects of this locus on plasma cholesterol levels (Boerwinkle *et al.*, 1987). Three major apo-E alleles are polymorphic worldwide, and have statistically significant average effects on cholesterol levels (i.e., the genotypic values differ) (Figure 8.1). The additive genotypic variance at this locus accounts for about 5–10% of total cholesterol variation in this French population.

Table 8.1. *Effects of RFLP haplotypes on triglycerides in 89 English subjects*

No. of individuals	Possible haplotype	No. of each haplotype						
		010	011	000	110	001	111	100
45	010/010	2	0	0	0	0	0	0
7	010/011	1	1	0	0	0	0	0
10	010/000	1	0	1	0	0	0	0
12	010/110	1	0	0	1	0	0	0
1	011/001	0	1	0	0	1	0	0
2	000/000	0	0	2	0	0	0	0
1	110/110	0	0	0	2	0	0	0
3	010/001 or 011/000	0.67	0.33	0.33	0	0.67	0	0
6	010/100 or 000/110	0.56	0	0.44	0.44	0	0	0.56
2	111/010 or 110/011	0.23	0.77	0	0.77	0	0.23	0
Haplotype frequencies		0.70	0.06	0.10	0.10	0.02	0.003	0.02
Allelic triglyceride effect		−0.06	0.45	−0.07	0.08	0.05	0.77	0.52

Note: Haplotype phases are assumed equally likely for ambiguous haplotypes. Phenotype is log-transformed triglyceride levels.
Source: Templeton *et al.*, 1988.

A locus can be 'measured' in a variety of ways. In fact, until we identify the sequence variants associated with the causal effects, there is no way of knowing what aspects of the sequence variation among sampled haplotypes at the locus are important. Electrophoretic or single RFLP variants have often been used in measured genotype studies, but more extensive haplotype data can be used. Such an analysis has been done of three RFLP cut-sites at the apo-A1 locus on chromosome 11, in a sample of 89 English subjects (Templeton *et al.*, 1988). Each cut-site was scored 1 (0) for presence (absence), yielding haplotypes 000 (all sites absent) to 111 (all present); fractional frequencies were assigned to haplotypes that were ambiguous in the observed (diploid) data, by weighting the probability of each possible haplotype for that individual based on the relative probabilities of the parental gametic phases.

Triglyceride levels were log-transformed to normalize their distribution, because the statistical significance tests in analysis of variance assume normally distributed variables. Table 8.1 shows the average triglyceride effects of the haplotypes, comparing these to the haplotype 010 as a baseline. Mean triglyceride level differences were observed for haplotypes 011, 111, and 100, although only the 011 effect was significant in this sample.

The top down* (G ← P) *approach: random markers in populations

It is conceptually straightforward to extend this kind of association analysis to random marker loci, to identify regions of the genome in which a true QTL may lie. One of the first attempts to find QTLs in humans was a test of 12 polymorphic genetic marker loci (blood groups, serum proteins, red cell enzymes) for effects on serum cholesterol, in a population sample of individuals from Tecumseh, Michigan (Sing and Orr, 1976). Each locus generated a separate measured-genotype test. Four had significant, approximately additive, effects. The *A* allele of the ABO system and the 'non-secretor' genotype were associated with higher cholesterol values. This finding has been corroborated by other studies (e.g., Sing *et al.*, 1988), but physiological hypotheses to explain the associations, involving gene × diet interaction have been inconclusive (e.g., Berg, 1983). However, it is now known that two of the tested markers are linked to genes involved in lipid physiology: haptoglobin to LCAT and CETP, and secretor to the LDL receptor and several apolipoprotein genes (Appendix 2.1).

The effects of these genotypes on cholesterol are of the lesser kind generally associated with 'polygenes'. Alleles with only little standardized effect (Chapter 6) are difficult to find in segregation analysis in family data, when the genotypes are unknown and must be inferred, but these lesser components of the causal spectrum can be documented if the locus involved is already known. We see in Chapter 11 how such loci can be found systematically, using reverse genetics.

Linkage analysis for quantitative traits

Linkage analysis with genetic markers in population data

Positive results from measured genotype studies using candidate loci in population data have usually been interpreted as implicating the tested gene, on the grounds that its 'candidate' status was already based on prior physiological knowledge. However, such statistical associations might be due to linkage disequilibrium, non-genetic causes, or even just sampling effects. Even prior suspicions should be treated with caution, because we know that the genome is characterized by clusters of functionally related genes, of which the candidate locus may be but one. Because of the widespread use of measured genotypes, it is worth showing how linkage disequilibrium and causal association are confounded.

Consider an unidentified diallelic *true* QTL with genotypic values μ_{DD}, μ_{Dd}, and μ_{dd}, and a random sample of individuals typed for some measured candidate or marker locus M. Assuming that polygenic and environmental variance are uncorrelated with either the QTL or the

Table 8.2. *Two-locus genotype frequencies with linkage disequilibrium, and their effect on a quantitative trait*

		QTL			
		DD	*Dd*	*dd*	
Marker	Means:	μ_{DD}	μ_{Dd}	μ_{dd}	Total
MM		γ_{DM}^2	$2\gamma_{DM}\gamma_{dM}$	γ_{dM}^2	p_M^2
Mm		$2\gamma_{DM}\gamma_{Dm}$	$2(\gamma_{DM}\gamma_{dm} + \gamma_{Dm}\gamma_{dM})$	$2\gamma_{dM}\gamma_{dm}$	$2p_M p_m$
mm		γ_{Dm}^2	$2\gamma_{Dm}\gamma_{dm}$	γ_{dm}^2	p_m^2
Total		p_D^2	$2p_D p_d$	p_d^2	1.0

Note: γ, defined in Chapter 4, is the two-locus haplotype frequency: $\gamma_{\mathrm{DM}} = p_D p_M + \delta$; $\gamma_{\mathrm{Dm}} = p_D p_m - \delta$; $\gamma_{\mathrm{dM}} = p_d p_M - \delta$; $\gamma_{\mathrm{dm}} = p_d p_m + \delta$.

candidate locus, how well will the marker locus represent the true effects of the linked trait locus?

Table 8.2 shows the two-locus (marker-QTL) genotype frequencies in terms of the corresponding haplotype frequencies, which are functions of the allele frequencies and (unknown) linkage disequilibrium, δ. As given in Chapter 4, the frequency of haplotype *ij*, where *i* and *j* are alleles at the two loci, respectively, is $\gamma_{ij} = p_i p_j + \delta$.

From the table, the mean phenotype in individuals with a given marker type can be computed. For example,

$$\mu_{MM} = [\gamma_{DM}^2 \mu_{DD} + 2\gamma_{DM}\gamma_{dM}\mu_{Dd} + \gamma_{dM}^2 \mu_{dd}]/p_M^2, \qquad [8.1]$$

with corresponding formulas for the other marker genotypes. These are functions of the allele frequencies (p), linkage disequilibrium (δ), and the genotypic values at the true D locus (μ). Because the numerator in [8.1] is based on population frequencies, the denominator 'normalizes' the computation to refer only to the mean among *MM* genotypes, which is what we want to know.

We can see quantitatively how these compare to the true trait genotype values over the range of possible disequilibrium values, under some simplifying assumptions, as shown in Figure 8.2. The example here assumes no dominance, $\mu_{DD} = 1, \mu_{dd} = -1$, and $p_D = p_M = 0.9$ and 0.5 in the two panels, respectively. The maximum spread of the vertical axis represents the true effect, $\mu_{DD} - \mu_{dd} = 2$. When $\delta = 0$ the marker and QTL are segregating independently and the phenotype means for each *marker* genotype equal the population mean, i.e., $\mu_{MM} = \mu_{Mm} = \mu_{mm} = \mu = \Sigma P_g \mu_g$ (the last sum is over trait-locus genotypes), so that in equilibrium even a closely linked marker shows no effect on the trait. With

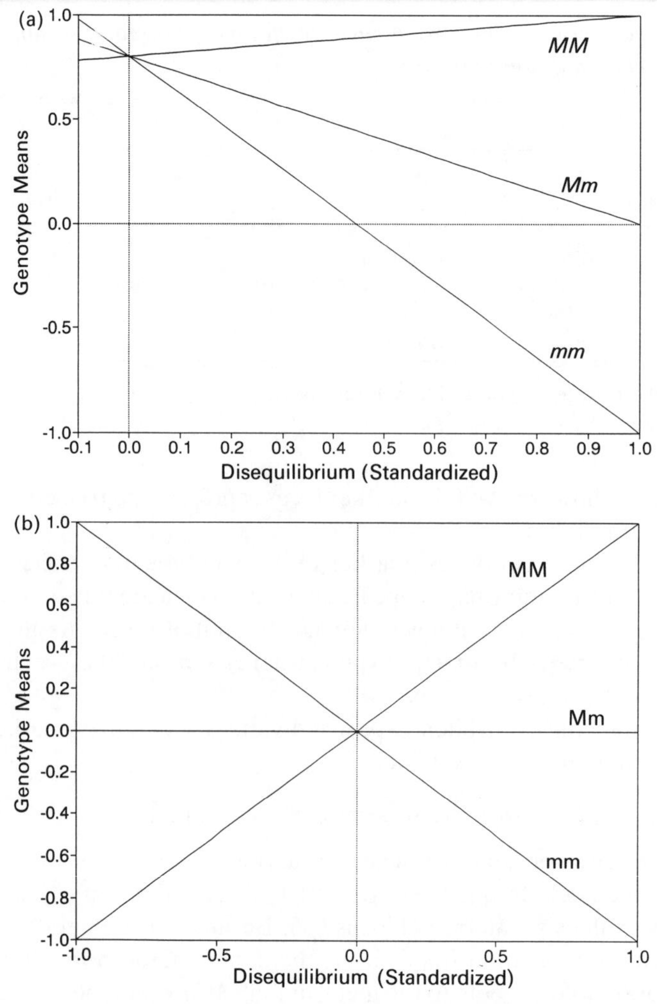

Figure 8.2. (a) and (b) Effect of linkage disequilibrium on phenotypic effects from a measured marker locus. The linkage disequilibrium, δ, has been standardized to have absolute value <1 (Hedrick, 1985). Additive model, no dominance, $\mu_{DD} = 1, \mu_{Dd} = 0, \mu_{dd} = -1$. (a) $p = 0.9$, $q = 0.1$ for trait and marker loci; (b) $p = q = 0.5$ for both loci.

maximum disequilibrium the marker genotype means are the same as those at the trait locus (e.g., $\mu_{MM} = \mu_{DD}$). Intermediate disequilibrium can generate almost any kind of pattern (Figure 8.2(b) is symmetric only because $p = 0.5$).

Linkage mapping approaches add a recombination parameter to the parameters of the measured genotype approach in populations. This allows a better estimate to be made of the true effect of a QTL near a marker locus (although whether this is simply due to the added parameter is something that must be tested, e.g. by likelihood ratio criteria). Linkage approaches will thus tend to give higher estimates of the QTL effects than measured genotype methods.

Linkage disequilibrium also affects the estimated variances. The variance explained by the marker genotypes in the sample will be less than that at the true locus. Further, the observed variances will differ among marker genotypes, contrary to the usual assumption that σ^2_{PG+E} is the same for all genotypes at a QTL (even if that assumption is in fact true at the D locus), perhaps leading us falsely to assume that there is a $G \times E$ interaction if we assume M is the true QTL. Such effects have been observed in experimental studies (Doebley and Stec, 1991; Edwards *et al.*, 1987; Lander and Botstein, 1989; Weller, 1986; Weller *et al.*, 1988).

Linkage analysis in family data

The message is that a marker locus with statistically significant measured genotype effects cannot be equated to a QTL without some further work. Typically, that involves linkage mapping in family data. A trait is expressed in terms of the standard mixed model. However, we now have two-locus oligogenotypes to consider, the marker genotypes which are known and the QTL genotypes which are inferred.

It is difficult to present the full likelihood computations in any simple way, and computers do the work. However, the components of the likelihood are really straightforward extensions of the methods given in Chapters 6 and 7. For founders, the unconditional genotype probabilities are the known marker genotype probabilities based on HWE expectations, and a summation cycle over all possible major-locus genotypes, plus an integral over all possible polygenotypes. For *non*-founders the marker type probabilities are just the mendelian transmission probabilities from Chapter 5, and the likelihood sums over all possible parental QTL genotypes as in earlier chapters, as well as polygenotypes; however, now the offspring QTL genotype probabilities, given their own and their parents' marker genotypes, depend on segregation at the inferred QTL (as in standard segregation analysis) modified by the possibility of recombination with probability θ between marker and QTL. As in Chapter 6, for each individual in the family, given his/her inferred QTL genotype, the penetrance probability density is $f(\phi - \mu_G|\sigma^2_{PG+E})$.

The parameters of the genetic model itself are sometimes known from prior segregation analysis, perhaps on the same pedigrees, or can be

Table 8.3. *Basis of intercross mixed model with marker genotype: likelihood of observing phenotype ϕ_o for each possible offspring genotype*

Recombination distance: θ.
Mating: $DM/dm \times DM/dm$.

	Paternal gametes			
Maternal gametes	DM $(1-\theta)/2$	Dm $\theta/2$	dM $\theta/2$	dm $(1-\theta)/2$
DM $(1-\theta)/2$	$(1-\theta)^2 f(\phi_o \mid DD)/4$	$\theta(1-\theta)f(\phi_o \mid DD)/4$	$\theta(1-\theta)f(\phi_o \mid Dd)/4$	$(1-\theta)^2 f(\phi_o \mid Dd)/4$
Dm $\theta/2$	$\theta(1-\theta)f(\phi_o \mid DD)/4$	$\theta^2 f(\phi_o \mid DD)/4$	$\theta^2 f(\phi_o \mid Dd)/4$	$\theta(1-\theta)f(\phi_o \mid Dd)/4$
dM $\theta/2$	$\theta(1-\theta)f(\phi_o \mid Dd)/4$	$\theta^2 f(\phi_o \mid Dd)/4$	$\theta^2 f(\phi_o \mid dd)/4$	$\theta(1-\theta)f(\phi_o \mid dd)/4$
dm $(1-\theta)/2$	$(1-\theta)^2 f(\phi_o \mid Dd)/4$	$\theta(1-\theta)f(\phi_o \mid Dd)/4$	$\theta(1-\theta)f(\phi_o \mid dd)/4$	$(1-\theta)^2 f(\phi_o \mid dd)/4$

Note: $f(\phi \mid G)$ is the normal density for phenotype ϕ given major locus genotype G; that is, $\mathrm{Nor}(\mu_G, \sigma^2_{PG+E})$; for example, $f(\phi \mid DD) = (1/\sigma\sqrt{2\pi}) \exp[-(\phi - \mu_{DD}/\sigma)^2/2]$.

estimated as part of the linkage analysis. In the linkage analysis the value of θ is of primary interest, and, if it is significantly less than $\frac{1}{2}$ by a likelihood ratio test, we assume physical linkage (synteny) at recombination distance θ_{MLE} between the marker and QTL.

To date, linkage mapping for a quantitative trait with a marker in family data has rarely been done. The reason seems to be that, if measured genotype effects are found, it is assumed that the tested candidate locus is the true QTL. But as shown above, this may not be accurate. We can see how linkage is taken into account by a simple example, offspring from a phase-known double intercross mating, $DM/dm \times DM/dm$. In a spirit similar to that of Table 7.2, Table 8.3 shows the offspring genotype frequencies, times the probability of observing offspring phenotype ϕ_o given the corresponding offspring genotype. Table 8.3 assumes recombination with probability θ between loci M and D. The likelihood of a given offspring phenotype with marker genotype M_o, is

$$\mathscr{L} = \sum_{o} \Pr(o \mid M_o)\Omega_o(\phi_o) = \sum_{o} \Pr(o \mid M_o) f(\phi_o - \mu_o). \qquad [8.2]$$

The sum is over all the possible offspring QTL genotypes as before. The probabilities of the offspring genotypes are determined by the parental and offspring marker genotypes and, given each possible parental QTL genotype, by the recombination fraction θ. The likelihood for a set of offspring is $\mathcal{L} \propto \Pi\, \mathcal{L}_i$.

For example, from Table 8.3, an *mm* offspring can have QTL genotype *DD* if there is a recombination on both parental *DM* chromosomes, creating *Dm* haplotypes, each event having probability θ, and these are both passed on to the offspring, each with probability $\frac{1}{2}$. The *mm* offspring could be *Dd* and *dd* at the QTL, with similarly evaluated probabilities. Overall, the likelihood is

$$\mathcal{L}(\phi_o) = (\tfrac{1}{4})[\theta^2 f(\phi_o|\mu_{DD}, \sigma^2) + 2\theta(1-\theta)f(\phi_o|\mu_{Dd}, \sigma^2) + (1-\theta)^2 f(\phi_o|\mu_{dd}, \sigma^2)], \quad [8.3]$$

where the penetrances are the normal densities for the phenotype given the QTL genotype. A likelihood ratio can test $\theta_{\mathrm{MLE}} < \frac{1}{2}$ (linkage) relative to $\theta = \frac{1}{2}$. In applying the full mixed model in families, polygenic variance would be taken into account separately from environmental variance.

Multipoint extensions and alternative approaches

Multipoint mapping analysis can be done for a quantitative trait in a straightforward extension of the likelihood, as discussed in Chapter 7. The idea of interval mapping over the whole genome as an 'ultimate' form of multipoint mapping was introduced in Chapter 7. Unfortunately, we do not have adequate data to implement this method on humans as yet (except for selected chromosome regions), and the only existing examples of whole-genone interval mapping are from agricultural applications to quantitative traits in various strains of tomato, for which an interval map has been developed (Paterson *et al.*, 1988).

The idea is roughly as outlined above for a single marker, using the mixed model and additive effects at the QTL. For example, consider the data relative to a specific interval *M–N*, in the chromosome map, where θ_{MN} is known. If a QTL is in this interval, it will have recombination values θ_{MD} and θ_{DN} relative to these markers. The probability of a given marker genotype in an offspring is a function of these recombination values; assuming that the phenotype depends on variation in the QTL, likelihoods for the offspring phenotypes can be constructed from the observed phenotypes and marker genotypes as described above.

For example, for a double backcross, *MmDdNn* × *mmddnn*, an *Mm–Nn* offspring will have the *dd* genotype with probability $\theta_{MD}\,\theta_{DN}$. The probability density of its phenotype ϕ is $f(\phi|\mu_{\mathrm{dd}}, \sigma^2)$, and so on. In the Lander and Botstein model, the genotypic mean is an additive function of

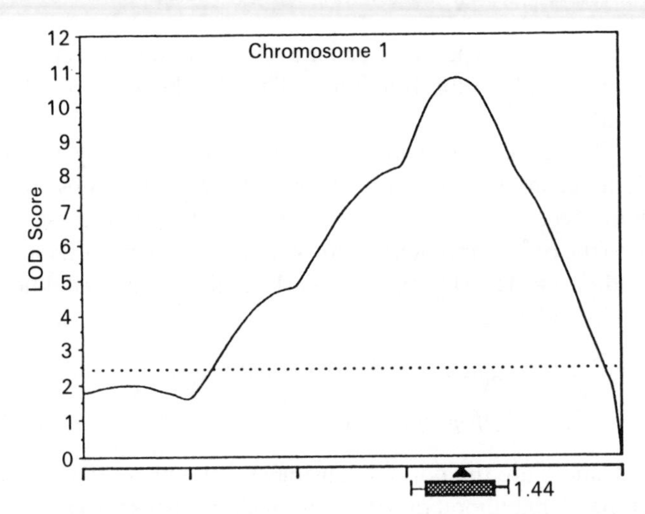

Figure 8.3. Interval mapping for a simulated QTL on a specific chromosome spanned by evenly spaced informative markers in offspring of an intercross. The position of the QTL is shown by a peak in LOD score. (From Lander and Botstein, 1989.)

the number of doses of the *D* allele, with background variance estimated from the data.

The location score maps that are generated by this method can be illustrated by a simulated example of a hypothetical 100 cM chromosome containing one true QTL and typed for a perfect 20 cM marker map (Lander and Botstein, 1989). Two hundred and fifty progeny were simulated from a backcross mating involving this chromosome. Figure 8.3 shows the LOD score for a QTL at all possible locations along the chromosome. The QTL is clearly mapped by the sharp LOD peak. The small dips in the LOD right above each marker locus itself correspond to the anti-modes in location score diagrams, discussed in Chapter 7, which arise when at least one recombination has occurred between a trait and that marker. Here is why: let *M* be the marker allele, with parental types *MD*/*md* and *mmdd*, and suppose the true trait-locus genotype means μ_{Dd} and μ_{dd} are quite different and the distributions hardly overlap. A nonrecombinant child can have marker type *Mm* but phenotype near μ_{dd}; such a phenotype has low likelihood (probability) under the assumption that $\theta_{MD} = 0$, since it would have to arise from the tail of the *DD* phenotype distribution i.e., where the probability is very low.

QTL screening is a powerful way to detect genes affecting complex traits. When a QTL is identified by measured genotype studies with a candidate locus, it should be followed up by linkage studies, to confirm whether the measured locus is the QTL or is merely linked to it, but few if any such follow-up tests have yet been done in humans.

Regressive models for quantitative traits

Chapter 6 described work by Bonney and others to simplify the mixed model by relaxing the constraining assumptions about polygenes and environmental effects in families. The method has been applied also to quantitative traits, and extended to include the use of marker loci as well as unmeasured QTLs (Bonney *et al.*, 1988b). The transmission probabilities for the marker and inferred QTL genotypes are expressed as given above and in Chapter 6, but the other effects are expressed in the form of general regressions, as in Chapter 6. Regressive models are gaining in popularity because they are easier, because most investigators do not really care about the details of polygenic patterns, and because they are available in many of the commonly used segregation and linkage computer packages.

The relative-pair method for quantitative traits

A regressive-type model has been developed for data on pairs of relatives. Unlike the situation for qualitative traits as discussed in Chapter 7, it makes less sense to talk of 'affected' pairs for a quantitative trait, for which the penetrance function has a different meaning and depends on a variety of genetic and non-genetic factors. *All* relative pairs of a given type, not just affecteds, are therefore used. The purpose is to find a relationship between the phenotype and the level of marker genotype sharing among the pairs.

One model regresses the squared difference between the phenotypes of the relatives i and j, $(\phi_i - \phi_j)^2$, on the fraction of genes they share at marker loci. The latter can be expressed in terms of recombination between the marker(s) and a true disease-related QTL; the effects at the QTL are assumed to be additive and there is provision for normally distributed non-genetic effects (Amos and Elston, 1989; Amos *et al.*, 1989; Haseman and Elston, 1972). The relationship between the phenotype difference and marker type sharing depends on recombination if the QTL and marker are linked.

This model is less sensitive to the assumed details of the genetic model, and although the method is formally less powerful than a complete likelihood approach (when the specified model is true), the computations are quicker and easier and the required data easier to collect, and the difference in power often is small (Demenais and Amos, 1989). The method has been extended to account for age of onset, e.g., of chronic diseases (Dawson *et al.*, 1990).

This method will be especially problematic if the important allele at the QTL is rare, or if the QTL has small standardized effect on the trait. Efficiency could be improved by approximating the 'affected' pair method by selecting pairs of siblings with trait values far out on the phenotype distribution, who may be more likely to share important QTL alleles (Carey and Williamson, 1991), but I know of no current examples of this strategy.

Mapping polygenic effects in pedigrees

We can even use these methods to 'map' polygenes that we do not try to identify individually! The phenotypic difference between relatives can be expressed as a function of the fraction of a given chromosome that the two relatives share i.b.d. The latter is determined by marker types in the relatives and the recombination events that could have occurred since they shared common ancestry. The phenotypic covariance between pairs of relatives can then be regressed on the fraction of a chromosome region shared. This accounts statistically for fractions of the phenotypic variance controlled by polygenes located in the shared region (Goldgar, 1990). For some purposes it may be enough to know that some gene(s) in the region have major effect on the trait; a hunt could then be undertaken to find the individual loci that are responsible.

Conclusion

Perhaps because their causation seems simple, we are ready to assume possible single-locus causation for qualitative traits. We seem to have been just as ready to assume that genetic influences on *quantitative* traits are polygenic and individually not identifiable or mappable. We may owe this attitude to the influence of Fisher (1918), showing that many small, additive, mendelian effects can produce a quantitative phenotype distribution. On the basis of those ideas, agricultural and experimental breeding has had a splendid century of success in selecting for desired quantitative traits by empirical breeding based on the theory of aggregate polygenic causation.

However, QTLs can now be identified and in principle it seems that we will always do better by understanding specific causes. Despite the complications that will be found, it is comforting to know that methods are available to make systematic genetic sense of quantitative traits – even by reverse genetics – when we know little or even nothing about the biology of the trait.

APPENDIX 8.1: **A note on statistical power for linkage studies**

It is somewhat difficult to determine the number of families needed to establish linkage between a marker locus and alleles at a QTL. However, we can get an idea of the sample size requirements by looking at optimal conditions: a marker with maximal *PIC*, very tightly linked to a major locus, nuclear families with large sibships, and a QTL with large standardized effect (e.g., $\mu_{DD} - \mu_{dd} \geq 3\sigma_{\phi}$). Samples of about 50 such families will detect linkage with about 80% power (Demenais and Bonney, 1989; Ott, 1991). As these conditions are relaxed sample size requirements quickly rise to the order of 100. Fewer multigeneration families would be needed to achieve the same power. Most QTLs have a standardized effect far less than 3σ. In such situations, the power of a study can be greatly improved by ascertainment from the tail of the phenotype distribution, as mentioned above, to raise the frequency of the rare high-effect QTL alleles (Carey and Williamson, 1991).

Part III *Evolution: the time dimension in populations*

9 *Genes over time and space*

In the domesticated animal all variations have an equal chance of continuance; and those which would decidely render a wild animal unable to compete . . . are no disadvantage whatever . . .

A. R. Wallace, 'On the Tendency of Varieties to Depart Indefinitely from the Original Type' (1858)

Survival of the luckiest.

M. Kimura (1989)

We have seen how to infer that genes have various effects on a trait, but how does the spectrum of genetic effects arise, and how is the variation distributed over space and time? This chapter provides a brief travelog of some of the relevant concepts of population genetics, the theoretical basis of evolutionary biology {Crow, 1986; Hartl and Clark, 1989; Hedrick, 1985; Li and Graur, 1990; Nei, 1987}. These concepts are used in subsequent chapters.

Life is fundamentally stochastic: the fate of a new mutation

(Nearly) each new mutation is unique

For most of this century it was thought that only a few alleles could exist at a locus, of which new copies arose via *recurrent mutation* at some rate, generally estimated to be about 10^{-5} per locus per generation, providing a continual supply of a given allele. This view was based on phenotype data, but at the sequence level the probability that the same mutation will recur is very small. Think of a *cistron* (coding stretch) of, say, a thousand nucleotides. Three thousand different non-synonymous (nucleotide-changing) point mutations are possible. But the mutation rate *per nucleotide* seems to be no more than 10^{-8} per meiosis, so that even if it is true that *some* mutation arises in the gene every 10^5 meioses, the exact change will rarely be seen twice. Of course, true recurrent mutations occur especially in the sequence hot-spots such as CpG runs (Chapter 1).

The idea that new mutations are unique in the population in which they arise is known as the *infinite alleles* or *unique mutation* model, and is at the core of modern population genetics.

Genetic drift: allele frequency change due to random factors

Mendelian segregation in sexual reproduction makes genetic variation *fundamentally* probabilistic. Every generation, every copy of every allele on every chromosome is subject to this sampling process if it is to survive to the next generation. In this sense, the future experience of any *specific* allele (i.e., on a single specific chromosome) is the same as that for a new mutation, say α. Initially, the mutation has frequency $1/2N$, where N is the diploid population size. What happens to α over time?

Since it is uniquely labeled, the person carrying α is a heterozygote, say $\mathbf{A}\alpha$, and will participate in an $\mathbf{A}\alpha \times \mathbf{AA}$ mating, where $\mathbf{A}$ denotes all other alleles at the locus. The probability that an offspring will bear the α allele – the segregation proportion – is $\frac{1}{2}$. Assuming for simplicity that the mating produces s offspring, the binomial distribution [5.4] specifies the probability that r of these have the α allele. Of particular interest is the probability that the marked allele will be lost, which is

$$\Pr(r = 0; s) = C(0; s)(\tfrac{1}{2})^0(\tfrac{1}{2})^{(s-0)} = \tfrac{1}{2}^s. \qquad [9.1]$$

This has nothing to do with 'selection' against α, but only with the chance aspect of mendelian transmission.

In a steady-state human population $s = 2$ on average, and a new mutation immediately becomes extinct with probability $\frac{1}{4}$. The binomial distribution also specifies how likely the other possible α-allele frequencies (e.g., values of r from 0 to s) are in the F_1 generation, and this can be iterated over subsequent generations. For example, for the t-th generation, if the frequency of descendants of α is p_t, the probability that the frequency will be p_{t+1} in the next generation (i.e., $r = 2Np_{t+1}$ copies) is

$$\Pr(p_{t+1} | p_t) = C(r; 2N)\, p_t^r(1 - p_t)^{2N-r}. \qquad [9.2]$$

An allele frequency 'meanders' between 0 and 1 in a way that can be described only probabilistically – that is, the exact 'sojourn' that a given allele will take cannot be specified in advance.

This process of allele frequency change is called *random genetic drift*. Equation [9.2] shows that each generation there is a new opportunity for extinction. Since the unique mutation model assumes that the same mutation will not arise again, the only way for the set of descendants of α to escape eventual extinction is for them to become fixed! The probability that this will happen to an allele is its current frequency, p, and the probability that it will be lost is $1 - p$. Thus, a new mutation has probability $1 - (1/2N)$ of becoming extinct, and in fact most mutations are quickly lost.

This is a continual process, as new mutations arise and drift in or out, and the *rate* at which they (i.e., descendent copies) become fixed, the *substitution rate*, equals μ, the mutation rate per locus per generation. Although this does not depend on the population size, the overall probability of fixation for a new allele ($1/2N$) is greater in small populations (small N) – such as those in which humans have spent most of their evolutionary history.

The basic calibrator of drift is the *effective population size*, N_e. The effective population size is the size of a homogeneous population of equally reproductive individuals that would generate the same rate of genetic drift as is observed in the real population of total census size N. A real human population is structured in many ways; it has individuals of different sexes and ages, geographical and social subdivisions. These factors affect the rate of genetic drift. N_e must usually be estimated indirectly or assumed; typically $N_e \approx N/2$ to $N/3$ for humans.

For the mutations that do eventually become fixed, the average *fixation time* is about $4N_e$ generations. For the remainder that eventually become lost, the average *survival time* is about $(2N_e/N) \ln (2N)$ generations. For example, in a population of $N = 10\ 000$, far larger than the local mating groups in which many of today's alleles originally arose, the mean survival time before all copies are lost, is only about 10 generations, approximately 250 years.

In addition to the effects of mendelian transmission, genetic drift is produced by the many other stochastic elements of our life-experience: mortality, failure to find mates, and variation in family size, including historical demographic bottlenecks and other fluctuations in population size. The result is a steady flux of new variation into, and existing variation out of, the population, like the River Ganges, ever flowing, ever changing, always the same, never the same.

The evolutionary impact of genetic drift at a locus

How realistic is this model of evolution by genetic drift? Alleles evolving only by genetic drift are said to be *selectively neutral* relative to each other. A locus evolving in this way will approach a steady-state, or equilibrium, allele frequency distribution, a balance between mutation and drift. Such an ideal equilibrium may never be reached exactly in nature, but the deviation may not be too great.

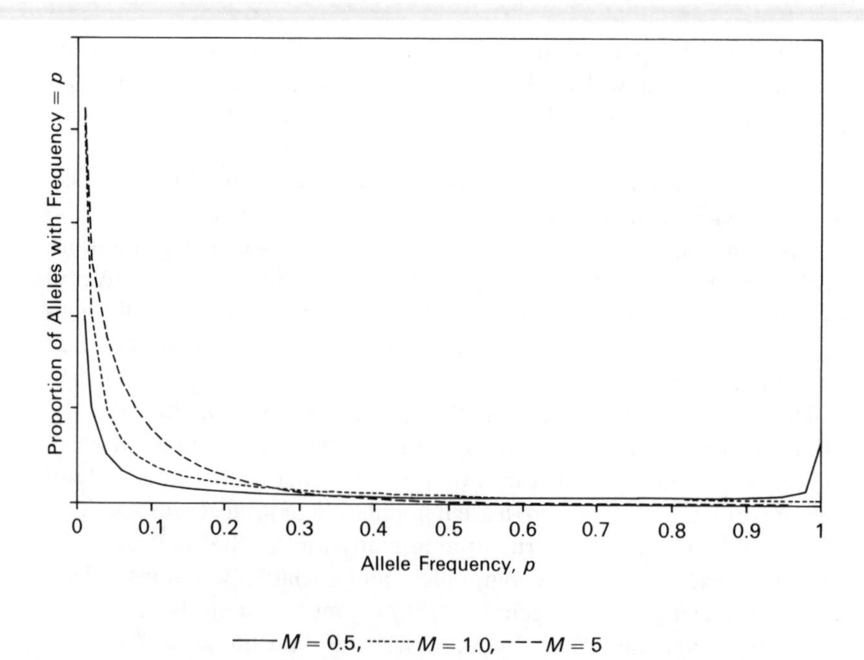

Figure 9.1. Expected distribution of allele frequencies under various assumed values for $M = 4N_e\mu$.

One simple approximate formula for the relative proportion of a sample of selectively neutral alleles that will have frequency p at equilibrium is

$$\Pr(p) = M(1-p)^{M-1}/p \qquad [9.3]$$

where $M = 4N_e\mu$ {Crow and Kimura, 1970; Gale, 1990; Nei, 1987}. As shown in Figure 9.2, most alleles will have very low frequency; drift allows only a few alleles to escape from low frequency, and of these even fewer reach high frequency. We can also predict the average *number* of different alleles in samples of a given size from a locus evolving strictly by genetic drift with mutation rate μ (Figure 9.3). This statistic has many important applications.

Data from a wide variety of loci and species are reasonably consistent with these predictions, as shown in Figure 9.2, expressed in terms of the heterozygosity $H = 1 - \Sigma_i p_i^2$. This does not imply that every mutation is selectively neutral, but does suggest that *most* variation segregating at any given time is approximately so.

Actually, there is evidence from the allele frequency distributions in large human populations, such as industrialized nations, that there are

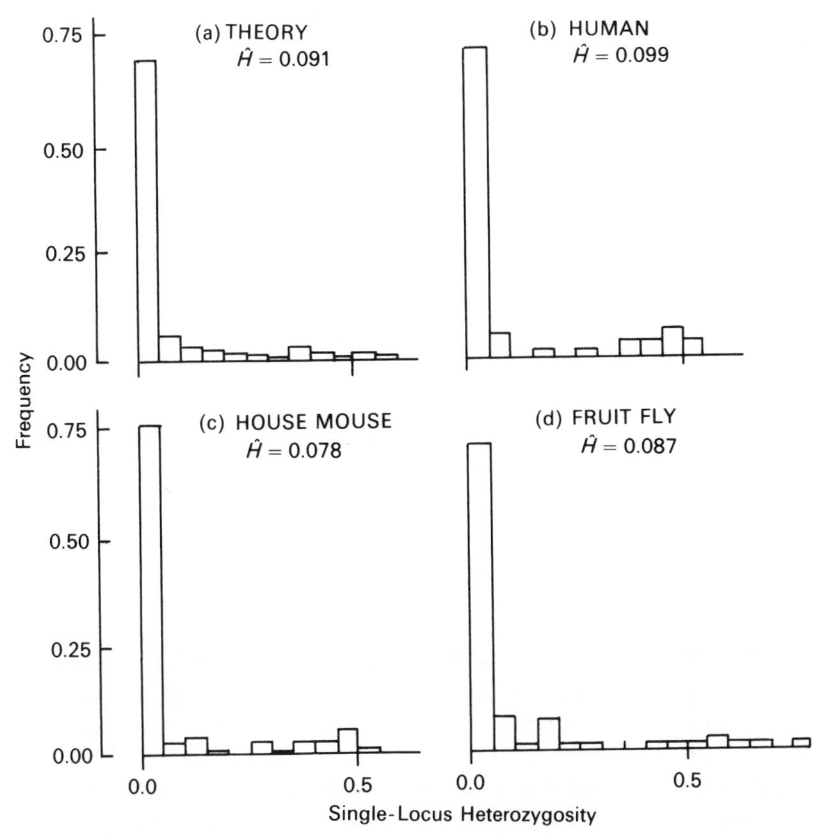

Figure 9.2. Theoretical and observed distributions of locus heterozygosity under the neutral model. (From *Molecular Evolutionary Genetics* by M. Nei, 1987, Columbia University Press, New York. Reprinted by permission from the publisher.)

too many rare alleles relative to expectations. One explanation is that today's populations have not grown gradually from a single, coherent population over the past few millennia, but rather were formed by the amalgamation of small, previously isolated tribal populations (Chakraborty *et al.*, 1988).

Of course, no one thinks that selective neutrality applies uniformly across the genome. At the DNA level, there is plenty of heterogeneity in the pattern of variation (i.e., substitution rate per nucleotide), reflecting drift in some regions and natural selection in others {Li and Graur, 1990; Nei, 1987}. The more functionally important the region of DNA, the

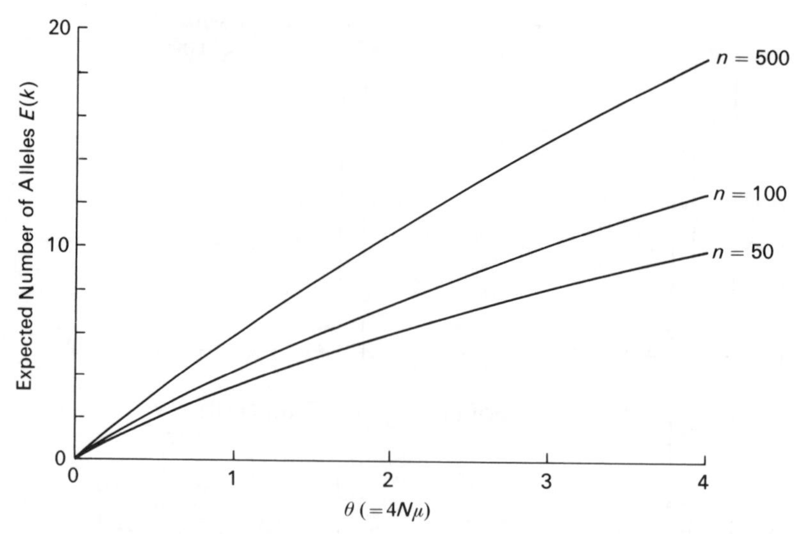

Figure 9.3. Relationship between $4N\mu$ and the number of alleles in a sample of neutrally evolving alleles. (From Hartl and Clark, 1989.)

fewer the substitutions that accumulate per unit time, and the proportionately higher the fraction of synonymous (not affecting the amino acid code), or 'semi-synonymous' (e.g., changing from one amino acid to another with similar chemical properties) substitutions. Codons for the important binding sites of a protein are highly sequence-conserving over evolutionary time. Non-functional regions such as introns and pseudogenes are most variable and fit the theory best.

Subtle patterns of this kind are pervasively distributed across the genome, found in all of our tens of thousands of genes involving hundreds of thousands of codons, billions of nucleotides. How can such patterns be maintained? We can only speculate. Perhaps the machinery for cell division and differentiation is screened by the need to undergo successful spermatogenesis, while housekeeping genes, those responsible for normal cell physiology are tested for years in ovarian follicles, because only long-term healthy oocytes can produce viable eggs. At present, we do not know.

More on the distribution of genetic variation at a locus

Even in random mating populations evolution generates identity by descent

Empirically, at a given stretch of one or a few nucleotides, most individuals in a population are homozygotes, but most are heterozygous over every 1000 bases or so; that is, for segments as long as a functional gene. We can see why this is so from the formula for the equilibrium heterozygosity at a selectively neutral locus:

$$H = 4N_e\mu/(4N_e\mu + 1). \quad [9.4]$$

The homozygosity is the complement of this, often denoted F because under the unique mutation model both alleles in a homozygote are i.b.d. from some original mutation. The amount of variability in a gene region depends on the compound parameter $N_e\mu$. For a typical human N_e, of the order of 10^4 or fewer for most human populations during our evolution, the mutation rate over a short stretch of nucleotides will be very small (e.g., 10^{-8}), and H will be correspondingly small. Drift fixes alleles much faster than mutation generates new ones. But over a length as long as 1 kb the mutation rate rises to a total of around 10^{-5} per generation, and heterozygosity is much higher.

This also implies that over substantial stretches of DNA our traditional concepts of allele frequency are somewhat obsolete – at the DNA level, nearly each allele at a locus is unique (except perhaps in close relatives), i.e., has a 'frequency' of about $1/2N$. However, specific mutations, for example single nucleotide changes that lead to disease, can have non-trivial frequencies.

A note on inbreeding and homozygosity

This is a good point to mention some of the subtleties and confusion that results from considering homozygosity. Within a pedigree, individuals who share a common ancestor can be homozygous i.b.d. from that ancestor, with a probability that depends on the nature of the pedigree paths that connect them (Chapter 6). This probability is called the *inbreeding coefficient*. An individual can also be homozygous due to i.b.s.; that is, not from the common ancestor in the pedigree. If there is an average F of inbreeding in a population, i.e., an excess probability of homozygosity above that expected from random mating alone, the total fraction of each type of homozygote in the population will increase by Fpq, e.g., to $p^2(1 - F) + Fp$. The frequency of heterozygotes correspondingly declines by $2Fpq$.

We are often interested in finding high levels of homozygosity. Inbreeding generates this, as does drift in small populations [9.4]. However, there is an important difference. Drift increases the frequency of randomly selected alleles at a locus, which therefore rise in frequency at the expense of the other alleles at the locus. Inbreeding, in the sense of non-random mating, however, elevates the frequency of *all* homozygous *genotypes* at the locus without changing the allele frequencies themselves. Population subdivision has similar effects (pedigrees are population subdivisions); assortative mating, the choice of mates like (unlike) oneself, raises (lowers) the frequency of homozygotes at loci related to the mate choice. This is a subtle and potentially confusing area, but is of major evolutionary importance {Hartl and Clark, 1989; Jacquard, 1974}.

Religious or other population isolates are of interest to geneticists for several reasons. First, isolates may be founded by a small number of individuals who non-randomly sample the alleles of their source population. This *founder effect* is a form of drift that can by chance raise some otherwise rare alleles to appreciable frequency. Subsequently, the smaller N_e in the isolate leads to more rapid drift effects, which can also raise a random selection of alleles to much higher frequency than previously. Finally, isolates experience high frequencies of consanguineous matings, in the pedigree sense, either by social rules or simply because a high fraction of the available mates are also relatives, e.g., cousins of various types.

The founder effect and drift can thus make some random set of otherwise rare alleles – for example, disease alleles – much more common in an isolate than in the larger population. By raising the frequency of homozygotes, inbreeding makes recessive diseases more common, for those recessive alleles that are in the isolate.

Gene clones, and the trail of history

Sequences that share common ancestry in the population constitute a *clone* of homologous alleles i.b.d. from that ancestral chromosome. At every locus some allelic variants, or sequences within the gene, will be very old, high up in the branching structure of the existing sequence variants at the locus, and widely dispersed in the species. Nonetheless, until a few thousand years ago (and even presently in some parts of the world) human populations were typically very small, and most mutations arise and remain in one localized area. Thus, while some RFLPs or haplotype frameworks are found species-wide, the details of many of the alleles that have arisen in and around them will often be highly local.

Among the members of a clone of alleles defined by a specific shared sequence, other mutations will arise around that sequence as the time since their common ancestry accumulates. The longer the sequence persists without itself being lost to drift, the more disparate will be the surrounding sequences among its copies in the population, and the more geographically dispersed the clone can become. This is a *nested* process,

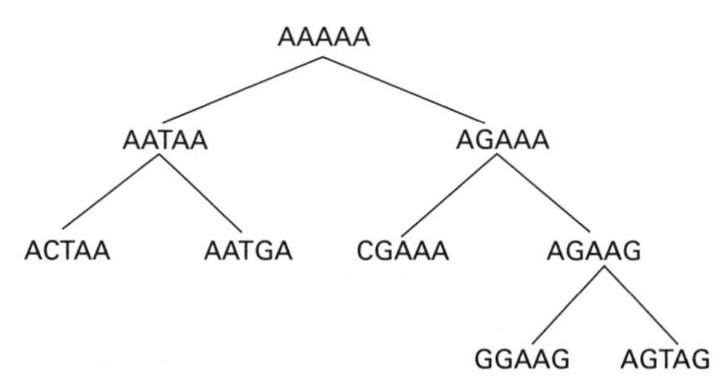

Figure 9.4. Hypothetical sequence cladogram.

in that each new mutation arises on a given haplotype background, and some of its descendent copies will, in turn, be modified by subsequent mutation. At any given time, a series of sequences in a population constitute a *hierarchically* structured *clade* of sequences, or alleles, that share some aspects but are diverse for subsequent mutations. These sequences can be arranged into a *cladogram* that represents the history of these events. The historically true cladogram cannot always be reconstructed unambiguously from a sample of haplotypes, but can often be approximated, and we can also sometimes reconstruct the original allele from which all alleles in the sample may have descended. This ancestral sequence is known as the *coalescent*. A simple hypothetical example is shown in Figure 9.4, where the coalescent is the sequence AAAAA.

If a population remains intact, the average coalescent time is the substitution time for neutral alleles; the population consists of an ever-changing set of coalescents as one is established and then in turn is replaced by mutation and drift {Hartl and Clark, 1989; Li and Graur, 1990; Nei, 1987}. But if a population splits, the coalescent time for alleles sampled *between* the descendent populations will remain the same as the absolute (historic) coalescent time at the time of splitting; drift can fix an ever-changing set of haplotypes in each population, but their most recent common joint ancestor will still be that at the time of splitting. Various computer programs exist for constructing cladograms from a set of sequences.

What can be done for alleles among individuals within a population can also be done for sequences of the same gene between species, and even among related genes within an individual. Chapter 2 provided diagrams of the sequence relationships among globin and apolipoprotein genes that have arisen due to mutation and drift since the genes were created by duplication events. The more similar the sequence, the more recent we infer that the duplication event occurred.

A corollary of all this is that cases of a given disease that arise in a specific population may be *clonal*; that is, due to alleles that constitute a clone of descendants of a specific pathogenic mutation. In fact, even when the evolutionary process generates *phenotypic convergence* in different populations (i.e., similar phenotypes), unless the populations derive from a recent common ancestral population, these phenotypes will be due to *genotypic divergence* (i.e., to different mutations) in each population.

Most new mutations are quickly lost to drift, or at least geographically isolated. Even true recurrent mutations will only rarely survive drift, and those that do can often be distinguished from each other because these arise at different times and the haplotypes on which they occur (surrounding sequence patterns sometimes called frameworks) are different and locally specific. Most new mutations remain localized.

Isolated populations become genetically divergent with time

Two sequences (copies of an allele) that start out identical become genetically divergent over time due to the accumulation of different mutations. With mutation rate μ per generation the probability of gene identity in two sequences is $(1 - \mu)^2$ after one generation, and by extension, after t generations is

$$I = (1 - \mu)^{2t} \approx e^{-2\mu t}, \qquad [9.5]$$

which declines exponentially with time. We say that the *genetic distance* between the sequences increases (Nei, 1987). Genetic distance can also be measured in terms of allele frequency differences between populations that derive from some common ancestral population; allele frequency distances also accumulate exponentially with time if the populations remain isolated (Roychoudhury and Nei, 1988). Cladograms exist because of this divergence; their structure can often tell us about the history.

Gene flow: allele frequency change due to the movement of people

If mutation and genetic drift cause divergent genetic patterns over space and time, mating between populations causes *gene flow* which makes the populations genetically more alike. Gene flow is important in the context of this book because: (1) it is useful to understand the ethnic and geographical distribution of disease genes; (2) recent long-distance movements of people have transported disparate genes to new areas rather rapidly (e.g., Europeans and Africans moving to the New World);

(3) *admixture* between populations can generate useful linkage disequilibrium between marker and disease loci; and (4) admixture between formerly disparate populations creates opportunities for new genotypes to arise and to be expressed in new environments. Such occurrences may aid in understanding the genetic basis of phenotypes in a way that is difficult to do in either parental population.

Gene flow has quantitatively simple effects. If two parental populations have allele frequencies p_1 and p_2, and a new population is formed by the admixture of a fraction m of genes from population 1, the allele frequency in the *hybrid* population is

$$p_H = mp_1 + (1 - m)p_2, \qquad [9.6]$$

a simple mixing-proportion equation. The amount of admixture can be expressed in terms of the allele frequencies by rearranging [9.6] as

$$m = (p_H - p_2)/(p_1 - p_2). \qquad [9.7]$$

The most informative allele frequencies are those for 'private' variants, i.e., mutations found in only one of the source populations, especially if they are fixed in that population (Chakraborty *et al.*, 1991).

The arrangement of genes: linkage relationships over time

Evolution changes the relationships between alleles at different loci. Every new mutation arises on one specific chromosome, thus creating linkage disequilibrium between the new mutation and alleles at linked loci (since the mutant cannot, at first, be found on all possible haplotypes in the population).

Because of recombination, linkage disequilibrium decays over time, and eventually recombination removes any association among the alleles. Linkage disequilibrium reduces by an amount θ per generation. If the disequilibrium initially is δ_0, at a late time t it is

$$\delta_t = \delta_0(1 - \theta)^t. \qquad [9.8]$$

This is a constant exponential rate of decay, which means that the value of δ drops off rapidly at first, then more gradually approaches zero over time.

Various factors such as drift and recent admixture that disrupt genetic equilibrium can also produce allelic association, even among alleles at loci that are not syntenic (physically linked). Equation [9.8] shows that linkage disequilibrium decays much more rapidly among loci that are unlinked ($\theta = \frac{1}{2}$) compared to closely linked loci ($\theta \ll \frac{1}{2}$). This is of course the basis of linkage inference in family data and can also be useful in mapping genes in admixed populations (Chakraborty and Weiss, 1988). There is no association between markers and trait loci if the two are in linkage equilibrium.

As seen in the previous chapter, the strength of effect and the degree of linkage disequilibrium are confounded. Indeed, at very short DNA intervals recombination may be about as rare as new mutations, and seemingly inconsistent relationships between several linked sequences may arise, i.e., the mapping function breaks down – disequilibrium is not proportional to recombination distance. We find this when several very closely linked RFLPs are analyzed relative to a specific coding locus, such as in some apolipoprotein regions that have been studied, and this can affect our understanding of genotype–phenotype relationships (e.g., Haviland *et al.*, 1991; Zerba *et al.*, 1991). At such short intervals disequilibrium is not proportional to recombination distance.

The sojourn of a mutation affected by selection

Some genotypes affect their bearers' ability to survive and reproduce; that is, affect the *net maternity function*,

$$NMF_g(t) = S_g(t)m_g(t), \qquad [9.9]$$

where $S_g(t)$ is the survival probability to age t, and $m_g(t)$ is the age-specific probability of reproducing, both specific to genotype. The sum of this function over the lifespan is known as the *net reproduction rate*, $R_{0,g} = \Sigma NMF_g(t)$. Two genotypes whose R_0 values differ have different *darwinian fitness*. The fitness of a genotype is often expressed in terms of its *fitness coefficient*, w_g, that is scaled so that the genotype with the highest fitness is assigned $w_g = 1$. Thus, $w_g = R_{0,g}/R_{0,\max}$. Actually, this is a simplification because relative fitness may also depend on the shape of the NMF as well as R_0 (Crow and Kimura, 1970) {Fisher, 1930}, but those details need not concern us here. If two genotypes have the same fitness they are selectively neutral relative to each other. Except for severely deleterious disease-related mutations how genotypes achieve their fitness is usually a mystery.

The fitness disadvantage of a genotype is the *selection coefficient* against it, $s_g = 1 - w_g$. The fittest genotype thus has $s = 0$; a lethal genotype has $s = 1$. For a diallelic locus, with allele frequency p, the frequency in the next generation is

$$p' = [p^2 w_{11} + p(1-p)w_{12}]/[\Sigma P_g w_g], \qquad [9.10]$$

where the denominator accounts for the total amount of selection against all genotypes at the locus, and normalizes the allele frequency relative to surviving genotypes in the postselection population. The same selection regime can alternatively be expressed in terms of selection coefficients for rather than against specific alleles, relative to a coefficient of zero for the least favored allele.

A deleterious mutation with high penetrance (e.g., dominant) will be eliminated from the population rather quickly, because its fitness-harming effects are manifest in most individuals who carry it. In terms of [9.10], if the fitness coefficients are less than 1 for all genotypes with the allele the change will be rather rapid.

A similarly harmful mutation that has low penetrance will persist for a somewhat longer time, because its effects may not be substantially detectable except in some genotype combinations; for example, if a is recessive, $w_{AA} = w_{Aa} = 1$; even if $w_{aa} = 0$, there is no selection against the allele until it reaches a frequency high enough for recessive homozygotes to appear. Selection is always relatively slow at eliminating deleterious recessive alleles, and in fact drift may often finish the job.

Selection for or against a specific allele is known as *directional selection*, because it systematically changes the allele frequencies in one direction, ultimately removing variation in the population by substituting one allele for the others at the locus. However, there are circumstances in which selection favors intermediate allele frequencies in which a set of alleles persist in the population. This is called *stabilizing selection* and can arise if the selective values of alleles vary over time, for example in seasonal environments, or are allele-frequency dependent. The best-known type of stabilizing selection is *balanced polymorphism* (*heterosis*, heterozygote advantage, balancing selection), in which the heterozygote has a fitness higher than that of any homozygote. This leads to a stable equilibrium, with $p_{eq} = s_{aa}/(s_{AA} + s_{aa})$, thus preserving variation in the population. The malaria-associated hemoglobinopathies, such as sickle cell hemoglobin (HbS), are still the best examples in humans. In a malarial environment, the homozygote for 'normal' hemoglobin is at risk for fitness-reducing effects of malaria, while the abnormal homozygote is at risk of serious anemia. The heterozygote has little anemia and is protected from malaria. However, even in a strongly selecting malarial environment, the equilibrium allele frequency is only $p \approx 0.2$ for HbS.

A definition of selection

Alleles affected by selection are subject also to genetic drift. For one thing the selection coefficient can usually be estimated only with great difficulty, by comparing the production of finite samples of each genotype in the offspring. Fitness is the net result of the age-specific pattern of

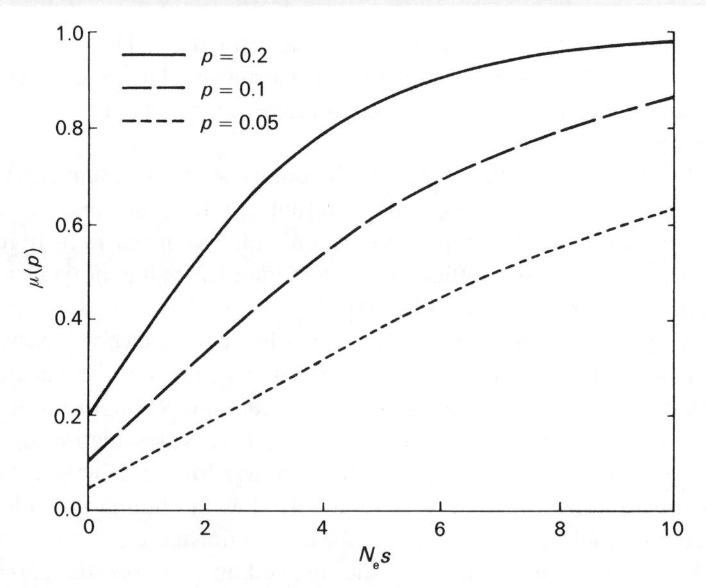

Figure 9.5. Probability of fixation as a function of selection and N_e, for purely additive fitness. (From Hedrick, 1985.)

mortality and fertility which (except for lethals) must be compared among all genotypes at a locus, in large enough samples to average over the entire range of irrelevant factors – sources of environmental and polygenic variance – that produce random fitness differences among individuals with the same genotype at the locus. Thus, the ultimate fate of a selected allele is partially probabilistic, and so is the concept of 'fitness' itself.

With favorable *allelic selection*, in which fitness relative to an allele is proportional to the number of copies in a genotype (e.g., relative fitnesses of $1 + \zeta$, $1 + \zeta/2$, and 1), the probability that a favored allele with current frequency p will eventually become fixed is approximately

$$\Pr(p \rightarrow 1 \text{ as } t \rightarrow \infty) \approx \frac{1 - \exp(-4N_e \zeta p)}{1 - \exp(-4N_e \zeta)}. \qquad [9.11]$$

Since $\exp(x) \approx 1 - x$ when x is small, as $\zeta \rightarrow 0$, i.e., for alleles that are nearly neutral, [9.11] converges to p; that is, as stated earlier the probability of fixation of a neutral allele is its current frequency. Essentially, whether there is selection or not may not generally be a clear-cut property of an allele, but can be defined only in terms of an allele frequency change relative to that expected by drift; that is, strictly probabilistically.

If there is selection the probability of fixation rises and is determined by the product $N_e\zeta$ (see e.g., Gale, 1990; Hedrick, 1985). If this product is much less than 1, the fixation probability approaches p, the expectation for a strictly neutral allele, and drift drives the system, but if $N_e\zeta$ is much greater than 1, selection drives the system.

There are complications other than those due to drift. Many genes are *pleiotropic*; that is, affect more than one phenotype. A gene may be expressed in various tissues with different effects. In experimental breeding, pleiotropy shows up as a correlation among traits in their response to selection on some index trait. An allele that may not be well suited to one of these biological systems (e.g., produces a particular disease), may be advantageous for other functions. The advantage or disadvantage may depend on local environments, or even on the frequency of the allele itself (the latter may apply, for example, to genes related to immune resistance; the more common the allele, the more selective pressure that allele puts on a pathogen to mutate to some other, newly virulent form).

Multilocus phenotypes: buffered systems evolving by phenotypic selection

For complex quantitative characters it is more difficult to write down a simple evolutionary theory for the allele frequencies at a specific locus. We can, however, attempt to understand how such traits evolve and thus to explain the genetic architecture observed for such traits. The starting point is to have some understanding of the general pattern of natural selection that applies to quantitative characters. Of course, in principle, selection can work on the phenotype distribution in any number of ways, but a typical pattern seems to be as follows.

At any point in time, the phenotype distribution represents a basically 'adaptive' pattern (i.e., it has evolved). Although most phenotypes in the population must function satisfactorily, a few at the extremes of their distribution are usually disadvantageous. If phenotypes at one extreme are more heavily affected than those at the other, the mean of the distribution will move over time in the relatively favored direction, as seen in Chapter 6 (response to selection). This is *directional* selection for a quantitative trait; because of allelic equivalency, however, the effect here is *not* referable to specific alleles.

For most traits most of the time, it seems that only rather slow and subtle changes are being wrought; that is, selection is largely of a *stabilizing* type. It may trim away the deleterious, usually rare, extreme phenotypes but pay rather little attention to the rest of the distribution.

Even if selection against the extremes is quite strong (some may even be lethal), the distribution does not change very much. This view is consistent with our historical difficulty in detecting selection within the normal phenotype range. Stabilizing selection reduces genetic variation somewhat, but the loss is opposed by the input of new mutational variation in each generation, at all the loci involved. Eventually an approximate balance is established that preserves genetic variance.

For example, there is still ample phenotypic and genotypic variation for human stature, brain size, and birth weight. These traits evolved slowly towards greater mean values over the past few million years, but that change basically stopped a few tens of thousands of years ago. The two-fold increase in human brain size during our evolution from ape-like ancestors presumably was produced by a combination of slow directional selection favoring larger brains, and stabilizing selection against cranial sizes at birth that would have impaired successful delivery. We do not know whether this evolution occurred smoothly and gradually, or by rapid, local bursts of intense selection (the fossil record does not allow us to resolve this question).

Most phenotype distributions appear to be in this kind of quasi-equilibrium, with high levels of genetic variability. This could explain why for many traits such as blood pressure (Chapter 6), the heritability is similar in different populations: by selection at the edges of the distribution, a roughly similar balance between selection and new mutation exists in each population. Of course, we must expect to find that the *specific* alleles differ, once we have been able to identify the loci involved.

However, experimental models for complex traits indicate that there can be a wide range of fitnesses among alleles in the experimental lines (see e.g., Clark, 1991). Traits directly related to fitness per se (e.g., clutch size in birds) manifest little heritability, presumably because selection has efficiently exhausted genetic variability (Mousseau and Roff, 1987).

The genetic causal spectrum of complex traits

It is of interest to develop as many generalities as we can about the genetic architecture (causal spectrum) of complex traits and how it evolves. The basic parameters we need to understand are the distribution of allelic effects, the number of loci, and the relationships among genotype, phenotype, and fitness. Attempts have been made to develop a general mathematical theory for the evolution of quantitative traits {Barton and Turelli, 1987, 1989; Bulmer, 1989; Hartl and Clark, 1989; Lande, 1987; Lynch, 1988; Turelli, 1987}. This theory is complex, and has run into difficulties; in the following, I present a basic characterization of some of the main points, based on the available data and some recent work on the problem by Adam Connor and myself (Connor, 1992; Connor *et al.*, unpublished results).

The distribution of allelic effects

There is no formal theory for the distribution of allelic effects on a biological phenotype. However, we can imagine a general process that might produce varying allelic effects over evolutionary time. Consider a given selectively neutral allele, say *A0*, associated with mean effect μ_{A0}. Over time, a tree (clade) of descendent alleles will develop (if the allele is not lost to drift), as mutations accumulate randomly in its various branches. Suppose for simplicity that each new mutation randomly adds or subtracts some fixed amount, d, to the phenotype. Now, suppose we look at two current descendent alleles, *A1* and *A2*, that have some fixed coalescent time, t, back to *A0*.

According to Figure 9.6, the first descendent allele will have accumulated a number $X1^{+}(t)$ of mutations that add, and $X1^{-}(t)$ that subtract amount d from its effect relative to μ_{A0}, and similarly for the second allele. If these occur with mutation rate μ and the positive and negative effects are equally likely (i.e., a Poisson process model), the distribution of the difference between the phenotypic effects of two alleles with known coalescent time t will approximate a normal distribution centered at μ_{A0}, as mutations make the phenotypes 'diffuse' about their origin. (The assumption of fixed effect d is not critical to this result.)

Actually, the coalescent time varies stochastically among random pairs of alleles because of the effects of random genetic drift on the alleles over time. The difference in allelic effects can be measured by their *standardized effect*, ξ, defined in Chapter 6, which takes into account their frequency. The distribution of standardized absolute effects (not considering sign), among random pairs of alleles will be *exponential*; that

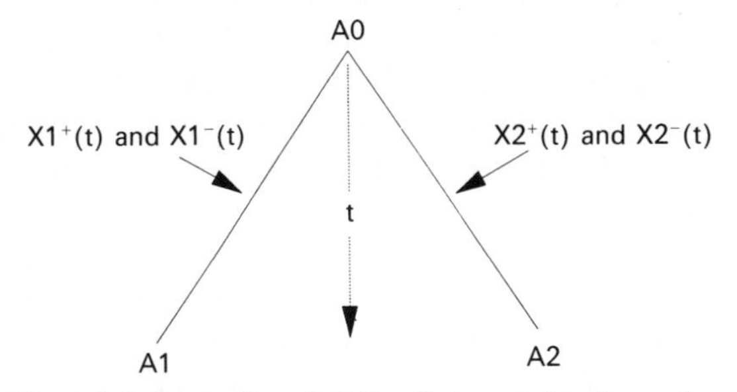

Figure 9.6. Generation of allelic effects over time by random neutral mutations.

is, the density of a given size of effect, x, will be $\Pr(x) = \lambda \exp(-\lambda x)$. If we take into account whether alleles raise or lower the phenotype, the distribution is a symmetrical exponential about the origin, known as a *Laplace* distribution.

A natural population has many alleles, not just two, and the distribution of allelic effects at a 'wild' locus will be more complex; the reason is that genealogies between pairs of alleles will differ in the length of time back to their coalescent (e.g., some may coalesce in the previous generation, others may go back to the common coalescent for all alleles at the locus and so on). The resulting distribution of allelic effects will have rapidly declining shape, more centered and with thinner tails, somewhat between a normal and a Laplace distribution.

We cannot easily test the latter model with existing data, but data from experimental crosses between inbred strains of plants and animals do allow us to test the two-allele model. Inbred lines are entirely homozygous; at each locus, wild-type alleles have been fixed by the person(s) creating the line. Apart from alleles whose effect was relevant to the development of the strain (e.g., some strains have been developed by selection as well as inbreeding), the alleles that become fixed by inbreeding are a random selection from the original wild-type alleles. Crosses between two such lines (e.g. $AA \times aa$) yield diallelic loci suitable for testing the reasonableness of the model just developed.

Most of the existing data are from studies of quantitative traits in inbred maize (corn) (Edwards *et al.*, 1987; Stuber *et al.*, 1987), tomatoes (Paterson *et al.*, 1988, 1990, 1991; Weller, 1986; Weller *et al.*, 1988), and *Drosophila* (fruit flies) (Shrimpton and Robertson, 1988a,b). The studies by Paterson *et al.* used formal interval mapping as discussed in Chapter 8; the maize studies have used measured genotype approaches, whereas the studies by Shrimpton and Robertson worked with engineered chromosomes. Deferring for the moment the question of how representative such data might be of wild populations, since each locus provides only one 'replicate' of pair-wise allelic effects, we assume that the loci affecting a given trait have the same evolutionary dynamics, i.e., represent the same distribution as would pairs of natural alleles at a single locus, and further, that this process is sufficiently generic that we can pool across traits and even across species.

Figure 9.7 provides an example of the allelic effects at loci affecting traits (here, in tomatoes, from Stuber *et al.* (1987)), and Figure 9.8 is a 'probability plot' showing the fit of such effects to an exponential distribution (a perfect fit would place the points all on the line). Clearly, the model is a good approximation. This is one of the better-fitting data

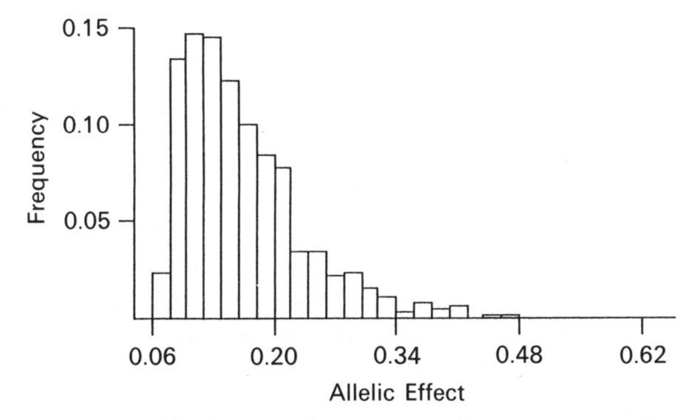

Figure 9.7. Distribution of pairwise allelic effects in experimental data, after standardization of strength of effect in terms of total phenotypic variance. (Data from Stuber *et al.*, 1987.)

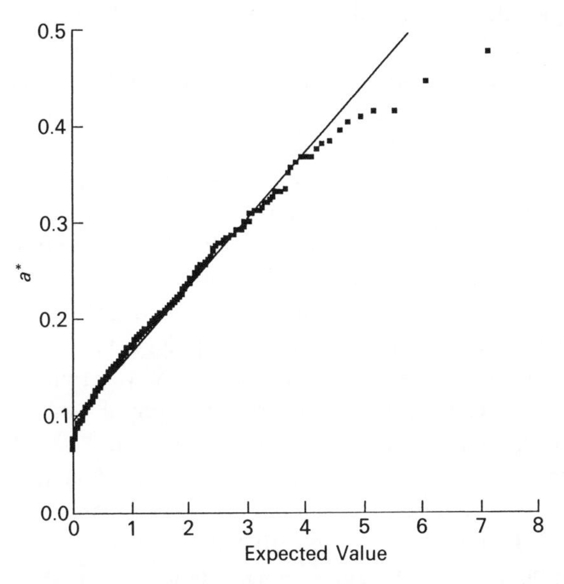

Figure 9.8. Probability plot showing fit of allelic effects distribution to the exponential distribution (perfect fit would place all points on the line). (Based on data from Figure 9.7.)

sets, but of the others referred to above, all provide a reasonable fit. There is usually a clear deficit of very small effects, which, though theoretically predicted, will be too small to generate statistically significant effects even with substantial sample sizes. Note that even among the effects that were detectable, most are small, $\xi < 0.5–1\ \sigma$ (Shrimpton and Robertson, 1988b; Weller *et al.*, 1988).

How well might these data represent the natural loci that affect quantitative traits? The data may underestimate very small effects for the sample size reasons just mentioned, and might overestimate the frequency of larger effects because, as noted above, the distribution of allelic effects at a locus with multiple alleles will be more centered than the distribution of a diallelic system. Also, the published data involve strains chosen because they differed for the tested traits – characteristics such as fruit weight or pH – perhaps further biasing the results towards larger effects. On the other hand, alleles with deleterious effects on viability (especially recessive deleterious alleles) would not be included.

Alleles relevant to the commercially relevant traits for which these strains may have been developed may have been strongly favored or disfavored. To this extent, current allelic variation is not representative of the frequency and effects in the wild-type ancestors. However, most allelic effects are small, including the ones the breeders had to work with, so that phenotypic selection probably does not specifically over-represent alleles with large effect. In any case, there are comparable data involving wild strains, and these show similar allelic effects distributions, suggesting that what we see in the experimental settings is not very misleading (Doebley and Stec, 1991; Doebley *et al.*, 1990; Paterson *et al.*, 1991).

The best systematic human data are on serum cholesterol levels (Sing and Boerwinkle, 1987). These show generally the pattern seen above (Figure 9.9). About half the phenotypic variation can be accounted for genetically, 25% by allelic variation at the few candidate and random marker loci that have been specifically tested, and 25% to residual correlation among relatives (i.e., aggregate polygenic effects). The remaining half appears to be due to environmental variance. These effects vary by age, sex, environmental exposures, population, etc.

It makes sense that alleles with strong effect are usually rare. For one thing, since we measure genotypic effects relative to the population mean (Chapter 6), there is a natural *confounding* of allele frequency and strength of effect. By definition, a large effect is one that differs substantially from the population mean. But if such an allele were common, most genotypes would contain it, and the mean would be dominated by that allele – whose effect would thus no longer be large relative to the mean.

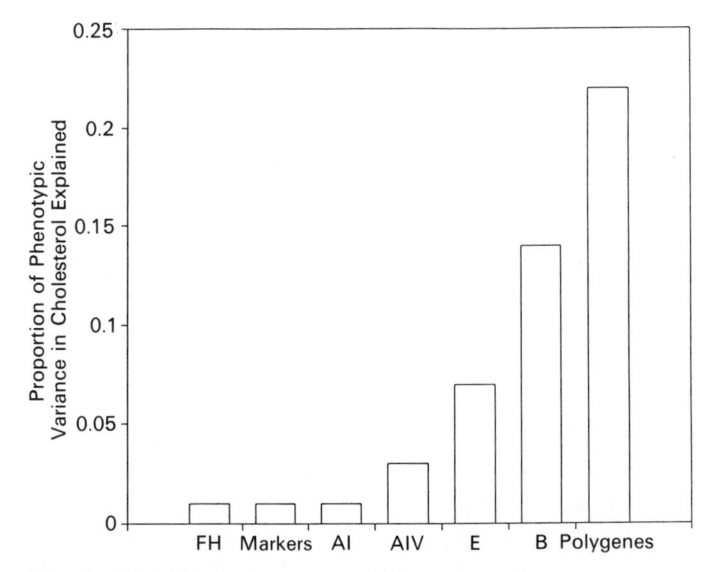

Figure 9.9. Distribution of known allelic effects on cholesterol. Only a sampling of existing data is included, and non-additive effects not included. FH, familial hypercholesterolemia; AI, AIV, E, B, apo-lipoproteins. (Replotted and modified from Sing and Boerwinkle, 1987.)

The kind of distribution described above also makes sense at the DNA level, where we know that a high fraction of mutations are synonymous, and that many others exchange amino acids with similar chemical characteristics, or affect relatively unimportant parts of coded proteins. The biochemical complexity of most biological systems may tend to dampen the effects of new alleles {Dean *et al.*, 1988; Hartl and Clark, 1989}. Dominance and epistasis probably have only local or temporary distorting effects on this picture. It is the additive effects that respond to selection (Chapter 6). Epistatic effects may be only transitory in the evolution of complex traits, depending on population history (Goodnight, 1988).

In sum, the picture of the distribution of additive allelic effects is probably a fair characterization of the genetic control of many quantitative traits.

Phenotypes, genotypes, and fitness

The fitness of a genotype depends on its phenotype distribution and the pattern of selection:

$$w_g = \Sigma_\phi \, w_\phi \, \Omega_g(\phi) \qquad [9.12]$$

(an integral for a continuous phenotype), where w_ϕ is the fitness of phenotype ϕ, scaled so that the most fit phenotype has $w = 1$. In general, fitness may not be symmetric about μ_g, so that as a rule $w_g \neq w_{\mu g}$, the average fitness of a genotype is *not* equal to the fitness of its associated genotypic value.

To quantify the fitness effects of genotypes related to a quantitative trait is difficult, but the problem can be explored with one of the best sets of reasonably relevant available data, namely the effect of serum cholesterol levels on the risk of coronary heart disease (CHD), particularly heart attacks. This is a leading cause of adult death in Western society, and should provide a useful case in point, at least for the genetic architecture of traits related to chronic disease.

The relationship between cholesterol and CHD has been observed in a long-term study in Framingham, Massachusetts (Anderson *et al.*, 1991). We have attempted to translate this relationship into approximate fitness effects (Weeks *et al.*, unpublished results). The phenotype we used was the *cholesterol ratio* (*CR*), between HDL and total cholesterol (TC), i.e., CR = TC/HDL; this ratio is known to be related to CHD risk. To estimate fitness effects, we computed the *NMF* specific to each CR value. From the Framingham data we computed $h\text{-}CHD_{\mathrm{CR}}(t)$, the age-specific hazard of CHD associated with each CR value. There is a monotonically increasing, approximately proportional, relationship between CR and its associated CHD hazard.

For a given CR value, the survivorship schedule depends on this hazard plus the hazard for competing (non-CHD) risks: $h_{\mathrm{CR}} = h\text{-}Other(t) + h\text{-}CHD_{\mathrm{CR}}(t)$; we used this to compute the CR-specific survivorship schedule $S_{\mathrm{CR}}(t)$ from [3.1]. Assuming that CR does not affect fertility, we used the US age-specific fertility schedule, $m(t)$, to determine $NMF_{\mathrm{CR}} = \Sigma S_{\mathrm{CR}}(t)m(t)$. The resulting fitness values are plotted in Figure 9.10. This shows that although high cholesterol is positively associated with the CHD hazard, for most cholesterol values there is very little fitness effect.

Equation [9.11] showed that key to the evolution of allele frequency is the value of $N_e \zeta$. Figure 9.10 shows the effects of selection for phenotypes rather than genotypes, expressed in terms of their fitness relative to a maximum fitness scaled to $w = 1$, for those with low cholesterol values. It is clear that the fitness among most genotypes in the population (the figure includes those with mean effects ranging up to 5σ above the optimum) differ by only about 0.001. These cholesterol levels and hazards were measured in an industrialized-nation population, but during most of the time that the causal alleles evolved the appropriate N_e has probably been of the order of 10^3–10^4. Thus, the vast majority of alleles

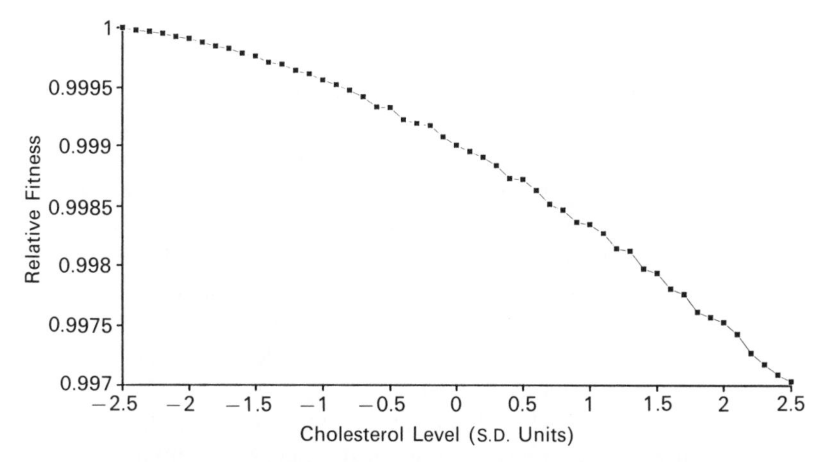

Figure 9.10. Fitness effects of cholesterol levels based on analysis of Framingham data on cholesterol levels and coronary heart disease. The CR scale is in units of standard deviation from the population mean.

responsible for cholesterol values are likely to have evolved largely as neutral alleles; that is, by genetic drift. If anything, it is only with the recent emergence of high-fat Western diets that the cholesterol level distribution has been markedly shifted towards the right, so there is much greater variation in fitness now than heretofore.

The upshot of this section is important: a nearly-neutral model will describe the causal spectrum of important human disease-related phenotypes reasonably well.

How many loci?

How many loci make a contribution to the variation of a typical quantitative trait? From a physiological viewpoint, one would guess that a few loci are specific in their function to a given phenotype, and that variation at many others, like generic 'housekeeping' genes, affect the trait only slightly. The number of loci is also affected by how finely we define the trait. Diseases such as PKU or Tay–Sachs disease that are defined essentially in terms of specific enzyme activities may involve many alleles at those specific loci but at few others. More crudely defined traits, however, will involve more loci, depending on how many pathways there are to similar phenotypes. A good example, is mental retardation. If that were our trait, we would find a great many loci, among them being the loci for PKU and Tay–Sachs disease!

From a purely statistical point of view the number of loci that affect a trait depends on how small an effect one wishes to count, and that in turn depends on sample size; larger samples can lead to the detection of smaller effects. With infinite allele models such as those used here, there is a continuum of alleles at a locus, so by definition the larger the sample size the more alleles will be detectable, ad infinitum. This is, of course, only an approximation to reality, but, the data are consistent with those approximate predictions. The agricultural studies cited above, which spanned large fractions of the genomes of the tested species, typically measured test progeny in the hundreds, and about half or more of the marker-trait tests were statistically significant, identifying about 5–10 substantial QTL associations per trait (Edwards *et al.*, 1987; Paterson *et al.*, 1988; Stuber *et al.*, 1987; Weller *et al.*, 1988). Classical biometrical and other indirect approaches had previously suggested similar numbers for most traits (e.g., Barton and Turelli, 1989; Robertson, 1967; Thompson and Mascie-Taylor, 1985; Thompson and Thoday, 1979; Wright, 1968).

QTLs are mapped only as regions between markers, and subsequent work is required actually to find the gene itself. A number of QTL regions have detectable effects in a variety of different strains that have been tested, and some of the loci themselves have been identified and shown physiologically to be related to the tested traits, confirming their statistical causal role (Paterson *et al.*, 1990). But effects at other QTLs have 'disappeared' when the same plants were raised in different environments (Paterson *et al.*, 1991). Thus, we must be careful about inferring that variation at a locus always affects a given trait. These are early days in our attempts to understand the causal spectrum, and we should be neither surprised nor discouraged to see that the major generalities that have emerged are littered with detail and exceptions. But typically there will be only a few alleles having moderate population-level phenotypic effects, and many loci with small effects. Large-effect alleles will be rare. These effects may vary among populations or individuals.

Gene duplication, complexity, and the evolution of dominance for quantitative characters

How does the complex control of a biological trait evolve? I can try to illustrate my own view of this subject with a somewhat speculative scenario based on what we know about genetic evolution and the effects of phenotypic, rather than genotypic, selection. For example, we can construct a Kiplingesque Just So story about the Lipoprotein River, referring back to Appendix 2.1. Suppose at some time in the distant past, a gene arose, call it apo-A0, whose coded protein structure bound lipids in a way that made them more easily transportable in the aqueous environment of the primitive circulatory system. Perhaps the original gene coded for the basic apolipoprotein 11-mer. With this gene a new and better method of lipid circulation arose, but the system is vulnerable to mutation: a diploid individual has two copies of the gene, so deleterious mutations can have a detectable effect.

The system may be improved by gene duplication, being initially made more efficient by duplications of the 11-mer to make a longer gene product. Subsequent duplication of the whole gene yields a system that has *redundancy*; even with just two proto-apolipoprotein (!) loci, a diploid individual has four copies rather than two, and a new mutation is less likely to do serious harm. There is a direct precedent for this kind of protection in the duplicated α-globin genes compared to the single-copy β-globin gene (Chapter 10).

Additional duplication events further increase buffering by redundancy. But since selection works on phenotypes not genotypes, the system can also become *anastamotic*, i.e., have several pathways to similar phenotypes, as mutations arising in these genes lead gradually to diverging, specialized functions, with homeostatic and compensating feedback mechanisms among them (Appendix 2.1; Taylor *et al.*, 1987). Lipoprotein particles can take on a larger distribution of sizes and hence functional effects; mutational drift among these alleles could, as now, create small variations in the size and function of particles of a given type, while mean differences between the different particle types could remain larger as their functions diverge due to selection. This complexity protects most genotypes, allowing most individuals' phenotypes to stay within a satisfactory 'fitness envelope'. For example, not all mammals have all the apolipoprotein genes that we have, and humans can function reasonably well with a number of known hypolipidemias, e.g., due to inactivated apo-B.

The evolution of dominance in complex traits

This whole process has the effect of the *evolution of dominance* for wild-type alleles relative to new mutations, a phenomenon long of interest to evolutionary geneticists {Dean *et al.*, 1988; Fisher, 1930; Hartl and Clark, 1989; Wright, 1977}. Mutational effects are *buffered* by the complexity that already exists. For a given level of stabilizing selection, the additive contribution of many alleles to a trait reduces the intensity of selection against specific alleles (Kimura, 1981), even more so if, as our model developed above suggests, most alleles have only tiny effects to begin with, and can evolve neutrally, at no fitness 'cost' to the organism. Of course, there will always be a few mutations with very large effects that are strongly selected for or (usually) against.

Stochastic self-similarity: is the causal spectrum 'fractal'?

The continual flow of mutations through a population leads to the kinds of standing-wave, or steady-state, allele-frequency spectrum discussed in this chapter that seems to be found quite widely (though with stochastic, or sampling deviation from the expected pattern). This has an interesting implication. A new mutation has only its own single frequency and effect. Over time, however, a clade of descendent alleles will develop that are clonal for the original mutation. The basic evolutionary parameters remaining constant, the topology of this tree, or clade, will be the same (except for stochastic variation) as that in the original population in which the mutation first arose. So, after suitably more time has passed will subclades of this clade be defined by some new mutation. The distribution of allelic effects will have similar characteristics, within each succeeding subclade, standardized to its new mean.

In this way, evolution generates a causal spectrum, with nested, cladistically structured branches that have similar shape (e.g., roughly exponential allelic effects and frequency distributions). This regeneration of similar structures is repeated *ad infinitum*, breaking down only for recently generated events that have not had time to substantiate their structure. This probabilistic self-similarity is reminiscent of the fractal geometry found in such traits as the branching structure of the blood vessels, bronchial tree, and so on (e.g., see the popular treatment by Gleick {1987}).

We can call this property of the causal spectrum of biological phenotypes *stochastic self-similarity* (*SSS*). I think approximate SSS applies widely in nature, for example within and across loci affecting a given trait, to the number of QTLs of given relative effect. The same thing is going on at all loci that affect the trait – and among all traits. I think that conceptually similar characteristics apply to the tree of regulatory control, and the role of gene duplication in the control of traits and the structure of the genome itself.

Conclusion

I have painted an evolutionary picture in which phenotype distributions, like the genome and its regulatory mechanisms themselves, are composed of whole series of independent (except for linkage relationships) overlapping cladistic structures of allelic effects. We might describe the spectrum of genetic causal effects figuratively as being like a Bonsai tree, which has fully tree-like branching structure, but whose overall size and shape are carefully pruned by natural selection.

Diseases are the extremes of the distributions of essentially any phenotype. We define diseases as 'abnormal' because we do not like them, but natural selection may not make the same distinction. The evidence suggests that a substantial fraction of alleles, even at disease-related loci, are selectively nearly neutral. This is especially true for chronic diseases that strike later in life, whose genetic architecture probably is largely the product of neutral evolution, with a small component of mutation-stabilizing selection balance. For complex traits, the observation that heritabilities are often similar across populations (see e.g., Table 6.5) may reflect a natural mutation–selection balance among alleles at the causal loci.

Certainly there are clear-cut selectively deleterious alleles, especially those leading to the severe diseases of childhood. Such alleles are systematically removed, but even so, *within* the class of deleterious alleles there will be a steady through-put of mutations *among which* there may be no selective difference; that is, which evolve neutrally relative to each other.

In sum, the genetic causal spectrum of typical complex quantitative traits may have evolved largely by selectively neutral genetic drift. This makes intuitive sense, and to the extent that it is accurate the neutral model usefully provides us with an expectation for the amount and pattern of genetic variation at QTLs for disease-related phenotypes. Further, the idea that biological traits have SSS is speculative, but may help us to understand at least the *structure* of the causal spectrum in a rather generic way even if, as seems likely, it will be difficult to enumerate all effects, which will vary among populations and among traits.

The contingent, stochastic nature of genetic evolution also leaves a history. This historical trail is important in many ways, in helping us to identify genes affecting a trait, understand their origins, and to refine our ability to do genetic screening and counseling in different populations. The remainder of this book is devoted largely to using and documenting these evolutionary points of view.

10 *Reconstructing history: the footprints of evolution*

And, departing, leave behind us
Footprints on the sands of time.

H. W. Longfellow, 'A Psalm of Life' (1839)

It may seem peculiar that genetic diseases, being deleterious and presumably selected against, exist at all. Yet, a number of genetic diseases, especially severely deleterious recessive conditions, have substantial incidence at birth. This chapter and the next illustrate the impact of evolution on the frequency and distribution of genetic disease.

What is responsible for the observed frequency of disease?

How often do mutations introduce new copies of disease-producing alleles?

A classical problem in human genetics has been to estimate how often new mutants leading to a given disease are produced. Interestingly, although the loci responsible vary tremendously in length, physiology, etc., a large number of studies, using very different methods, have reached similar estimates. For alleles related to both recessive and dominant qualitative traits, new mutations occur about every 100 000 meioses (i.e., $\mu \approx 10^{-5}$) {Vogel and Motulsky, 1986}. The range of these estimates is about 100-fold, which roughly encompasses the range of size of coding regions of the genes involved, so that overall the rate per nucleotide per generation is consistently around 10^{-7} to 10^{-9}.

What maintains the frequency of disease-related alleles?

Other important questions, perhaps not asked as often as they should be, are: what factors maintain the observed frequency of alleles that are deleterious? is it selection? do the alleles affect fitness?

For many disease-related alleles, fitness effects are very difficult to measure directly, for reasons discussed in Chapter 9. Perhaps the most

Table 10.1. *Variation in frequency for the major apo-E alleles*

Population	N	ε_2	ε_3	ε_4
US Blacks*	194	0.034	0.706	0.260
Nigerian Blacks*	365	0.027	0.672	0.296
Michigan, USA	223	0.073	0.806	0.121
Washington, DC, USA	74	0.095	0.756	0.149
Boston, USA	152	0.130	0.750	0.120
Ottowa, Canada	102	0.078	0.770	0.152
Grampian, Scotland	400	0.080	0.770	0.150
Finland*	615	0.041	0.733	0.227
Holland*	2000	0.082	0.751	0.168
Nancy, France	223	0.130	0.742	0.128
Munster, Germany	1000	0.078	0.783	0.139
Marburg, Germany	1031	0.077	0.773	0.150
Christchurch, NZ	426	0.120	0.720	0.160
Chinese*	196	0.084	0.852	0.064
Japanese	110	0.023	0.891	0.086
Japanese*	319	0.081	0.849	0.067
Amerindians	107	0.000	0.816	0.184
Mayans	135	0.000	0.911	0.089
Mexican-Americans	963	0.039	0.859	0.102
New Guineans*	110	0.146	0.486	0.368
Samoans	67	0.045	0.798	0.157
Means (weighted)	8812	0.071	0.772	0.157

Note: N = sample size.
Source: Samoans, Crews *et al.*, 1991; Mayans, Kamboh *et al.*, 1991; Mexican-Americans, Hanis *et al.*, 1992; others from Kamboh *et al.*, 1991 (*), or Boerwinkle *et al.*, 1987 (unmarked), where original references are given.

useful, if indirect, data would be the distribution of allele frequencies in different populations. Selective constraints should take the form of a restricted amount of frequency variation among populations. Table 10.1 provides a recently studied example, allele frequencies at the apo-E locus. There is variation, but the ε_3 allele is always much more common than the others, and ε_2 almost always much the rarest. Unfortunately, without knowing more about the evolutionary dynamics of these populations (e.g., N_e), it is difficult to determine whether this is what we would expect on the basis of drift alone, e.g., for ancient alleles geographically correlated frequencies related to the historical relationships among the sampled populations . . . or is it the effect of selection?

Many diseases are caused by mutations that are clearly deleterious, but is their frequency greater than would be expected under mutation–selection equilibrium?

Cystic fibrosis

CF is a recessive disorder characterized by serious respiratory problems due to the dysfunction of the mucus membrane lining the lungs, as well as pancreatic deficiency. CF is the most common recessive disorder in Caucasian populations, with allele frequency of about 0.022, an incidence of about 1 per 2000 births, and over 4% ($2pq$) of the population being carriers. Reported prevalences of CF in other populations are generally lower. The disease is produced by mutations at the *CFTR* locus, a membrane-bound ion-transport gene product (Chapter 7). The disease is so severe that CF homozygotes rarely survive to reproduce successfully, so that $s_{aa} \approx 1$. For a recessive trait in mutation–selection balance, the loss to selection in each generation equals the input of new mutational variation; that is, $\mu = sq^2$ {see e.g., Hedrick, 1985}. This would imply that the CF mutation rate in Caucasians is $\mu = 4.8 \times 10^{-4}$, considerably higher than that generally observed in humans, and also inconsistent with the variability in allele frequency worldwide (since μ should be the same in different populations). Yet, selection against CF homozygotes is so severe that genetic drift seems unlikely to explain the prevalence of CF in Europe.

To maintain the *CFTR*CF* allele frequency by balanced polymorphism, the normal homozygote would require $s_{AA} \approx 0.02$ against it relative to heterozygotes. But what is the heterozygote's advantage? Hypotheses related to fertility advantage (Jorde and Lathrop, 1988) or diarrheal disease (Romeo *et al.*, 1989) have been suggested but appear to have been excluded. A hypothesis that remains is that CF heterozygotes are resistant to endemic infectious disease(s) that have been common in Europe, such as influenza, typhus, bubonic plague, syphillis, or tuberculosis (TB). *CFTR*CF* carriers may secrete large amounts of mucopolysaccharide molecules in their pulmonary mucus, which may help them to resist invasion of the mucosa by the TB bacterium, but which causes severe lung damage in homozygotes (Meindl, 1987); this idea has not yet been proven.

Tay–Sachs disease

Tay–Sachs disease (TSD) is a serious recessive disease with aggregate allele frequency of about 0.017 in the Ashkenazi Jewish population, whose ancestors emigrated from the Middle East to Europe about 1000 years ago. Roughly one birth in 3600 is affected, 10 times the risk in most other populations. TSD is a GM_2 gangliosidosis, a disorder caused by the toxic accumulation of gangliosides, or sugar-containing lipid molecules, in the central nervous system. Normally, these gangliosides are degraded in lysosomes by the removal of their sugars by the degradation enzyme β-hexosaminidase-A (HexA). The TSD locus, on chromosome 15, codes for one of the subunits of this enzyme. Variation in HexA concentrations are associated with phenotypic variation, the most serious of which leads to infant forms of TSD that involve severe mental retardation, paralysis, blindness, and early death.

The major TSD mutations in the Ashkenazi Jews include an exon 11 insertion, and a splice-junction mutation in intron 12, both of which lead to severe infantile TSD, and a Gly → Ser substitution in codon 269 in exon 7 leading to a milder form of the disease in adults (Grebner and Tomczak, 1991; Meyerowitz, 1988; Navon and Proia, 1989; Navon *et al.*, 1990). Half of all Ashkenazi patients are exon 11 homozygotes, 25% are exon 11/intron 12 heterozygotes, both severely affected, but most of the remainder have milder, adult-onset disease. As we will see for other classical 'recessive' diseases, many TSD cases are actually heterozygotes.

Is the elevated frequency of TSD in Ashkenazi Jews due to founder effect or to balancing selection? Or, as has been suggested, could there have been heterozygote resistance to the tuberculosis that was endemic in urban European areas inhabited by Jews (see e.g., Chakravarti and Chakraborty, 1978; Paw *et al.*, 1990; Rotter and Diamond, 1987; Wagener *et al.*, 1978)?

The probability density of allele frequencies, p, at equilibrium, under recessive lethal selection, mutation, and drift {Crow and Kimura, 1970; Gale, 1990}, is approximately

$$\Pr(p) = \frac{2(2N_e)^{(M/2)}}{\Gamma(M/2)} p^{M-1} e^{-2N_e p^2}, \qquad [10.1]$$

where $M = 4N_e\mu$, and $\Gamma()$ denotes the gamma function from statistics. (This differs from [9.3] in considering only two allele classes rather than an unlimited number of identified alleles.) An appropriate test of the drift model is the probability that the (aggregate) Ashkenazi TSD allele frequency would be its observed frequency p or higher, which from [10.1] is

$$\Pr(p \geq p_{obs}) = \int_{p_{obs}}^{1} \Pr(p)\, dp. \qquad [10.2]$$

Table 10.2. *Percentage HexA allele frequencies in Ashkenazi Jewish and other European populations*

	Jewish $N = 304$		
Mutation	Unconditional	Conditional	Non-Jewish $N = 79$
Exon 11	222 (73)	—	12 (15)
Intron 12	40 (13)	(48)	0 (0)
Exon 7	9 (3)	(11)	2 (3)
Other	33 (11)	(41)	65 (82)

Note: Conditional distribution is that after removing exon 11 (see the text).
Source: Modified from Grebner and Tomczak, 1991, and Paw *et al.*, 1990.

The size of the founding Ashkenazi population is unknown, but in the range 1000 to 5000 the probability in [10.2] ranges between 0.03 and 0.007 (Chakravarti and Chakraborty, 1978; Wagener *et al.*, 1978). If the former size were true, it would not be surprising, given the hundreds of diseases there are, that at least one would be found that by chance had a frequency as high as that of TSD, but it would be more surprising if the second population size assumption were correct – in which case, one might suspect balancing selection.

Can the allele frequency distribution help to resolve this? Table 10.2 shows the data. The founder effect would typically increase only one allele frequency at a locus; the Ashkenazi population seems to have at least two abnormal *HexA* alleles at relatively high frequency, suggesting that selection may have favored these alleles although both have serious phenotypic and fitness effects; however, this idea would have to be tested by comparison with the TSD mutation distributions in other populations, especially in Europe, for which adequate data are not yet available.

Malaria-associated hemoglobinopathies

The malaria-associated hemoglobinopathies are among the best-understood human genetic traits {Antonarakis *et al.*, 1985; Gelehrter and Collins, 1990; Higgs *et al.*, 1989; Livingstone, 1986, 1989; Vogel and Motulsky, 1986; Weatherall, 1991}. As described in Chapter 2, the constituents of hemoglobin are coded for by the α-globin cluster on chromosome 11 and the β-globin cluster on chromosome 16. Hemoglobin is found in red blood cells, where it is required for disposing of CO_2 and acquiring O_2 in the lungs for supply to all tissues of the body. Hemoglobins made dysfunctional by mutation lead to various forms of *anemia*. Usage varies, but here I use the term *thalassemia* for mutations that inactivate a globin gene, and allele designations for those that alter the functional efficiency of the gene product (the official allele designations are arcane, so I use familiar ones in this book).

Across the tropical areas of the Old World that have had endemic malaria caused by the parasite *Plasmodium falciparum* (and others), a number of thalassemia deletions as well as high-frequency point mutations have been found in the globin clusters. The best-known is that for sickle-cell hemoglobin, *HbS*, a GAG $\rightarrow$ GTG mutation causing a Glu $\rightarrow$ Val substitution in codon 6 of the β-globin gene. However, there are several other β-globin mutations (*HbC*, *-D*, etc.). In homozygotes, these alleles produce anemias of various kinds, and most if not all appear to involve resistance to malaria, although the pathophysiology of globin-based malarial resistance is only partly known.

The frequencies of these alleles are far too high to be explained by recurrent mutation, and once it became clear that *HbS* conferred resistance to malaria in heterozygotes, even though causing lethal anemia in homozygotes, the explanation of the observed frequencies by balancing selection was established. In malarial environments such as are found commonly in West Africa, the equilibrium *HbS* frequency (Chapter 9) is $q_{eq} = s_{AA}/(s_{AA} + s_{aa}) \approx 0.2/(0.2 + 0.8) = 0.2$, based on estimates of the selective coefficients.

Other than for *HbS* itself, the bulk of the evidence for malarial selection for anemia mutations is still indirect. For example, there is a correlation between the prevalence of malaria and of α-thalassemia deletions along a gradient of altitude in Papua New Guinea (PNG), as shown in Table 10.3 (Flint *et al.*, 1986). That this is not due to drift is shown by an absence of such correlations for alleles at random marker loci in the same data (not shown). The inference extends along the nearby Melanesian island chain of Vanuatu, which was probably settled from PNG, in which thalassemia prevalence is concordant with clines of malaria prevalence and/or the first known date of endemic malaria.

Malarial resistance can also be conferred by mutations at other loci, and these also are common in traditionally malarial areas. These include the X-linked gene for the enzyme glucose-6-phosphate dehydrogenase (G6PDH), and the Duffy blood-group locus, at which a mutant common in Africa protects against malaria caused by the parasite *P. vivax*.

Table 10.3. *Incidence of α-thalassemia chromosomes by region in Papua New Guinea and the Pacific Islands*

Place	No.	Genotypes			Total
		αα/αα	*α−/αα*	*α−/α−*	%*α*−
Papua New Guinea					
Highlands	539	502	36	1	4
Coastal areas	188	84	61	43	39
Pacific Islands					
North Solomon Is.	30	9	15	6	45
Vanuatu Is.					
Espiritu Santo	178	72	75	31	38
Maewo	169	80	67	22	33
Pentecost	149	80	56	13	27
Tanna	224	149	69	6	18
Futuna	56	51	4	1	6
Aneityum	66	56	10	0	8
New Caledonia Is.	84	74	10	0	6

Source: Modified from Flint *et al.*, 1986.

Clonal and cladistic analysis of qualitative diseases

Clonal disease was defined in the last chapter as cases caused by the effects of alleles descended from some common original mutation. Diseases may often be clonal in isolates or individual families, but even in the larger general population a disease may be produced by only a small number of clones of mutations. Since these will usually be specific to each population, their identification is important for accurate fetal diagnosis and genetic counseling. Also, we can sometimes reconstruct population history from the distribution of such mutations.

Phenylketonuria: a world-wide story

PKU is a recessive defect in the metabolism of the amino acid phenylalanine (Phe), especially the conversion of Phe to tyrosine (Tyr). The accumulation of Phe disrupts many metabolic pathways that depend on it. In the central nervous system (CNS), lack of Tyr prevents adequate production of catecholamines, neurotransmitters needed in normal CNS development, leading to potentially severe mental retardation {Scriver *et al.*, 1988}. Untreated, the fitness of a severe PKU homozygote is nearly zero. The responsible gene is on chromosome 12 and codes for the enzyme phenylalanine hydroxylase (PAH) (see e.g., DiLella and Woo, 1987), expressed actively in the liver. In Caucasians, the incidence of

PKU at birth ranges from about 1 per 4500 in Ireland to 1 per 16 000 births in Switzerland and most other populations (there are other local hot-spots of high frequency) (see e.g., Konecki and Lichter-Konecki, 1991; Wang *et al.*, 1989).

Genetic variation in the *PAH* region has been characterized by a standard classification system based on eight restriction enzyme recognition sites. More than 50 haplotypes (sometimes called RFLP frameworks) have been identified using these enzymes; for example, haplotype 1 is +−−−−+−−, haplotype 2 is +−−−−+++, and so on (most are tabled by Rey *et al.* (1988)).

Figure 10.1 shows the distribution of normal and abnormal haplotype frequencies in various European populations for the most common haplotypes (in the figure, frequency bars for each haplotype correspond (left to right) to the country symbol keys (bottom to top)). Most European PKU mutations are found on haplotypes 1 to 4. For example, in Danish data a PKU-producing GT → AT mutation at the *PAH* intron/exon 12 splice junction is found on haplotype 3. The mutation reaches its highest frequency there and is thus presumed originally to have arisen on that haplotype. A North European haplotype 2 mutation is in codon 408, and is also found in French Canadians (John *et al.*, 1990) and other Europeans (Okano *et al.*, 1990; Rey *et al.*, 1988). The mutation associated with haplotype 1 is a Glu → Lys mutation at codon 280 (Okano *et al.*, 1990). The haplotype 3 mutation accounts for about 38% of PKU chromosomes in northwest Europe, and this along with mutations on haplotypes 1, 2, and 4 account for 90% of cases.

As is to be expected if mutations arise randomly on different haplotypes, the majority of PKU mutations are found on the normal haplotypes that are most common in the same population. It is also on these common normal haplotypes that more than one PKU mutation has been identified; for example, two are on haplotype 1, four on haplotype 4. This increases the overall association between PKU and these haplotypes (see e.g., Konecki and Lichter-Konecki, 1991). The geographic *clines* (gradients) of PKU haplotype frequencies are what we would expect of mutations that arose in one area, and are diffusing through other populations from that original source. Thus, across most of Europe the great majority of PKU alleles are descendants of a few original mutations, with many of the remainder being derived by recombination with these, plus a scattering of other recent, rare mutations (Daiger *et al.*, 1989a; Hertzberg *et al.*, 1989).

The evidence suggests that most of these mutations have arisen only once, but there are occasional possible exceptions. For example, the

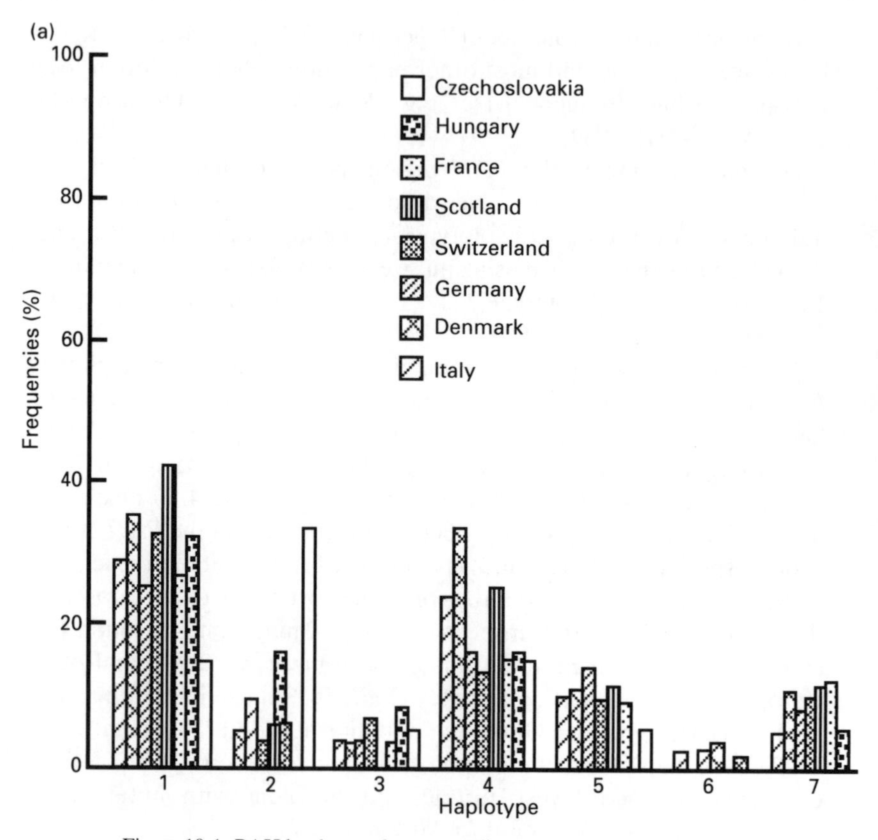

Figure 10.1. PAH haplotype frequency distributions in Europeans. The frequency of each haplotype, defined as in the text, is shown for each country for (a) normal, and (b) abnormal chromosomes. (Data from Konecki and Lichter-Konecki, 1991 and sources cited therein.)

codon 280 mutation usually associated with haplotype 1 occurs also in Mediterranean French patients and is found on haplotype 38 in North Africa (Lyonnet *et al.*, 1989). This has been interpreted as a possible recurrent mutation because: (1) the site is a CpG potential mutational hot-spot, which may make recurrent mutation more likely than the recombination events that would have been required to introduce the original mutation onto haplotype 38 and (2) there is little if any overlap in haplotypes affected with this mutation in different populations (Okano *et al.*, 1990).

The PKU story in Asia is similar. The disease is about as common in Chinese and Japanese as it is in Europeans, but the haplotype diversity of the first two, based on the standard coding system, is less than that of

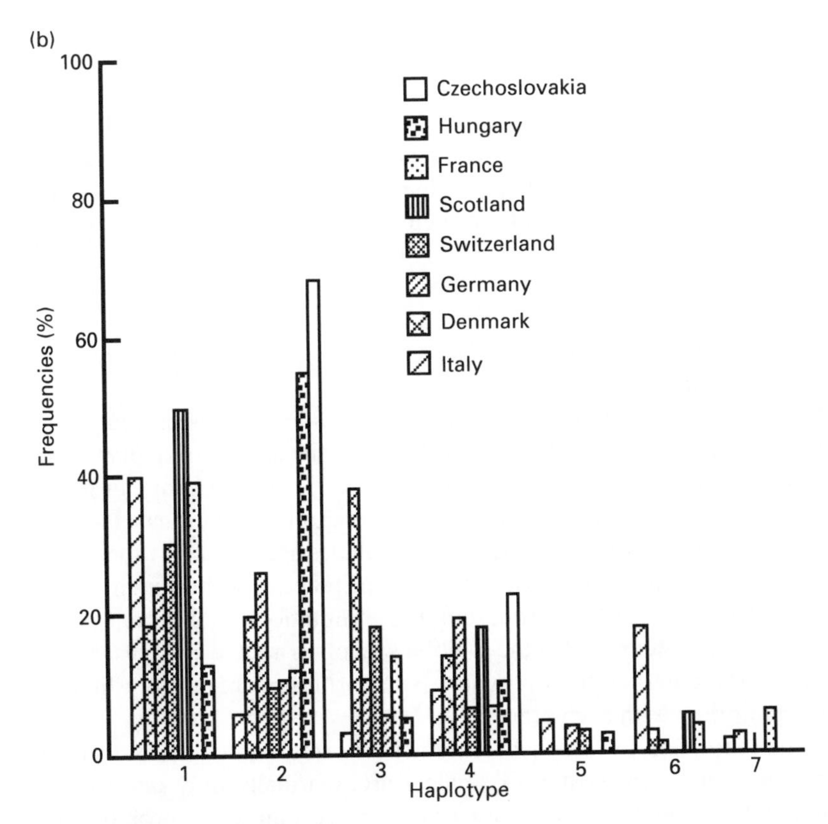

Figure 10.1. *Continued.*

Europeans. Haplotype 4 has a prevalence of about 80% in Asia, which, along with its worldwide distribution, suggests that it is an ancient haplotype (Daiger *et al.*, 1989b). A number of unique, geographically restricted, PKU mutations are found in Asia (Wang *et al.*, 1989, 1991a). The geographical patterns of these mutations are useful in understanding the population history of the region. For example, the same Asian haplotypes are found in the island populations of the Pacific, which confirms evidence from archeology, linguistics, globin, and HLA genes, that this area was originally settled from southeast Asia (Hertzberg *et al.*, 1989), unless these ancestors replaced an exterminated population of prior inhabitants, which seems unlikely. And the distribution of PKU mutations on haplotype 4 in China and Japan confirm the northern-Asian origins of the present population (Wang *et al.*, 1991b).

The evidence suggests to me that PKU mutations are not being maintained by balancing selection. At the sequence level, most of the mutations appear on just one haplotype, or else on haplotypes that can be derived from it by recombination. Most are rare and scattered across many haplotype backgrounds. They seem to be recent. At least, there do not seem to be any *clades* of PAH haplotypes all of whose members share a mutation; that is, the mutations seem to have arisen more recently than the haplotype divergence.

Malaria resistance and the hemoglobinopathies: how many mutations?

A debate has long raged over how many unique sickle-cell mutations there are in the world today. HbS is found at substantial frequency from Africa across to India, and at low frequency in some European populations and elsewhere. Is this due to recurrent mutations aided by selection in many different regions? *HbS* alleles are defined by the codon 6 point mutation, so that evidence for recurrence must be sought in the haplotype frameworks surrounding those mutations in different populations. The system used in the β-globins involves a set of 5–12 restriction sites (varying among studies), whose polymorphisms are in strong linkage disequilibrium with each other.

The (aboriginal) world distribution of abnormal hemoglobins is shown in Figure 10.2. In most geographical areas, usually only one or two mutant hemoglobins are common (plus some recombinant chromosomes that have transferred an original mutant to another haplotype) (Antonarakis *et al.*, 1985). The distribution can generally be related to historical population movements and the local history of endemic malaria, but it is not clear how completely clonal the pattern is (Antonarakis *et al.*, 1985; Livingstone, 1986, 1989; Higgs *et al.*, 1989; Vogel and Motulsky, 1986).

Most *HbS* mutations are found on one of three RFLP haplotypes in Africa, one predominating in each local area. If in fact there is only one original African *HbS* mutation, the clone of descendants must include several recombinants and gene-conversion chromosomes to generate the haplotype clades observed in different geographical regions. With such a scenario, the original mutation may have arisen in the Middle East and spread from there, carried by Arab traders into Central Africa, and then diffusing westward, always preceded by the spread of malaria. To the east, the *HbS* allele spread to India where intermediate frequencies are now widespread. Computer simulations have been used to argue in support of these ideas (Livingstone, 1986, 1989).

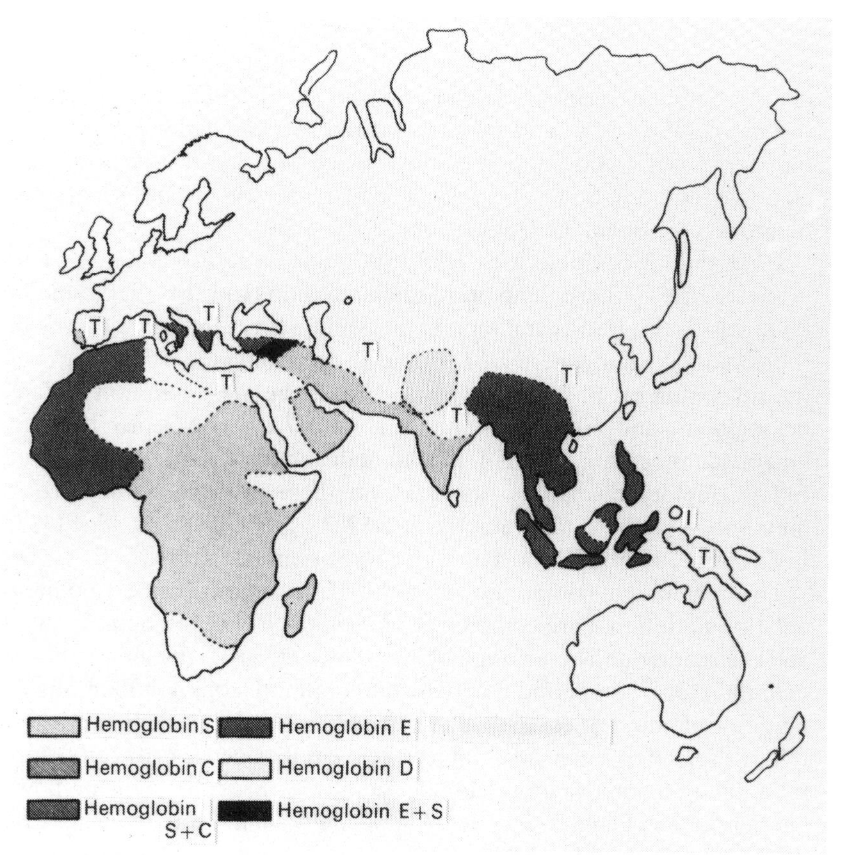

Figure 10.2. Distribution of malaria-related globin mutations, T, thalassemia. (From Vogel and Motulsky, 1986.)

However, some authors prefer a multiple-origin explanation (Antonarakis *et al.*, 1985; Kan and Dozy, 1980). Sequences 5′ flanking and intragenic to the β-globin gene suggest only a single origin of Arab–Indian *HbS*, to some authors four African patterns seem more likely to be due to recurrent mutation than to recombination or gene conversion (Trabuchet *et al.*, 1991). In this latter interpretation, rather than spreading from the Mediterranean into Africa, a 'Benin' *HbS* mutation has spread from an origin in Africa outward to the Mediterranean.

The severity of malarial selection gives *HbS* a strong advantage that may make it a 'predatory' allele that can displace the other currently available mutations at the locus. This would allow *HbS* to spread rapidly to areas where malaria has become endemic. Malaria is now much more widespread in Africa than is *HbS*, suggesting that *HbS* is a relatively recent mutation, currently expanding at the expense of prior thalassemia and *HbC* mutations in Africa and eastward (Livingstone, 1989). *HbC* is common in West Africa on an otherwise rare African haplotype, and in a number of isolated tribal populations, suggesting that it may be a prior adaptive mutation now being displaced by *HbS*, which confers a greater heterozygote advantage (Livingstone, 1986, 1989).

HbO and -*E* compete with *HbS* in Middle Eastern populations and across to India, where the pattern gradually shifts towards that found in Southeast Asia. *HbS* is common in the Middle East, but more patchy in India and Pakistan, where *HbD* is found, especially in northern parts of the subcontinent. In Southeast Asia, *HbE* is the most common mutant hemoglobin, and *HbS* is essentially absent. *HbE* apparently has two independent centers of origin, in Cambodia and in the Assam region of India. Haplotype analysis suggests that these are independent *HbE* mutations (e.g., Antonarakis *et al.*, 1982, 1985; Deka *et al.*, 1987; Livingstone, 1986); *HbE* occurs as a rare variant in Europe.

The dynamics of competition in multi-allelic systems (be they point or deletion mutations) are complex. Each genotype (i.e., all homozygotes and each compound heterozygote) has its own fitness coefficient. In many malarial areas, compound heterozygotes are more common than single-mutation homozygotes. The allele-frequency dynamics depend on the relative selection intensities; under some circumstances drift can determine the outcome, favoring an allele that, by chance or by gene flow, happens to reach high frequency. Stable multi-allelic equilibria can exist if all homozygotes are disadvantageous, i.e., $s_{ii} > 0$ for all alleles i, and $s_{ii}s_{jj} > s_{ij}^2$ for all heterozygotes; if these criteria are not satisfied at least one allele will be eliminated by selection (see e.g., Vogel and Motulsky, 1986). The first condition may be met in malarial populations, but the second may require implausible selective coefficients. Estimating these coefficients is difficult.

For example, with regard to the high prevalence of *HbE* in Asia, it is not clear whether the homozygote has any selective disadvantage; some estimates are in the range $s_{EE} = 0.2$ to 0.3 (Vogel and Motulsky, 1986), but a direct study in Assam found no fertility disadvantage for *EE* homozygotes, or even for *EE* × *EE* matings (Deka, 1981; Deka *et al.*, 1987; Livingstone, 1986). In Asia most of the competition is from thalassemia mutations. Denoting these collectively by T, it seems that even if there *is* little selection against *HbE*, in some genotypes, it is unlikely that $s_{EE}s_{TT} > s_{ET}^2$ is generally true. Thus, a stable equilibrium among the common Asian hemoglobinopathy mutations is unlikely. Overall, the low selection (if any) against *HbE* suggests that it is simply advantageous in malarial environments, and may even be displacing the 'normal' *HbA* allele in some populations: *HbE* reaches frequencies of about 50% in parts of India and Nepal, where 75% of individuals carry at least one copy (e.g., Deka, 1981; Livingstone, 1986).

Evidence for malarial selection favoring α-thalassemias (in PNG) was described earlier (Flint *et al.*, 1986). This region contains two adjacent α-globin genes that have very similar sequence. The sequences surrounding these genes are also highly conserved, perhaps maintained by unequal crossovers and gene conversion. This may lead to frequent deletion and copy number mutations by misalignment during meiosis for example. In any case, many such mutations, as well as a number of gene-inactivating point mutations, have been found in the α-globin genes. α-Thalassemia mutations are common in the tropical parts of the world but not elsewhere, whereas non-pathogenic mutations are distributed worldwide. Selection affecting α-globin mutations appears to be weaker than for the β-globin mutations, because of the buffering provided by redundancy due to the gene duplication (Chapter 9); the result is less intense selective competition, and more mutational variability in the α-gene region. Nonetheless, each local area still has its own characteristic mutations (Flint *et al.*, 1986; Higgs *et al.*, 1989).

Finally, mutations at the X-linked enzyme G6PDH make red blood cells less hardy but can deprive the *P. falciparum* malaria parasite of nutrients it needs during the part of its life cycle that it must spend in host red cells. Thus, G6PDH deficiency leads to balancing selection in malarial environments, adding to the complex repertoire of adaptive responses to malaria.

Some thoughts on evolution occasioned by the globin gene story

It is common to hear persons opposed to the idea of evolution argue that adaptations such as flight cannot arise fully formed by the kinds of mutation–selection mechanism that are the core of evolutionary biological theory. The usual biologist's response is that no such saltation is needed, and that adaptation need only involve gradual improvements over considerable time periods.

One way this gradual change might occur is for new function to arise in a rather crude form initially, and be focused by selection thereafter. The evolution of buffered systems by gene duplication was discussed in the previous chapter, and is exemplified by the α-globin gene redundancy. The evolution of the malaria-related hemoglobinopathies may also constitute direct evidence for another mechanism.

The crudest form of mutational change is to inactivate a gene. While this is not very sophisticated, there are many mutational ways to accomplish this end; thus, such mutations, as a group, arise frequently. This may explain the plethora of different thalassemia (deletion and inactivation) mutations in malarial areas. Mutations with less dramatic effect, such as those slightly modifying protein function, will arise less frequently (since they have to be more 'precise' than simple inactivation), but may arise at a number of loci and provide natural selection with a more or less quantitative array of phenotypes with which to work. Thus, it is possible that thalassemia mutations were Nature's first adaptive response to malaria, but that these are being replaced by the more refined point mutations at the globin genes, such as *HbS* and *HbE*. The competitive confrontation between these two adaptive mutations, which would take place in the region that includes India if malaria were not being brought under control, would be very interesting to watch.

The adaptive responses to malaria exemplify another evolutionary principle referred to in the previous chapter: phenotypic convergence in different populations (i.e., 'malaria resistance') is produced by genotypic divergence among those populations, owing to the unique nature of mutations.

Cladistic analysis for quantitative traits

An analog of this type of cladistic thinking applies to genes that affect quantitative traits. Chapter 9 discussed a model for the generation of allelic effects on a quantitative trait by the mutational processes that also produce the branching sequence structure of alleles in a population. Each new allele is associated with some slight 'displacement' of the phenotype. A clade of mutations that eventually descend from each surviving mutation will have a distribution of effects that is displaced in a similar direction. For example, if the mean associated with a given allele is μ and a mutation changes this to $\mu + x$, subsequent mutations in this clade will generate a distribution of effects centered statistically at $\mu + x$. Other clades in the population will have other mean effects, based on the mutations they share.

A population consists of a set of such clades. If these can be identified, and alleles grouped accordingly, it may be possible to explain the control of the phenotype more simply – that is, to reduce the 'dimensionality' of the causal spectrum of effects. Some preliminary attempts have been made to do this.

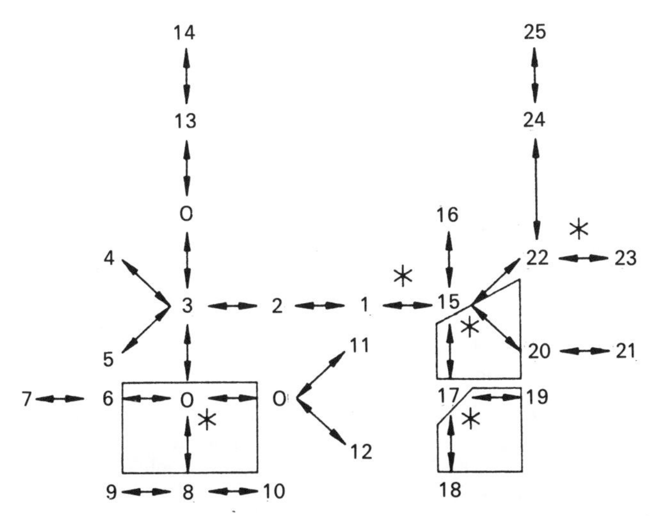

Figure 10.3. Cladistic analysis for effects of mutations on ADH levels in *Drosophila*. Each arrow represents a single haplotype change. Asterisks indicate positions of mutations that have statistically significant effects. Polygons show alleles that share significant mutational effects. (From Templeton *et al.*, 1987.)

Cladistic patterns of measured genotype effects

Figure 10.3 shows a cladogram for mutations at the *alcohol dehydrogenase* (*ADH*) locus in the fruit fly *Drosophila melanogaster* (Templeton *et al.*, 1987). The ADH concentration in animals with each of these mutations was determined. Each branch represents a single restriction site difference between haplotypes (branches labeled 0 are inferred intermediate haplotypes not found in the data). A nested analysis of variance of ADH level, beginning at the tips of the diagram and working towards the middle, identified branches whose alleles, collectively, accounted for a significant fraction of the variance in ADH levels. These are marked with an asterisk in the figure. Presumably the alleles in each clade share some important mutation along with the other sequence variation that defines the clade. A non-cladistic analysis might attribute most of this variation to unordered polygenes.

The only analysis of this kind to date in humans is a study of the effects of apo-AI haplotypes on triglyceride levels discussed in Chapter 8 (e.g., Table 8.1) (Templeton *et al.*, 1988). Although the sample was small and preliminary, cladistic analysis showed that the 010 to 011 mutation was

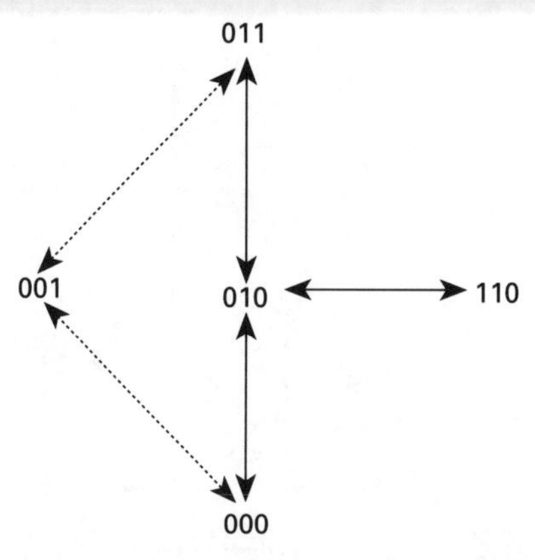

Figure 10.4. Cladogram for apo-AI/CIII/AIV gene region haplotypes and their effects on log-transformed triglyceride levels. Dashed lines indicate two possible sources for haplotype 001 (recombination or mutation). (From Templeton *et al.*, 1988.)

associated with a statistically significant fraction of the triglyceride (Figure 10.4) (Templeton *et al.*, 1988). The origin of haplotype 001 is ambiguous (e.g., whether owing to recombination or to two mutations) and was excluded from the analysis, so that in this example there is no nesting of haplotypes.

These are single-locus examples, but for complex traits (such as triglyceride levels), many loci will contribute and the effects of alleles at each of the loci will be cladistically structured – because evolution is occurring all the time, at all loci.

Genetic admixture: natural experiments with disparate genes

Different mutations arise in different populations, and the longer the populations have been isolated the more genetically different they become. Different clades of alleles will become fixed in each population. The mass-scale long-distance movement of human populations of the past few centuries has brought together such disparate populations. One need think only of the millions of American Blacks and Hispanics to realize how important such populations are. Admixture between such populations may constitute an informative 'natural experiment' that can approximate the effects of crosses between inbred strains. This can sometimes be useful to us in understanding the genetic basis of a disease or other trait.

Basic principles of admixture studies: prevalence and model fitting

Admixture at the population level

Equations [9.6] and [9.7] described the effects of admixture between two populations {Chakraborty, 1986}. The genotype frequencies in the hybrid population can be computed from the allele frequencies, for example, $P_{H,AA} = p^2_{H,A}$. If the allele frequencies in the parental populations are known, we can estimate the amount of admixture in the hybrid from a sample of n_i individuals of genotype i from the hybrid population, by maximizing the likelihood

$$\mathscr{L}(m) \propto \prod_i (P_{H,g_i})^{n_i}. \qquad [10.3]$$

The genotype probabilities can be replaced by expressions for the phenotype probabilities if there is dominance, and multiple source populations can also be accommodated by expanding this expression.

Drift will lead to different admixture estimates for different loci, even though all loci have the same population history. The average $m_{i,\text{MLE}}$ values over all loci i tested provide one estimate of m, or a single estimate can be derived from $\mathscr{L} = \Pi\, \mathscr{L}_j$, where each $\mathscr{L}_j$ is [10.3] for a specific locus. Complications such as linkage among the loci and heterogeneity of the locus-specific estimates can be addressed. It is important to use selectively neutral markers, since selection will change the allele frequencies and hence the apparent admixture at a locus, also generating interlocus heterogeneity in estimated admixture.

Genetic models and the prevalence of disease in an admixed population

The prevalence of a genetic trait in an admixed population, Prev_H, is computed from [4.9], using the hybrid allele frequency, based on the appropriate admixture expression. For a dominant disease at a diallelic locus, the prevalence in population i, $\text{Prev}_i = p_i^2 + 2p_i(1 - p_i)$ (Chapter 4), so

$$\begin{aligned}
\text{Prev}_H &= p_H^2 + 2p_H(1 - p_H) \\
&= [mp_1 + (1 - m)p_2]^2 + 2[mp_1 + (1 - m)\,p_2] \\
&\quad \times \{1 - [mp_1 + (1 - m)\,p_2]\} \\
&= m^2\,\text{Prev}_1 + (1 - m)^2\,\text{Prev}_2 + 2m(1 - m) \\
&\quad \times \{\surd[1 - (1 - \text{Prev}_1)(1 - \text{Prev}_2)]\}
\end{aligned} \qquad [10.4]$$

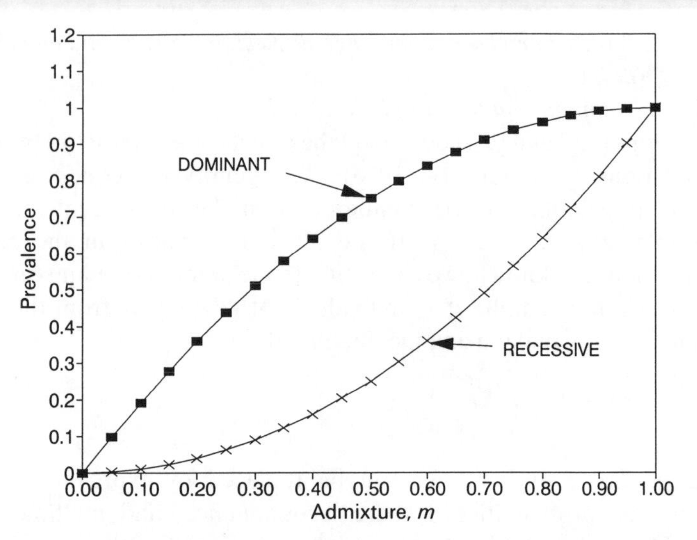

Figure 10.5. Relationship between admixture and prevalence of a disease caused by dominant and recessive genotypes.

for a di-hybrid population. The comparable expression for the recessive phenotype is

$$\text{Prev}_H = m^2\,\text{Prev}_1 + (1 - m)^2\,\text{Prev}_2 + 2m(1 - m) \times \surd(\text{Prev}_1\,\text{Prev}_2). \qquad [10.5]$$

Similar formulas for some other models were worked out by Chakraborty and Weiss (1986). In the simple case when a dominant allele is fixed in one source population and absent in the other, [10.4] and [10.5] become $1 - (1 - m)^2$ and m^2. These are plotted as a function of m in Figure 10.5.

Prevalence with variable onset age

For some diseases, the effect of a genotype is to alter the hazard function, $h_g(t)$. Assuming no differential loss to mortality among the affecteds, the age-specific prevalence (Chapter 3) of a pure population of genotype g is

$$\text{Prev}_g(t) = 1 - S_g(t) = 1 - \exp\left[-\int_0^t h_g(y)\right] \text{d}y. \qquad [10.6]$$

In a di-hybrid with allele frequency p_H, the age-specific prevalence, $\text{Prev}_H(t)$, is

$$\begin{aligned}\text{Prev}_H(t) &= \Sigma P_{H,g}\,\text{Prev}_g(t)\\ &= p_H^2\,\text{Prev}_{AA}(t) + 2p_H(1-p_H)\,\text{Prev}_{Aa}(t)\\ &\quad + (1-p_H)^2\,\text{Prev}_{aa}(t).\end{aligned} \qquad [10.7]$$

If there is differential mortality, the allele frequency will change with age, and this must be built into the equation.

Many epidemiological studies report cross-sectional rather than age-specific prevalence; if the *age distribution*, i.e., the fraction of individuals at age t, is c_t, then the cross-sectional prevalence (Chapter 3) in a hybrid population can be expressed using the above equations as:

$$\text{Prev}_H = \Sigma\, c_t\,\text{Prev}_H(t). \qquad [10.8]$$

Example: Amerindian diabetes

Adult onset NIDDM has become endemic in Amerindian populations, and populations admixed with them, in the last 50 years (Szathmary, 1990; Weiss *et al.*, 1984a, 1992). The relative uniformity of the pattern among tribal groups across North America suggests that the problem is due to a $G \times E$ interaction, and that NIDDM may be a clonal disease in Amerindians. One hypothesis is that some allele(s) arose early in the original settlement of the Americas, perhaps 20 000 or more years ago, that may have conferred a selective advantage, for example in storage of food energy in the form of fat as adaptation to an unpredictable arctic environment. The allele(s) may have become nearly fixed due to this advantage. Only after the advent of the high-caloric, less vigorous, modern lifestyles would such a gene lead to pathogenic obesity and diabetes as a sequel in today's Amerindians who have, in common, descendent copies of the original allele(s). This idea is a modification of a more general classic speculation about metabolically 'thrifty' genotypes (Neel, 1962).

The notion that Amerindian NIDDM is clonal, or at least genetic, can be tested at least in an exploratory way by admixture studies, as shown in Figure 10.6. The *expected* prevalence is plotted as a function of admixture and genetic model. To do this, I used [10.6] to [10.8] with age- and sex-specific NIDDM incidence rates reported for Amerindian populations by the US Indian Health Service, and for US Caucasians, ignoring the small amount of differential mortality that may occur. I computed cross-sectional prevalence for adults over age 25 years by assuming the published age distribution for Texas Mexican-Americans, which is sufficiently similar among all the populations for practical purposes.

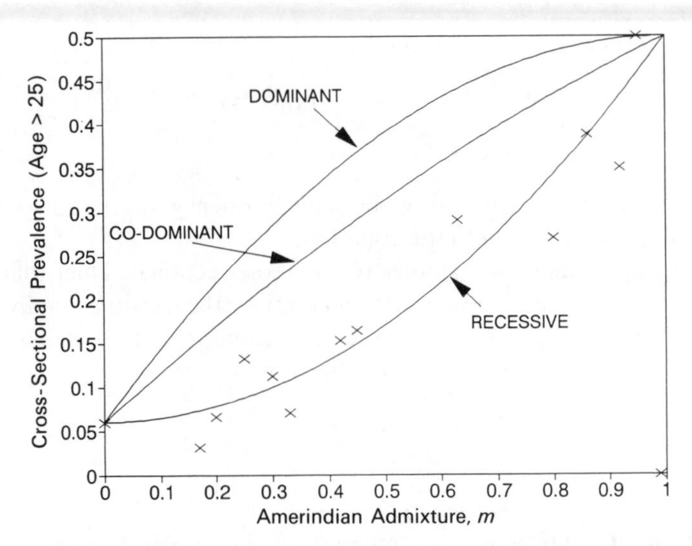

Figure 10.6. Relationship between cross-sectional prevalence of NIDDM in Amerindian-admixed populations. Plot suggests recessive etiology (see the text).

To account for the possibility of dominance, in the admixed population, of the hypothetical Amerindian (A) allele relative to its Caucasian (C) counterpart, I expressed dominance as the fraction, k, that the heterozygote hazard is of the difference between the hazards of the two homozygotes, averaging over all adult ages; that is, $h_{AC} = h_{CC} + k(h_{AA} - h_{CC})$. The figure shows prevalence–admixture relationships for fully dominant ($k = 1$), fully recessive ($k = 0$) and co-dominant models ($k = 0.5$).

Each point in the figure represents a cross-sectional NIDDM prevalence estimate from a population for which the amount of Amerindian admixture has been estimated from genetic marker or other data. Many of the populations shown are not 'Indian' by ethnic classification but are Hispanic populations that are admixed with Amerindians (e.g., Mexican-Americans). The data seem roughly consistent with a single-locus recessive model.

Admixture at the individual level: relationship to prevalence

Each *individual* in a hybrid population can be genotyped for loci whose allele frequencies are informative for admixture. From these genotypes it is possible to estimate each individual's m, the fraction of his/her genes from each population. Following the logic of population admixture, the likelihood for this m is

$$\mathcal{L}(m) = \Pi\,\Pi\, P_{H,g}^{\delta_{ij}}, \tag{10.9}$$

where the sum is over all genotypes at all the tested loci, δ_{ij} is an indicator variable equal to 1 if the individual has genotype j at locus i, and 0 otherwise, and the P is hybrid genotype probability expressed as functions of m as before. The value of m_{MLE} estimated from [10.9] applies only to the individual. These values can vary among individuals, each of whom has his/her own unique genealogy in the admixed population, with opportunities for drift and mendelian segregation to vary among loci.

Now it is natural to think that disease prevalence should be associated with such individual admixture estimates, but this is incorrect. Admixture is estimated from a number of informative marker loci; unless there is strong linkage disequilibrium between marker and disease alleles, based on different haplotypes in the source populations, independent segregation will quickly dissociate the alleles at the two loci. In any case, unless admixture is estimated from a large number of marker loci, the likelihood of close linkage of any random marker to the trait locus is small {Ott, 1991}. Even when such a marker is closely linked, the admixture estimate is the average over all tested marker loci, and the fraction of information contributed to that estimate by the single linked marker will generally be small. Thus, unless admixture has just occurred with little opportunity for independent assortment of marker and trait alleles, we cannot expect individual admixture to be very informative about a genetic model.

The example continued, and a tricky problem

Several studies have failed to find strong associations between individual admixture estimates and the prevalence of NIDDM in Amerindian admixed populations (Knowler *et al.*, 1988) or Mexican-Americans (e.g., Chakraborty *et al.*, 1986; Hanis *et al.*, 1986) despite population data such as in Figure 10.6 (Knowler *et al.*, 1983; Szathmary, 1990; Weiss *et al.*, 1984a, 1992). For example, there is a positive association between NIDDM prevalence and the self-reported fraction of Amerindian ancestry, in eighths, among the Pima (Knowler *et al.*, 1988). Table 3.4 showed a negative association between the prevalence of a European marker haplotype at the *Gm* locus ($Gm^{3;5,13,14}$), and the fraction of ancestry, again when people were *grouped* according to ancestry. However, as shown in Table 10.4, if we look within an ancestry group, for example in the self-reported 'pure-blood' group, there is clearly no association between individual presence of the European *Gm* admixture marker and NIDDM.

Table 10.4. *Association between diabetes (NIDDM) and presence of European* Gm *marker in self-reported full-blood Pima Amerindians*

$Gm^{3;5,13,14}$	Diabetes	No diabetes	Total
Present	10	7	17
Absent	1058	706	1764
Totals	1068	713	1781

Source: Knowler *et al.*, 1988.

Natural experiments

Besides formal admixture-based analysis, we may be able to take advantage of admixture as a natural experiment to create informative matings for a particular disease or other trait. In the Amerindian diabetes example, suppose Amerindians are nearly fixed for some susceptibility allele that is essentially absent in Caucasians. In Mexican-Americans, who have about 30% Amerindian ancestry, only $0.7^4 + 0.3^4 \approx 25\%$ of all matings will be of uninformative $DD \times DD$ or $dd \times dd$ at the disease locus, but the remaining 75% will segregate variation at the locus. Informative families may be rare in either parental source populations.

Similar arguments may pertain to problems such as the high prevalence of hypertension in American Blacks, although there is disagreement about the interpretation of the available data (e.g., Motulsky *et al.*, 1987; Ward, 1990). Recent studies in American Blacks and in the Caribbean have used skin color (Klag *et al.*, 1991), and genealogical and genetic admixture data (Darlu *et al.*, 1990) to argue that there are positive correlations between African admixture and blood pressure. The latter may be a multilocus trait, which will somewhat circumvent the individual-group admixture paradox; that is, an individual's admixture estimate may well also estimate the fraction of the group of blood pressure-related alleles he/she has from a given source population, by marking chromosome regions that carry relevant genes, as outlined in Chapter 8 (and see Goldgar, 1990). Recombination takes much longer to break up such associations than associations with a single causal locus. Physiological salt sensitivity has been suggested to be an explanation for this racial difference in blood pressure (e.g., Aviv and Gardner, 1989); however, environmental factors also are clearly important (Hutchinson, 1986; Klag *et al.*, 1991).

Admixed families may be useful for other forms of family analysis. For example, if NIDDM in Amerindians is largely clonal, and is due to a tractably small number of loci, then affected siblings from $AC \times AC$ matings (intercross matings between Amerindians and Caucasians) will all be homozygous for gene regions surrounding the disease loci (Chapter 7). If the trait is recessive, marker sharing among sets of such affected siblings may efficiently be able to identify candidate regions. If the trait is dominant, one would sample offspring from $AA \times AC$ backcross matings.

Conclusion: phenotypic convergence due to genotypic divergence

The picture drawn by example in this chapter is generic. We are finding essentially the same pattern wherever we have adequate data. This applies to the loci responsible for quantitative and single-locus traits. Similar phenotypes arise as a result of different mutations in different populations. Diseases are locally clonal, although cladistically structured diversity often modifies the 'parent' mutation of the clone. Many disease-related loci have levels of variation that balance the input of new pathogenic mutations and selection against their deleterious effects. Such mutations may appear phenotypically convergent (similar effects), especially if we define phenotypes only vaguely, but are usually genotypically divergent (i.e., at the sequence level).

The same general ideas apply even when there is balancing selection that specifically preserves variation. The phenotype 'resistance to malaria' has arisen independently across the malarial belt of the Old World, but the genetic basis of the phenotype is locally different, each area having its own clone(s) of mutations at diverse loci. Were malaria to persist, eventually one of these, perhaps *HbS*, might out-compete the others and spread worldwide; in the 'end' there would only be one or a few clades of descendants of original *HbS* mutations. But these will bear the trace of their history in their haplotype backgrounds. And there is no 'end' really, because new mutational refinements will create new sub-clones, which will in turn expand through the population, like ripples in a pond.

This does not require a 'fit' population to expand into a malarial area at the expense of its inhabitants. All we need is admixture between them (e.g., by peaceful mate exchange at their frontier), aided by common selection pressures (e.g., malaria in both places). This probably explains the pattern of *HbS* in Africa.

The position of disease-related alleles on the gene tree at the locus or loci concerned may indicate something of their history. If all deleterious mutations are near the ends of the branches, this would suggest that they are recent. But if there are older clades whose member alleles have some disease-related mutation in common, such a mutation may have been maintained by selection. However, an effective theory to discriminate between a mutation that is old by chance and one maintained by selection remains to be developed.

Cladistic analysis tells us about the shape or structure of the causal spectrum, but each area will differ in its mutation, selection, and drift history, and hence in its sequence details.

11 *Evolution generates heterogeneity*

All healthy families resemble each other. Each unhealthy family is unhealthy in its own way.

With apologies to Leo Tolstoy, *Anna Karenina* (1877)

The 'Rusty Rule'

The previous chapter focused on the evolutionary determinants of the frequency of disease-related alleles, and the trail and structure left by the unique history of mutations in each population. Evolution systematically generates variation, but is the amount of variation so great as to change our traditional notions that there is 'a' locus, or 'a' mutation that is responsible for 'a' disease in the majority of cases?

This chapter uses examples to characterize the level of heterogeneity that led me to introduce (Chapter 2) the 'Rusty Rule' that whatever can go wrong will go wrong – in some family, at some time. In fact, the amount of heterogeneity associated with most traits can lead to great difficulty in applying the segregation and linkage methods of Part II, a problem we are just beginning to face.

Etiological and phenotypic heterogeneity for qualitative traits

A number of the classic genetic diseases have now been studied at the DNA level. In each case, a similar story is told, one that is consistent with evolutionary genetics. It is the story of many different alleles, or even loci, associated with the same phenotype, or of what seemed originally to be a unitary phenotype decomposing before our eyes as subtle phenotypic differences associated with identifiable allelic differences are discovered.

Cystic fibrosis

As with TSD and PKU, most cases of CF are caused by mutations at a single locus, *CFTR*. Like the other diseases, there are several CF-related mutations at this locus (Kerem *et al.*, 1989) differing among populations.

Table 11.1. *Phenotypic variation in cystic fibrosis associated with allelic heterogeneity*

	Genotype	Frequency
Pancreatic insufficiency (PI)	*ΔF508/F508*	0.459
	ΔF508/S	0.331
	SS	0.060
Pancreatic sufficiency (PS)	*ΔF508/M*	0.106
	SM	0.038
	MM	0.006

Notes: *ΔF508* is the 3 bp deletion in codon 508; *S* and *M* are severe and mild mutations, respectively. Frequency values are predicted by a causation model described by Kerem *et al.* (1989).
Source: From Kerem *et al.*, 1989.

CF has two basic clinical forms, a severe 'pancreatic insufficiency' (CF-PI) form and the less severe 'pancreatic sufficiency' (CF-PS) form. The severity is related to the genotype (Table 11.1). About half of all cases are heterozygotes (we saw this earlier for the 'recessive' diseases TSD and PKU). The *ΔF508* mutation is consistently associated with severe disease, homozygotes being more severely affected than mixed heterozygotes (Kerem *et al.*, 1990). In fact, a full spectrum of alleles and effects at the *CFTR* locus is being documented (Tsui, 1992).

Thalassemias

Chapter 10 discussed the geographically patterned nature, and possible evolutionary explanations, of malaria-associated globin gene variation. Here, I briefly review that variation in the context of its phenotypic implications. What we find is qualitatively similar but quantitatively different for the ancestrally related α- and β-globin regions {Antonarakis *et al.*, 1985; Gelehrter and Collins, 1990; Higgs *et al.*, 1989; Weatherall, 1991}.

β-Globin mutations

The genes in the β-like globin cluster are differentially expressed during development, generally sequentially in their chromosome order (Figure 11.1(a)). The wide variety of β-globin deletion mutations that have been observed in this cluster and are associated with hemoglobin disease are shown below this. These variants exemplify the Rusty Rule, because

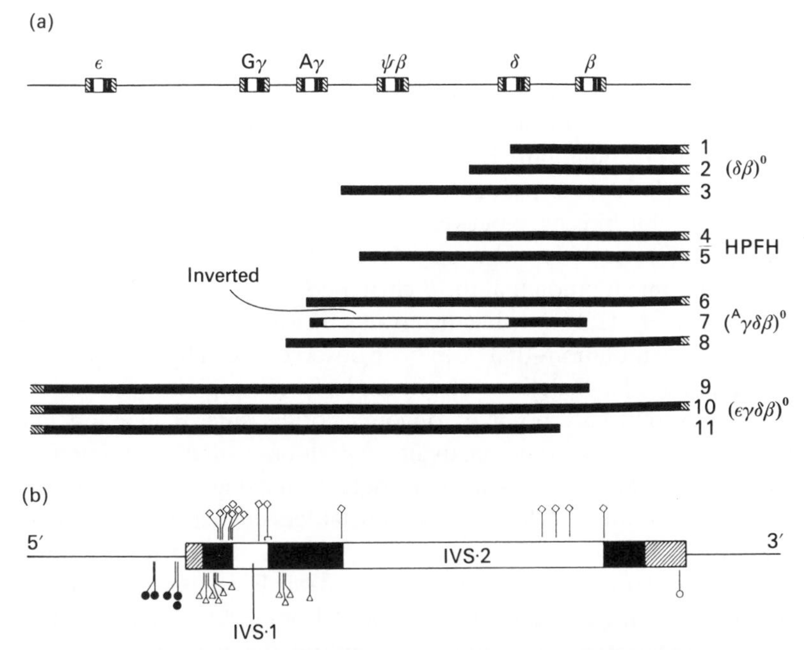

Figure 11.1. (a) β-Globin gene cluster arrangement showing location of deletion mutations. (b) Structure of β-globin gene, showing location of point mutations: filled circles, transcriptional mutations; open circles, RNA cleavage mutations; diamonds, RNA processing mutations; triangles, frameshift and nonsense mutations. (From (a) Weatherall, 1991; (b) Antonarakis *et al.*, 1985.)

essentially every step in β-gene expression is inactivated by one or more of them: promotor-site mutations leading to complete or depleted mRNA transcription, deleted intron–exon splice sites, added splice sites, activated cryptic splice sites, altered mRNA polyadenylation sites, altered translation initiation or termination sites, and frameshift or nonsense mutations yielding unstable or short-lived polypeptides. The many point mutations that result in coded, but abnormal, β-globin protein such as HbS were discussed earlier.

Some thalassemias result from deletion–fusion mutations bringing together, as one gene, parts of separate genes in the globin cluster (these are known as Lepore and Kenya thalassemia mutants). Another is a mutation in the upstream regulator region of the γ-globin gene, leading to its elevated expression at the expense of the normal β-globin. Hereditary persistence of fetal hemoglobin (HPFH) results if the δ and β genes are deleted (and can also result from regulatory region mutations).

α-Globin mutations

The α-globin region contains four pseudo-genes. In the embryo the ζ gene and the two function α-globin genes are expressed; although probably transcribed in small amounts, the θ gene is not an important constituent of hemoglobin (Higgs *et al.*, 1989). In adults, there is a 3:1 excess of α_2 relative to α_1 product, although these are biologically similar.

An important difference between this and the β-globin cluster is that the dual α-genes provide a measure of redundancy and hence phenotypic protection from mutation that the β-cluster does not provide. Figure 11.2 resembles Figure 11.1, showing the arrangements of this region and some of the deletion mutations that have been observed (see Higgs *et al.*, 1989; Weatherall, 1991). Deletions of one and of both α-globin genes have been observed. Many point or oligonucleotide (micro) deletion mutations are also known that inactivate an α-globin with effects similar to the deletions. Chromosomes with one deleted and one mutated α-globin gene have been found. There is a somewhat less diverse array of micro-mutations in the α-region than in the β-region; in particular, fewer regulatory mutations have been found to date. For unknown reasons, most of those affect the α_2-gene without leading to a compensatory increase in the α_1-expression and hence have relatively severe phenotypic effects.

The number of functional α-globin genes is the most important determinant of the phenotype of an individual relative to this region. A normal individual has four (two $\alpha\alpha$ chromosomes), but deletions can produce any combination of $--$, $\alpha-$, and $\alpha\alpha$ chromosomes, and the function of the remaining genes can be modified by mutation. The resulting phenotype is α-dose dependent, as shown in Table 11.2. On a single-deleted chromosome, the remaining α-globin gene is transcribed at a level between the normal α_1 and α_2 levels. $\alpha\alpha/\alpha-$ are essentially normal, while $\alpha-/\alpha-$ or $\alpha\alpha/--$ genotypes have mildly impaired hemoglobin function; $\alpha-/--$ genotypes are severely impaired and form a substantial amount of β-globin tetramers known as hemoglobin H. The genotype $--/--$ produces a lethal condition known as hydrops fetalis.

Collagen production and brittle-bone disease

Collagen is one of the most important and ubiquitous molecules in the body, a long structural protein used in connective tissue and bone {Byers, 1990, 1991; Gelehrter and Collins, 1990; Weatherall, 1991}. The collagen molecule comprises three procollagen subunits, which join to form a helical structure; each of the chains has a pro-peptide tag, which is cleaved after the molecules are secreted from the cells in which they are

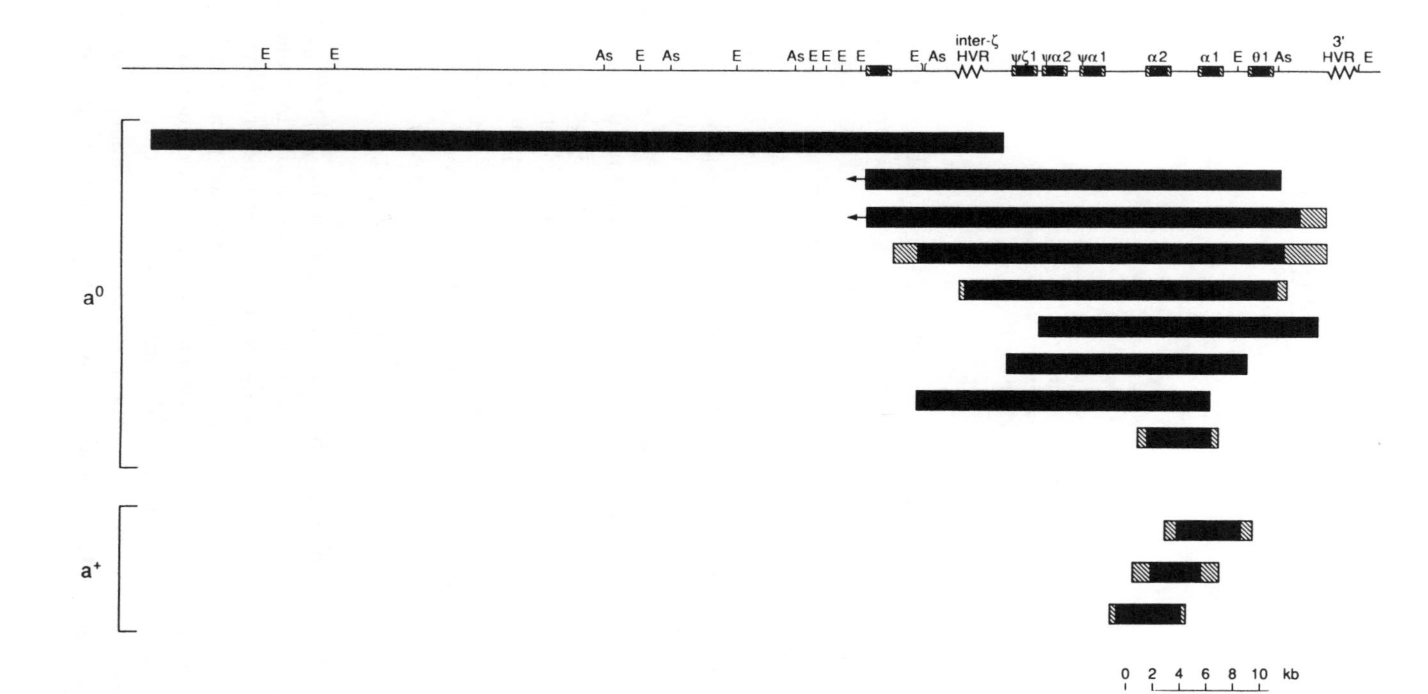

Figure 11.2. Known deletions that delete both α-globin genes. Each bar represents the deleted region. As, E, HVR, marker sites; ψ: pseudogene; a^0, a^+: double- and single-α inactivations, respectively. (From Weatherall, 1991.)

Table 11.2. *Phenotypic variation associated with α-globin genotypes*

Phenotype	Genotype
Normal	$\alpha\alpha/\alpha\alpha$
Clinically silent carrier (also called α-thal. 2)	$\alpha\alpha/\alpha-$
Heterozygous α-thalassemia trait (also called α-thal. 1)	$\alpha\alpha/--$
Homozygous α-thalassemia trait	$\alpha-/\alpha-$
Hemoglobin H disease	$\alpha-/--$
Hydrops fetalis with hemoglobin 'Barts'	$--/--$

Sources: Gelehrter and Collins, 1990; Higgs *et al.*, 1989.

produced. The subunit peptides each comprise large numbers of repeated amino acid triplets, typically consisting of (Gly-X-Y)$_n$, where n is about 3380, X is typically proline and Y the modified amino acid hydroxyproline (sometimes hydroxylysine). The repeat structure is critical to the functional stability of the molecule.

The corresponding collagen gene structure has an interesting history that relates both to the coded protein structure and to the potential effects of mutation. The current genes appear to have arisen from multiple duplications of nested sequences built upon an original 9 bp codon triplet, in a manner reminiscent of the apolipoprotein structure described in Appendix 2.1 (Runnegar, 1985). The primary subunit has been tandemly repeated many times, including the formation of a 54 bp secondary structure consisting of three of the primary tandem repeats; a 702 bp tertiary structure consisting of 13 secondary units was subsequently copied five times during the evolution of contemporaneous vertebrate collagen genes. These various subsequences have of course also been modified by point mutation. Collagen genes span 38 kb, with 51 exons.

Two $\alpha_1(1)$- and one $\alpha_2(2)$-chains (coded by separate collagen genes) unite to form type 1 collagen, used in bone, tendon, and skin. Collagen types 2 and 3 are homo-trimers of (different) α_1-type gene products, and are expressed in various connective tissues in the body; there are other minor collagens. The genes comprise a multigene family of more than 20 members scattered over at least 10 different chromosomes.

Various diseases involve collagen metabolism, including Marfan's syndrome, Ehlers–Danlos syndrome and perhaps even osteoporosis and

Table 11.3. *Collagen mutations and disease severity*

	Heterozygotes		Homozygotes	
Mutation	Phenotype	Severity	Phenotype	Severity
Usable α_1	$\frac{1}{4}$ normal $\frac{3}{4}$ α_1 abnormal	Severe	All α_1 abnormal	Lethal
Unusable α_1	$\frac{1}{2}$ normal $\frac{1}{2}$ α_1 missing	Mild	All α_1 abnormal	Early lethal
Usable α_2	$\frac{1}{2}$ normal $\frac{1}{2}$ α_2 abnormal	Mild	All α_2 abnormal	Lethal
Unusable α_2	$\frac{1}{2}$ normal $\frac{1}{2}$ α_2 abnormal[a]	Very mild	All α_2 abnormal[a]	Moderate

Notes: [a] Partial production of abnormal $\alpha_1\alpha_1\alpha_1$ collagen. 'Usable' mutations are coded peptides that are able to form mature collagen; 'Unusable' mutations cannot bind properly into triple helix.
Source: after Weatherall, 1991.

osteoarthritis. However, the major reflection of collagen mutations are the diverse diseases collectively known as *osteogenesis imperfecta* (*OI*), whose unifying characteristic is brittle bones.

OI is classified into four major clinical types (1 to 4), inherited with varying degrees of dominance and major allelic heterogeneity among and between them. The most common, OI type 1 is dominant, and is produced by inactivation of the α_1 (1) gene, or by mutations that lead to a gene product that cannot be incorporated into the triple helix of mature collagen. Heterozygotes for such mutations produce only 50% of the normal amount of α_1-chains, and hence of total type 1 collagen, leading to the brittle bones that are symptomatic of the condition.

OI type 2 is caused by code-changing in the collagen exons. Even though the collagen chain components are produced rather than absent, these disorders can be more serious than OI type 1, and are often lethal. When the collagen subunits have the wrong amino acids, triple helix formation is impaired and the resulting molecule does not function or is quickly degraded. Since the molecule requires a 2:1 α_1 to α_2 ratio, if half the α_1-chains are defective as in a heterozygote, $\frac{3}{4}$ of the resulting collagen might be defective. Some details of the genotype–phenotype relationship are given in Table 11.3.

Tens of mutations including exon rearrangements and deletions, and many point mutations have been identified in the collagen genes (e.g., Figure 11.3). Disease-associated point mutations often replace Gly at

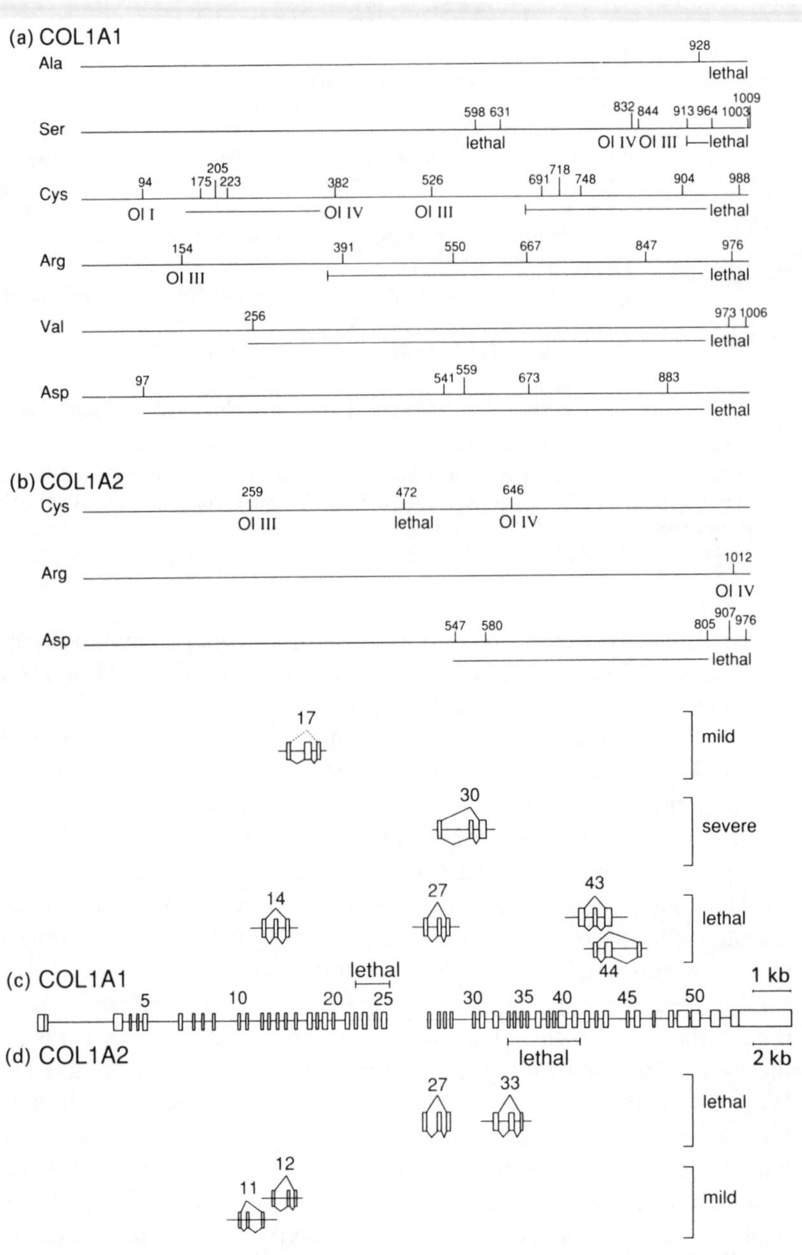

Figure 11.3. Diagram of mutations in type 1 collagen genes and their effects. (a) and (b) Mutations replacing glycines in the α_1 (a) and α_2 (b) proteins, the numbers referring to the amino acid position in the protein chain; (c) and (d) effects of rearrangement mutations in these genes, the numbers referring to the exons involved. (From Byers, 1990.)

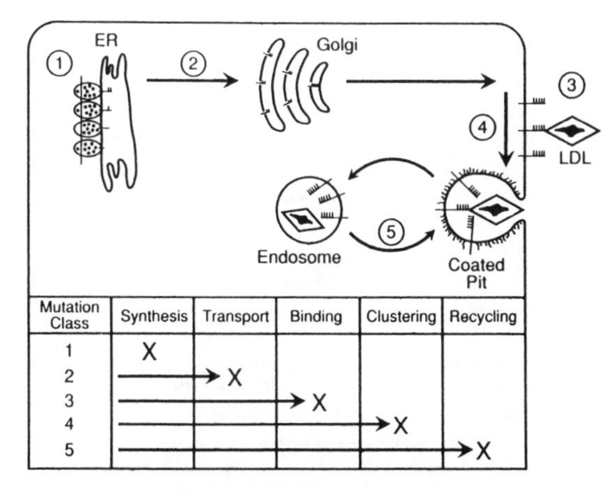

Figure 11.4. Places in its physiology where known mutations affect LDL receptor function. ER, endoplasmic reticulum; LDL, low density lipoprotein. (From Hobbs *et al.*, 1990. Reproduced, with permission, from the *Annual Review of Genetics* **24**, © 1990 by Annual Reviews Inc.)

various points in the repeat structure. The position of a mutation determines its phenotypic severity; mutations towards the high codon numbers are more severe or lethal, those in low-numbered codons produce less severe effects. Whole α_1 exon deletions have severe effects, generally regardless of which exon is involved.

The LDL receptor locus

Mutations at the *LDLR* locus show the consistency and replicability of the pattern seen in most loci that have been looked at in detail. The *LDLR* gene product resides on the surface of hepatic cells (among others) and is responsible for recognizing the apo-B100 and apo-E molecules on LDL particles and removing cholesterol from the blood into the liver. Even a single mutation may produce seriously elevated blood cholesterol levels. The array of mutations found in the *LDLR* locus exemplifies the Rusty Rule. As shown in Figure 11.4, these mutations affect all of the five different basic aspects of LDL receptor function: synthesis, transport of the protein to the Golgi apparatus (and thence to cell membrane), binding ability to recognize LDL particles, endocytosing action to bring cholesterol into the hepatocyte, and recycling of the protein {Gelehrter and Collins, 1990; Goldstein and Brown, 1988; Hobbs *et al.*, 1990}.

Over 150 different mutations have been identified at the LDL receptor locus. These include deletion mutations, shown schematically in Figure

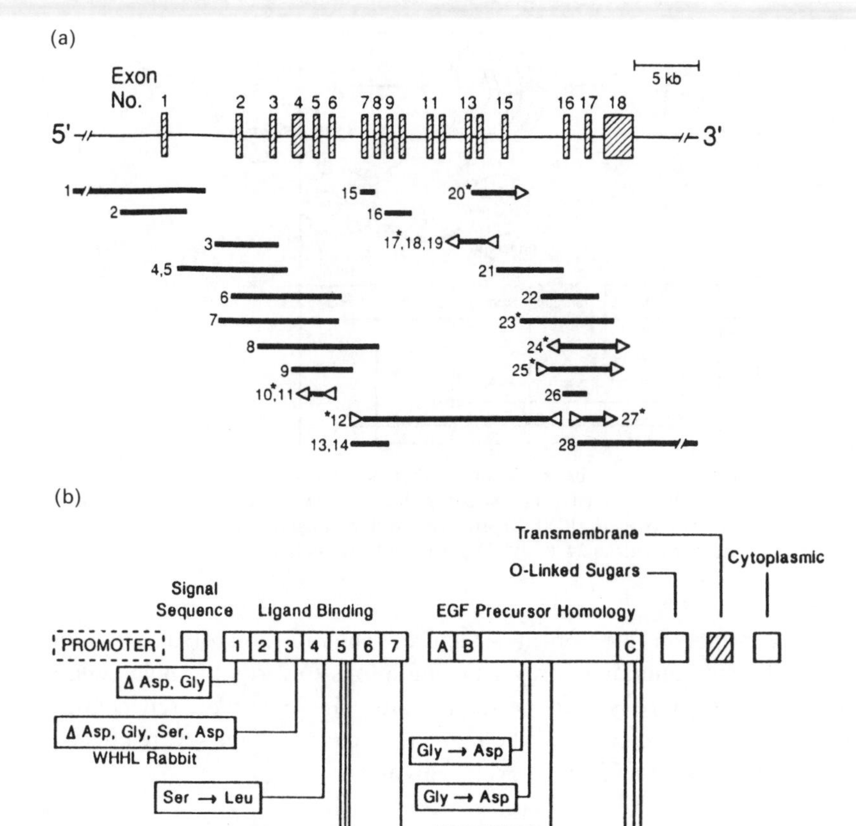

Figure 11.5. Known mutations in the *LDLR* locus: (a) deletions and (b) point mutations showing domains of the LDL receptor protein that are affected. (From Hobbs *et al.*, 1990. Reproduced, with permission, from the *Annual Review of Genetics* **24**, © 1990 by Annual Reviews Inc.)

11.5(a), and smaller mutations that affect LDL receptor transport within the cell, by causing misfolding of the protein, shown in Figure 11.5(b). These mutations are typically unique and geographically confined to specific populations.

LDL receptor defects cause *familial hypercholesterolemia* (*FH*), a co-dominant disorder leading to early heart attacks and other problems associated with greatly elevated circulating LDL-cholesterol. The degree of cholesterol elevation is gene-dose dependent. Deletion homozygotes have the greatest effects, but there is considerable variability, especially among heterozygotes, about 20% of whom have cholesterol that is in the normal range (Roy *et al.*, 1991). This is because LDL-cholesterol levels are affected by other loci as well (Hobbs *et al.*, 1989, 1990) and by dietary lipid levels. Interaction between FH and apo-E alleles has also been reported (Hopkins *et al.*, 1991).

Boundaries blurred: the single locus as a polygenic system

Even at a single locus, the many genotypes resulting from the array of mutations found in a given population can produce a nearly continuous phenotype distribution. That is, a single locus can itself act as a polygenic system (e.g., Humphries *et al.*, 1987).

This may apply to somewhat more complex systems, too. DS has long been known to be caused by triplication of chromosome 21, but not all DS patients have the same collection or severity of phenotypes. Why is this? The evidence is rapidly accumulating to show a mappable correlation between the regions of chromosome 21 that are aneuploid (abnormal copy numbers) in a given patient and his/her phenotypes (Korenberg, 1991).

PKU: PAH *mutations and their effects on phenylalanine hydroxylase concentration*

PKU results from errors in phenylalanine metabolism that can be produced by dysfunctional genes at various loci {Scriver *et al.*, 1988}; here we consider only the main locus, *PAH*. The allele frequencies given in Chapter 10 for European PKU mutations show that, with random mating, heterozygosity is about 85% – most PKU patients are actually heterozygotes. The nucleotide sequence at the carboxyl end of the *PAH* gene, which codes for its hydroxylation activity, is highly conserved. The substrate-recognition region, at the amino end of the protein, is less conserved.

The phenotypic effects of various PKU mutations have been quantified (Okano *et al.*, 1991). Among 206 Europeans typed, eight PKU alleles accounted for 64% of all PKU chromosomes in the sample. The effects of these alleles were estimated experimentally. Each mutation was generated by site-directed mutagenesis on cloned *PAH* genes, introduced with

Table 11.4. *Allelic effects of eight common* PAH *mutations*

Codon affected	Relative frequency	% normal PAH activity
243	0.2	<1
281	2.2	<1
408	20.1	<1
IVS12	25.5	<1
280	0.7	<3
158	4.1	10
261	3.9	30
414	7.3	50

Note: IVS12 is an intervening sequence (intron) in which the mutation is found. Relative frequency, among 412 chromosomes in 206 PKU patients from Denmark and Germany. Some of the mutations were discussed in Chapter 10.
Source: after Okano *et al.*, 1991, with other pertinent references therein.

an expression vector into cultured monkey kidney cells, and the PAH activity level estimated. The level of all mutants was quantified relative to the level of the normal *PAH* allele when tested by the same assay (recall that allelic effects are defined in relative terms). Table 11.4 shows these allelic effects.

Assuming purely additive effects, these experimentally determined allelic effects were used to predict the PAH activity of the diploid genotypes that could be formed from the alleles; that is, $Activity(i, j) = (\alpha_i + \alpha_j)/2$. The results are shown in Table 11.5. Study subjects with these same mutations were tested to determine whether their Phe levels corresponded to the predicted PAH levels. The expected inverse relationship between PAH activity and Phe concentration was found, with highly significant correlations of the order of 0.8 or more, confirming the additive model as an adequate approximation to the mutations' dose effects.

In introducing the idea of phenotype distributions associated with specific genotypes, Figure 4.1(a) and (b) showed how we sometimes put the underlying continuous phenotype distribution into categories such as 'normal', 'mildly affected', and 'severe'. In this vein, we can provide a quantitative genotypic basis for the clinical variability of PKU. This is shown in Table 11.6, and begins to explain the variability in clinical

Table 11.5. *Clinical course and* PAH *phenotypes*

Mutation	Clinical phenotype				
	243, 281 408 IVS12	280	158	261	414
243,281 408, IVS12	Severe	Severe	Severe	Severe and mild	Severe and mild
280	—	NA	NA	NA	NA
158	—	—	Severe	NA	NA
261	—	—	—	Mild	Mild
414	—	—	—	—	Mild

Notes: NA, genotype not present in sample studied. IVS, intervening sequence (intron). Lower triangle denoted '—' is omitted as it is a duplicate of upper triangle.
Source: Okano *et al.*, 1991.

course and response to dietary intervention that has long been known for PKU. The ability to determine haplotypes should provide better guidelines for diagnosis and treatment (Guttler *et al.*, 1987; Lyonnet *et al.*, 1989; Okano *et al.*, 1990).

We can use these data to look at the spectrum of allelic effects at this single locus (not including any variation due to different 'normal', or non-PKU mutant alleles). I have assumed that the 36% untested PKU haplotypes in the sample have the same distribution of allelic effects that the tested 64% did. Thus, the percentage relative frequency of each allelic effect is derived by re-scaling the frequencies in Table 11.4 by a factor 100/0.64 to obtain the percentage of all PKU alleles. These adjusted frequencies are shown in Table 11.6.

Because the aggregate of normal *PAH* alleles has a frequency 0.99 or thereabouts, most PAH variability in the population is associated with normal allele(s), and the aggregate PKU genotypes comprise only a small fraction of the distribution. Non-PKU hyperphenylalaninemia, which involves mild symptoms, is typically caused by compound heterozygosity, usually involving a PKU mutation and an otherwise 'normal' allele (Avigad *et al.*, 1991). Table 11.6 shows that the spectrum of effects, when restricted to the population of PKU chromosomes, looks much like the spectrum discussed in Chapter 9, a roughly exponential distribution of effects. The *PAH* distribution given here is conditional on removing

Table 11.6. *Predicted genotypic values (normal type) and relative genotype frequencies among PKU patients (bold type), based on allelic effects of eight* PAH *mutations*

		Predicted PAH activity (% normal)				
Allele	Normal	243,281 408, IVS12	280	158	261	414
(Normal	100	50	51.5	55	65	75)
243, 281		0	1.5	5	15	25
408, IVS12		**56.1**				
280			3.0	6.5	16.5	26.5
		1.7	**0.0+**			
158				10	20	30
		9.6	**0.1**	**0.4**		
261					30	40
		9.6	**0.1**	**0.7**	**0.4**	
414						50
		17.1	**0.2**	**1.5**	**1.4**	**1.3**

Notes: For computational details see text. IVS, intervening sequence (intron).
Source: Modified after Okano *et al.*, 1991. Uses data from Table 11.4.

the (common) 'normal' allele(s), and shows the self-similarity also discussed in Chapter 9.

Etiological heterogeneity for quantitative traits

Multiple simultaneous QTL control: the causal spectrum in action

Chapter 8 provided means to find a QTL, or major gene, associated with variation in a quantitative trait, and Chapter 9 suggested that many such genes may be involved in a given trait, having a spectrum of effects. But how are the genes in this spectrum identified? Of course, one could individually test a battery of candidate genes or marker polymorphism for measured genotype effects, as suggested for cholesterol in Chapter 9.

A better way, less subject to linkage disequilibrium errors, is to perform linkage studies using a complete marker map, the ultimate extension of reverse (i.e., **G**←**P**) genetics. The interval-mapping method developed by Lander and Botstein was described earlier with regard to qualitative (Chapter 7) and quantitative (Chapter 8) traits. These methods are rapidly becoming methods of choice in human family data as the marker map is becoming complete and polymorphic enough. However, a suitable map is nearly available in the form of widely dispersed VNTRs, dinucleotide repeats, and other hypervariable regions. Therefore, it is worth outlining briefly how the method works.

An example of QTL searching

Figure 8.3 illustrated schematically the use of a marker map to find QTLs for a quantitative trait, and Chapter 9 discussed the results of screening progeny of tomato backcrosses for QTLs related to several phenotypes (Paterson *et al.*, 1988, 1990, 1991). That work screened all tomato chromosomes, yielding maps such as Figure 8.3, identifying a total of about 5–10 QTLs per trait, typically finding one QTL per trait per chromosome.

Two recent studies have used similar methods to screen spontaneously hypertensive laboratory rats for loci related to blood pressure regulation. *Hypertension*, or elevated blood pressure, is a risk factor for stroke and other cardiovascular disease. Over the years, extensive efforts have been made to find genes with important effects on blood pressure, but despite consistent heritabilities of about 20–30% (Table 6.5) these efforts have largely been unsuccessful {Sing *et al.*, 1986; Turner *et al.*, 1989; Ward, 1990}. One reason seems to be that the trait is poorly defined, probably a collection of diverse underlying phenotypes that we measure collectively as 'blood pressure'; we would be similarly confused if we tried to find genes for the diffuse trait 'mental retardation', rather than studying specific loci such as those for PAH, TSD, DS and so on.

As will be seen in the next chapter, only one locus has yet been identified with strong effects on human blood pressure, with suggestive evidence from segregation analysis for another. Therefore, laboratory rats have been turned to for 'assistance'. A recent study searched for blood pressure QTLs using 181 mapped markers polymorphic in F_2 progeny of crosses between normotensive rats and a strain of rats that develop hypertension in response to dietary sodium (Hilbert *et al.*, 1991). A locus on rat chromosome 10 was identified that, based on rat–human linkage homologies, may be a good blood pressure QTL, since it is near the gene for *angiotensin conversion enzyme* (*ACE*), an enzyme known to be related to blood pressure physiology (though not particularly to hypertension). In the experimental cross, variation at this locus explained more than 20% of the variation in response to dietary salt loading. Another gene on the X-chromosome was identified that is consistent with previously known sex-specific blood-pressure response patterns in these rats. Using the same phenotype data and other markers, Jacob *et al.* (1991) have found evidence for an additional QTL on rat chromosome 18.

These studies show the power of using a marker map to screen for loci by reverse genetics, to identify genes involved and show their physiological function. Variation at the human homologs of these mapped rodent genes can now be studied in families or by measured genotype approaches.

The Rusty Rule for quantitative traits

Inbred strains represent unrealistic working material for reasons given in Chapter 9, but the usefulness of experimental crosses as model systems is clear. While human families that are 'crosses' of various types at specific loci can be found, the overall context of human heterozygosity will likely generate patterns different from those seen in experimental systems, but I think precedent shows that the information we gain from the latter will be applicable to humans.

Experimental back- or intercrosses distort reality in an important way. They raise all segregating allele frequencies to the same high value (e.g. $\frac{3}{4}$ and $\frac{1}{4}$ in an $AA \times Aa$ backcross). Thus the distribution of allelic effects seen in QTL maps of such material may not represent the relative contribution of the same (homologous) alleles in humans, given the typically inverse relationship between strength of effect and frequency. As noted in Chapter 9, the standardized effect of alleles at such diallelic loci will not be representative of those in a natural population with exponential allele frequency distributions.

However, alleles of strong effect can be artificially raised to high frequency by sampling human 'natural experiments'. Families with individuals admixed between distantly related populations may mimic experimental crosses at some loci. And probands with extreme phenotypes are likely to be heterozygotes (or even homozygotes) for rare alleles, and their immediate families will perforce have high frequencies of those same alleles.

Approaches to heterogeneity in family data

It is one thing to document etiological heterogeneity from known genes, and quite another to take care of the problem adequately using inferential methods such as segregation and linkage analysis in families or other samples, i.e., before any genes are actually identified. Otherwise, we obtain seriously misleading – especially false negative – results.

Etiological heterogeneity can involve environmental factors as well as genes. In a sample of families, only some may be segregating risk alleles, the others expressing 'sporadic' cases. It can be very important to be able to detect such risk heterogeneity and to identify the subset of families that is likely to be genetically interesting.

Permutation tests and other model-free family-ranking tests

The simplest tests for detecting such families are model-free tests that ask the simple question: are the families in a set different in their risk of a given disease? Essentially, a set of families has been collected and the phenotype (disease status) determined for each member. We want to know if all families are at equal risk. An exact test of this can be impossible to work out, since we do not have a good way to specify the significance of the deviation of the patterns of cases in a family from 'expectation'. A variety of empirical approaches have been developed to address this problem, relying on a method known as *permutation testing*.

For a given family structure (age, sex, and numbers of relatives, etc.), we either (1) estimate the risk in those families on the assumption that each member was drawn randomly from the population, or (2) assume that the expected risk in each family member equals the average risk of all persons of the same age, sex, etc. in the set of sampled families (i.e., pooling across families).

For each individual in a family the expected number of cases equals the prevalence in the population (model 1) or in the family data set (method 2) for a person of his/her age, sex, and so on. For example, if the prevalence is 50%, the person is assigned $\frac{1}{2}$ an expected case. The expected number of cases in the i-th family is $e_i = \Sigma\, e_{ij}$, the sum of the individuals' expected cases over all sampled members of the family. This number can be compared to the number of cases actually observed in those individuals, o_i, creating a risk index, $I_i = o_i - e_i$ for the family. It is usually not possible to specify the distribution of I values in a set of families, so that an exact significance test for a given family, or heterogeneity test for the set of families, cannot easily be formulated.

One approach to a significance test is to permute the individuals in the families, randomly exchanging persons (excluding probands) of the same age, sex, etc., and to recompute the I values for the rearranged set of families. This is done a large number, say n, of times, generating risk indexes $I_{1i}, I_{2i}, \ldots, I_{ni}$ for the i-th family. If the observed index, I_i, for that family is in the top α% of this set of randomly permuted index values, the family can be considered to have a significant excess of cases (and the converse for a deficit).

A test for homogeneity in risk would be that the observed *variance* of index values among families is not large compared to the set of n variances generated, one each, by the permutation runs. Another is that the rank of the observed family indexes, among their respective permutations, should have a uniform distribution; for example, if risk is randomly distributed among families, the observed index in 10% of the families will be in the top 10% of the permutation-based index values generated for the family, 20% will have index in the top 20% of simulated index values, and so on.

Tests of this type have been used to show that colorectal cancer clusters in families ascertained by a proband affected with colorectal cancer, but not in families ascertained via probands with other types of cancer (Lynch *et al.*, 1981, 1986), and to show that there is heterogeneity in breast cancer risk among families; some families are at excess, and others at reduced, risk (Schwartz *et al.*, 1988). With the fast computer power now ubiquitously available, permutation tests and a variety of other tests of similar spirit (e.g., 'bootstrapping', 'jacknifing') are widely used to determine significance levels in complex situations in genetic epidemiology.

Model-based tests in families

Segregation analysis programs normally model all families as having the same etiology. However, there are ways to use likelihood methods and the power of genetic models even when there is heterogeneity. One approach is to compute a desired likelihood ratio (e.g., comparing major gene v. sporadic models) separately for each sampled family. Normally the likelihood for each hypothesis would be computed jointly for the entire set of families by multiplying the likelihoods for each family, and the net likelihood ratio computed. Computer simulations of family data under various models, with heterogeneity, have shown that suitable cutoff points can be found such that most families whose likelihood ratio exceeds the cutoff will correctly be classified as segregating a major gene, while families below the cutoff will not.

For example, a cutoff of $\mathcal{L}_{\mathrm{MG}}/\mathcal{L}_{\mathrm{NTF}} \leq 1$ is suitable for discriminating major gene (MG) from no major gene (NTF) for simple quantitative traits (e.g., Beaty, 1980; Boehnke and Moll, 1989). This criterion has been used to suggest that only a subset of families segregate a major gene for apo-AI levels (Moll *et al.*, 1989), in the study discussed in Chapter 6. Table 11.7 shows the parameter values for the overall data and for these two subsets.

There are many potential problems with such generalized methods, and a number of families may be misclassified, especially if their likelihood ratio is near the cutoff value. In the apo-AI study the parameter values are very similar between the two subsets, despite significant differences in their likelihoods and the transmission models of the two

Table 11.7. *Comparison of parameter estimates for models in whole apo-AI family set and in etiological subsets identified by the likelihood cutoff method*

Parameter	Whole data general model	Non-Transmitted Factor subset	Single Major Locus subset
No. families	283	126	157
p	0.784	0.698	0.857
μ_{AA}	129.65	132.35	128.47
μ_{Aa}	143.55	132.47	143.58
μ_{aa}	182.27	171.82	189.20
σ	17.46	16.56	16.82
h^2	0.34	0.952	0.0*
$\tau_{AA,A}$	0.92	p^*	1.0*
$\tau_{Aa,A}$	0.73	p^*	0.5*
$\tau_{aa,A}$	0.26	p^*	0.0*

Note: * indicates values fixed by authors (based on the selection criterion for the subsets analyzed). For NTF model, see Chapter 6.
Source: Extracted from Moll *et al.*, 1989.

groups. At least, those families with likelihood ratios far above the cutoff should be those that, from the available data, are most likely to be segregating a major locus and it would be in those that linkage or other follow-up studies would be done. Regardless of their true etiology, the families near the cutoff simply do not provide the required information, due to the vagaries of, for example, incomplete penetrance or their age/sex structure.

Standard linkage methods can successfully identify map locations of diseases that are predominantly caused by segregation of alleles at a single locus, such as CF, HD, PKU, etc. In all families the critical phenotypic variation segregates 'properly' and will map to the same locus. However, if the same phenotype can be generated by variants at different loci in different families, a single linkage study may not generate sufficient power to identify all of them. Indeed, strange 'emergenetic' phenotypes can arise; for example, two albinos can marry and produce normal children, showing that they are homozygous at different loci. A single-locus linkage or segregation analysis would not pick this up.

Whenever possible the best way to approach such a problem is to classify the phenotypes accurately into homogeneous categories (or in some other way identify clonal cases), in which families are more likely to be homogeneous for segregation and linkage. When, as is typical, this is not possible, a parametric approach can be taken {Ott, 1991}. For example, a population can be assumed to be an 'admixture' of a fraction m of families segregating alleles at one locus at a recombination distance θ_m from some marker, and a remaining fraction $1 - m$ of families that are due to some cause unlinked to the marker locus.

The likelihood of this hypothesis can be compared with that under the assumption of etiological homogeneity; that is, that all families segregate at a locus at distance, say θ_1, from the marker. The first hypothesis has parameters θ_m and m, the second only θ_1. The likelihood for the set of families (e.g., product of the likelihoods for each family) can be computed for the two hypotheses; for the i-th family one likelihood is $\mathcal{L}_i(m, \theta_m) = m\mathcal{L}(\theta_m) + (1 - m)\mathcal{L}(\frac{1}{2})$ and the other is $\mathcal{L}(1, \theta_1) = \mathcal{L}(\theta_1)$. Comparing these likelihoods by

$$\chi_1^2 = 2 \ln [\lambda(m_{\mathrm{MLE}}, \theta_{m,\mathrm{MLE}}) - \lambda(1, \theta_{1,\mathrm{MLE}})]. \qquad [11.1]$$

yields a test for heterogeneity [Chakravarti *et al.*, 1987; Ott, 1991; Rao *et al.*, 1978; Risch, 1990a; White and Lalouel, 1987].

Admixture tests have low power to distinguish between uniform loose linkage and admixture, do not by themselves identify which families are segregating linked alleles and which are not, and do not account for the possibility that some families may segregate at more than one locus. An alternative approach, which in principle is not restricted to any admixture assumptions, is to ensure that each family collected is large enough, on its own, to generate a significant LOD score for a linked marker (Appendix 7.1). Lander and Botstein (1986a,b, 1989; Lander, 1988) discussed applications of their interval mapping approach for simultaneous searches across the genome for linkage that can vary among families [Martinez and Goldin, 1991].

At present, we are not close to being able effectively to identify more than one major locus per family, and even further from being able to map all the genes in the spectrum that have more than trivial effects on the trait or on risk. The variability in frequency and strength of effect are too difficult to disentangle in most observational family data. Current studies can identify one major locus per family (or that shared in common among a set of families) but it is likely that many of the inconsistencies and uncertainties in the current human disease linkage literature are due to various forms of heterogeneity.

Conclusion

Heterogeneity has only relatively recently been 'discovered' as a serious complicating problem in human genetics, but explains many common difficulties in genetic epidemiology. The level of complexity was not anticipated by genetic epidemiologists, or by classical evolutionary geneticists, who generally worked with two-allele recurrent mutation concepts, but it is fully predictable on the basis of modern population genetic theory. It is becoming an increasing challenge to sort through the complexity. Each family may be segregating alleles of substantial effect at one or more loci; that is, the trait may have multiple unilocus, or even multiple oligo-locus, etiology (not to mention polygenic and environmental effects). The genes affecting lipid levels (and hence risk of CHD) are the best-documented human example for a quantitative trait, but we have seen that even the classic 'simple' qualitative traits manifest comparable complications.

Families ascertained via extreme-value probands may tend to be individually more homogeneous, in a sense artificially raising the frequency of the major allele that generated the proband's phenotype (Carey and Williamson, 1991). These families will, however, differ among each other and may not accurately reflect the causal spectrum in the general population.

Cladistic analysis, by clustering under a single category (clade) many alleles of related effect, may help to reduce the dimensionality of the problem. Increasing reliance on animal models may also help, since the organization of the genome is conserved to a great extent even between humans and laboratory rodents, and the latter have been highly mapped. Experimental studies using marker maps, and controlled crosses among strains, have shown a potential for identifying genes that have similar effects in humans (e.g., Plomin *et al.*, 1991). This has been done, for example, with regard to diabetes (see Chapter 14).

Despite these advances, we should probably not aspire to identify every mutation that affects a trait. These are continually added and lost through mutation, selection, and drift, and vary among populations. However, we should be able systematically to identify the important loci and the characteristics of the causal spectrum as discussed in Chapter 9.

Heterogeneity presents its most formidable obstacle for poorly defined quantitative traits for which we as yet have no good candidate genes. An example may be the genetic basis of blood pressure. Genome QTL mapping will help, but vague or tertiary phenotypes may not yield easily to genetic understanding. The heterogeneity problem is worse when environmental or other complicating factors are important. These are the subject of the next chapter.

Part IV *Modification of the inherited genotype: the time dimension in individuals*

12 *Phenotype amplification by the environment*

Diverse forms of allometry

Allometry is a term that refers to the quantitative scaling relationships between biological traits, such as between metabolic rate and body size among species. The term can be extended to refer to systematic dose–response relationships between genes, environments, and phenotypes – that is, to *risk factors* (covariates or concomitants) that determine the penetrance (**G** → **P**) function.

A challenge to genetic inference is to identify those situations that magnify or clarify the phenotypic differences among genotypes. Exposure to environmental risk factors often leads to *phenotype amplification*; that is, to phenotypes that become increasingly divergent, and easier to identify, with exposure dose and duration (or age). Other factors, however, may obscure the effects of different genotypes. We saw in Chapter 6 how to include unmeasured risk factors, such as the 'environmental' variance, in terms of their aggregate effect on phenotypic variance. This chapter is concerned with the effects of those risk factors that we can identify and measure on individuals.

We usually have no specific biological model for the dose–response relationships of such variables, and usually parameterize those relationships with general epidemiological models such as those given in Chapter 3.

The complex map from genotype to phenotype

The complexities and levels at which genetic and environmental factors can affect a subsequent outcome phenotype are illustrated schematically in Figure 12.1 for the case of CHD; for example, cholesterol is one component of the atherosclerotic plaques that may lead to clotting reactions producing heart attacks.

Figure 12.2 shows the regular empirical relationship between the CR, defined in Chapter 9, and life expectancy (average years of life remaining) at each age, with CR expressed in terms of standard deviations away

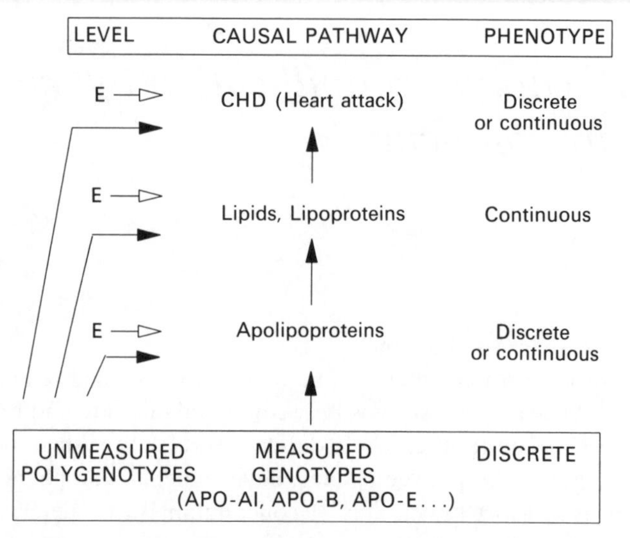

Figure 12.1. Levels of observation and determination for complex phenotype. E, environmental factors. (Modified from Sing *et al.*, 1988.)

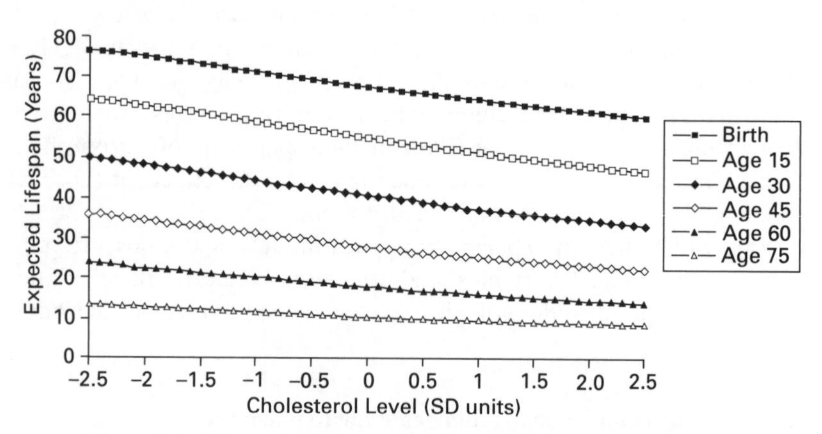

Figure 12.2. Schematic of relationship between genotypic effects, phenotype, and disease outcome for cholesterol. (S. Weeks *et al.*, unpublished results.)

from the mean CR in the population. For a specific CR value read down the chart from the top, which represents that CR's associated life expectancy to a newborn, to the bottom which gives the life expectancy of a 75-year old with that CR. This computation was the basis of the fitness computations given in Chapter 9.

CHD aggregates in families, and allelic variants of the apolipoproteins that transport lipids, affect the levels and characteristics of the lipoproteins. Yet, because the system is complex, redundant and anastamotic (Chapter 9), CHD risk does not segregate in a neat mendelian way (except for some severe *LDLR* mutations in FH). Generally, the CHD-predictive power of most specific lipid gene variants is rather low (complex systems evolve to be buffered against individual mutants). As illustrated in Figure 12.1, as each physiological level is crossed, the effects of additional genotypes and environmental factors are interposed.

For example, apo-E is a constituent of VLDL and HDL and hence is involved in cholesterol transport both to and from the liver. Allelic variation at the apo-E locus explains about 20% of the variation in the level of the apo-E product itself, but only 5% of the variation in plasma cholesterol level and 6% of the overall risk of CHD (Boerwinkle and Utermann, 1988; Davignon *et al.*, 1988). Diet, exercise, gender, smoking, and other risk factors amplify the phenotype relative to the primary effects of the locus itself.

Heart attacks are of primary interest to us as a society, but heart attacks are just a statistical outcome of the effects of elevated lipids and other risk factors, and there are no 'genes for heart attacks' per se. From a genetic point of view, we may learn more by studying the physiological basis of these risk factors than those of heart attack itself {Sing and Boerwinkle, 1987; Sing and Moll, 1989, 1990; Sing *et al.*, 1986}.

Risk factors for quantitative phenotypes

In order to model the effects of measured risk factors on quantitative traits we have to specify their *effects* on the phenotype. The basic methods were introduced in Chapter 3.

Additive risk factors

Consider a set of measured risk factors, $\boldsymbol{x} = \{x_i\}$, for a quantitative trait. Let the mean exposure to factor i in the population be μ_i. The effect of exposure x_{ij}, on an individual j's phenotype can be expressed by regression

$$\phi_j = G_j + PG_j + \Sigma \beta_i x_{ij} + E_j. \qquad [12.1]$$

For example, if the phenotype is affected by age, sex, and daily caloric intake we could write $\phi_j = G_j + PG_j + \beta_1 \times \mathit{age} + \beta_2 \times \mathit{sex} + \beta_3 \times \mathit{calories} + E_j$. As shown, these are linear relationships between exposure and phenotype, but higher-order effects can be included, for example by adding terms such as $\beta_4 \times \text{age}^2$ and so on. (Hereafter, for ease of reading, the subscript j will be omitted where the context is clear). Here, the PG and E terms are expressed in terms of deviations, as in earlier chapters, and represent unmeasured variables. The exposures can be re-scaled in terms of deviations from the mean exposures:

$$\phi = \mu + MG + PG + \Sigma \beta_i (x_i - \mu_i) + E. \quad [12.2]$$

In [12.2] the notation MG is used to denote that these are major-genotype deviations ($\mu_G - \mu$) rather than absolute effects. The unmeasured environmental effects (E) can be divided into terms representing stratum-specific unmeasured environments (e.g., sibship, nuclear family, occupation group) as well as exposures specific to each individual. Qualitative risk factors can be taken into account by dummy variables. Equation [12.2] simply extends the general models of Chapter 6.

When environmental effects are additive, i.e., unrelated to genotype, they provide no information about genetic models, and are only a source of extraneous statistical variation. In such instances, it is customary to remove the effects by *adjusting* the phenotype values before genetic analysis is done, as $\phi' = \phi - \Sigma \beta_i (x_i - \mu_i)$. The regression coefficients (βs) can be estimated from other population data or directly from the sample (see Chapter 6, Note 3).

Genotype-specific risk factors

When there is $G \times E$ interaction, the effects of some risk factor(s) differ among the major genotypes. The model can be revised to make the risk factor regression specific to those genotypes (see e.g., Moll *et al.*, 1984):

$$\phi = \mu + MG + PG + \Sigma \beta_{G,i} (x_i - \mu_i) + E. \quad [12.3]$$

Genotype-independent regression effects could also be included; for example, there may be general age or sex effects, plus additional effects that differ among genotypes. If $G \times E$ interaction relative to unmeasured environmental effects is suspected, the likelihood can contain separate environmental terms for each genotype, e.g., E_G.

If the major locus is an identified ('measured') locus, we can group persons according to their genotype and estimate these genotype-specific regression coefficients directly. When we must infer the major locus in segregation analysis, however, we have to build these regressions into the penetrance function and estimate their coefficients along with the other parameters of the model. This is statistically not very powerful as a rule.

Equation [12.3] is a reasonable model but may require estimation of so many parameters as to be impractical. With a diallelic major locus and just one risk covariate these parameters include μ, major genotype means

σ^2_{PG}, σ^2_E, $\beta_{DD,1}$,$\beta_{Dd,1}$, and $\beta_{dd,1}$, and more if transmission probabilities (τs) are to be estimated or if the risk factors affect the environmental variance, or that variance is divided into components, etc. While many genetic studies use adjusted levels of phenotypes for variables such as age and sex, there are few examples of serious efforts to include genotype-specific effects in segregation analysis, and indeed only a few have done so for measured genotypes.

Effects of age and sex in the lipid system

It is routine to age- and sex-adjust lipid levels before doing genetic analysis, but there is some evidence that these and other factors may have genotype-specific effects. For example, a segregation analysis study of cholesterol levels in a large Canadian family with FH found a two-fold difference in the regression coefficients on sex, and a three-fold difference in the residual environmental variance among genotypes at the LDL receptor (FH) locus (Moll *et al.*, 1984).

An environmental factor that has been studied for $G \times E$ interaction is the effect of smoking on the reverse cholesterol transport system introduced in Appendix 2.1. Genetic variants at, or in linkage disequilibrium with, *CETP* locus are associated with interindividual differences in plasma apo-AI levels in smokers but not in non-smokers (Kondo *et al.*, 1989). Similarly, structural variants at apo-H, another lipid transfer apolipoprotein, affect apo-AII levels in smokers but not non-smokers (Kaprio *et al.*, 1989). Apo-AII and apo-AI are constituents of HDL, which, like CETP, is involved in reverse cholesterol transport. Smoking is known to suppress HDL levels, and as a result may increase risk of heart disease.

An interesting application is the study of the effect of the apo-E locus on the *longitudinal changes* in cholesterol, triglycerides, and β-lipoproteins (the apo-B-containing lipoproteins, VLDL + IDL + LDL), and glucose over a 5-year period in 128 unrelated adults and 55 children in Nancy, France (Gueguen *et al.*, 1989). After adjusting for age, height, and other extraneous variables, the apo-E genotypes by themselves did not affect these longitudinal patterns. But the authors also looked at apo-E effects on longitudinal changes in weight as these affect changes in lipid levels. Among adults there were significant apo-E genotypic differences in weight-associated changes in triglyceride and β-lipoprotein levels (Table 12.1). In particular, the apo-ε_4 allele was associated with greater changes in these outcomes per unit increase in weight. This is consistent with reported increases in hypertriglyceridemia among obese individuals with the ε_4 allele (Fumeron *et al.*,1988), and may be related to apo-E

Table 12.1. *Apo-E polymorphisms and changes in weight in adults, as they affect changes in plasma levels of several physiological variables*

Change in:	N	$\beta_{\varepsilon 2}$	$\beta_{\varepsilon 4}$	Significance
Cholesterol	128	−0.002	0.042	NS
Triglycerides	89	−0.015	0.115	<0.001
β-Lipoproteins	125	−0.056	0.263	<0.05
Glucose	128	0.031	−0.031	NS

Note: N is sample size; NS, not significant.
Source: Gueguen *et al.*, 1989.

effects on VLDL-triglyceride clearance rates, and to associations between elevated triglyceride and obesity.

Effects of age and sex on blood pressure

A recent study of 278 pedigrees randomly ascertained in Minnesota has found evidence for a single locus that has sex- and age-dependent effects on systolic blood pressure (Perusse *et al.*, 1991; these authors, like many others, use the term gender rather than sex). This study found additive sex- and age-effects (e.g. systolic blood pressure increases with age in all genotypes), but also genotype-specific effects, as shown in Figure 12.3. The curvature in the *LL* group is due to the effects of an age^2 term in the regression.

The effects of phenotype amplification can be seen from the fact that blood pressure in children is a poor predictor of adult blood pressure. Among children, the effects of different genotypes will be difficult to detect even if a highly polymorphic major locus is segregating. For example, the distribution of blood pressure in a sample of French-Canadian adults suggested that the sample consisted of two sub-distributions that could in principle be referred to major genotypes, but there was no such bimodality in the offspring of the same adults (Rice *et al.*, 1990). Segregation analysis in these families suggested that there were major effects, but the segregation proportions were non-mendelian, probably because all the offspring in the families were under age 26 years; that is, the penetrance functions differ only at later adult ages.

What might the major blood pressure locus be? A number of gene products are known to affect blood pressure physiology, including hormones such as angiotensin (Chapter 11), that affect vascular tension or kidney filtration of blood. These affect immediate blood pressure

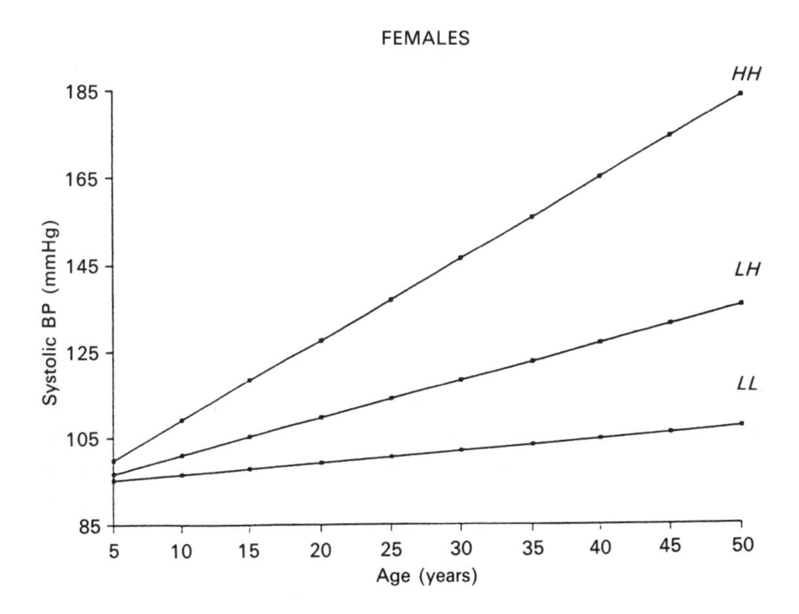

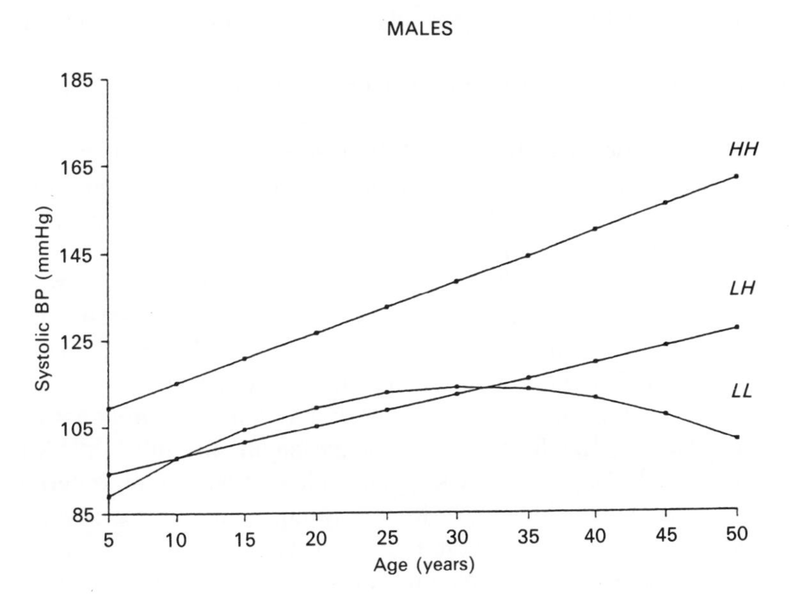

Figure 12.3. Genotype-specific age effects on systolic blood pressure (BP). (From Perusse *et al.*, 1991. © 1991 American Society of Human Genetics.)

homeostasis, but seem largely unrelated to the health issues involving long-term ambient blood pressure levels. Only a few loci that may affect human ambient blood pressure have been identified. One is the level of *sodium–lithium countertransport* (Na–Li CNT), an enzyme whose activity is assayed in red blood cells but presumably is important in kidney osmoregulation mechanisms. Although the locus itself has not been found or mapped, Na–Li CNT variation is apparently genetic and related to the risk of hypertension (Motulsky *et al.*, 1987; Ward, 1990).

The level of Na–Li CNT activity has a bimodal distribution in the population of the USA, as shown in Figure 12.4 (Turner *et al.*, 1989; see also Boerwinkle *et al.*, 1984; Canessa *et al.*, 1979). This varies little with age (i.e., is present before blood pressure bimodality is itself manifest in the population). Persons with essential hypertension have CNT means and variances like those of the upper mode (Boerwinkle *et al.*, 1984), and there is a relationship between blood pressure and CNT levels among persons in the upper mode (but not in the lower) that may help to explain the fact that a subset of that group develop hypertension (Boerwinkle *et al.*, 1986b; Turner *et al.*, 1989; Weder and Schork, 1989). CNT appears to be one of the phenotypes intermediate between primary genotypes and hypertension, analogous to the role played by lipoprotein levels in CHD risk.

Family studies have suggested that membership in these CNT models may be controlled by a single major locus (SML), in at least a fraction of the population (Boerwinkle *et al.*, 1984; Dadone *et al.*, 1984; Hasstedt *et al.*, 1988; Motulsky *et al.*, 1987), even though other families suggest that some other NTF is responsible (Boerwinkle *et al.*, 1986b; Rebbeck *et al.*, 1991). So the role of CNT is far from clear.

In this terminology, a NTF is some risk factor to which everyone is exposed with the same probability, say p – that is, the transmission probabilities are all equal to p in family members. The SML model specifies mendelian transmission probabilities. A recent study of the Minnesota pedigree data, referred to earlier, has found that providing for a sex-specific effect significantly improved the likelihood under both SML and NTF models (Table 12.2). Using the likelihood ratio cutoff criterion discussed at the end of Chapter 11 for separating families into etiological subsets, Rebbeck *et al.* suggest that at least 12 pedigrees may be segregating alleles at a SML. Whether this is Na–Li CNT or some other locus is not yet known.

An interesting aspect of this study, which is found in other genetic studies, is that the mean phenotypic effects estimated by maximum likelihood for the different NTF categories in some families are almost

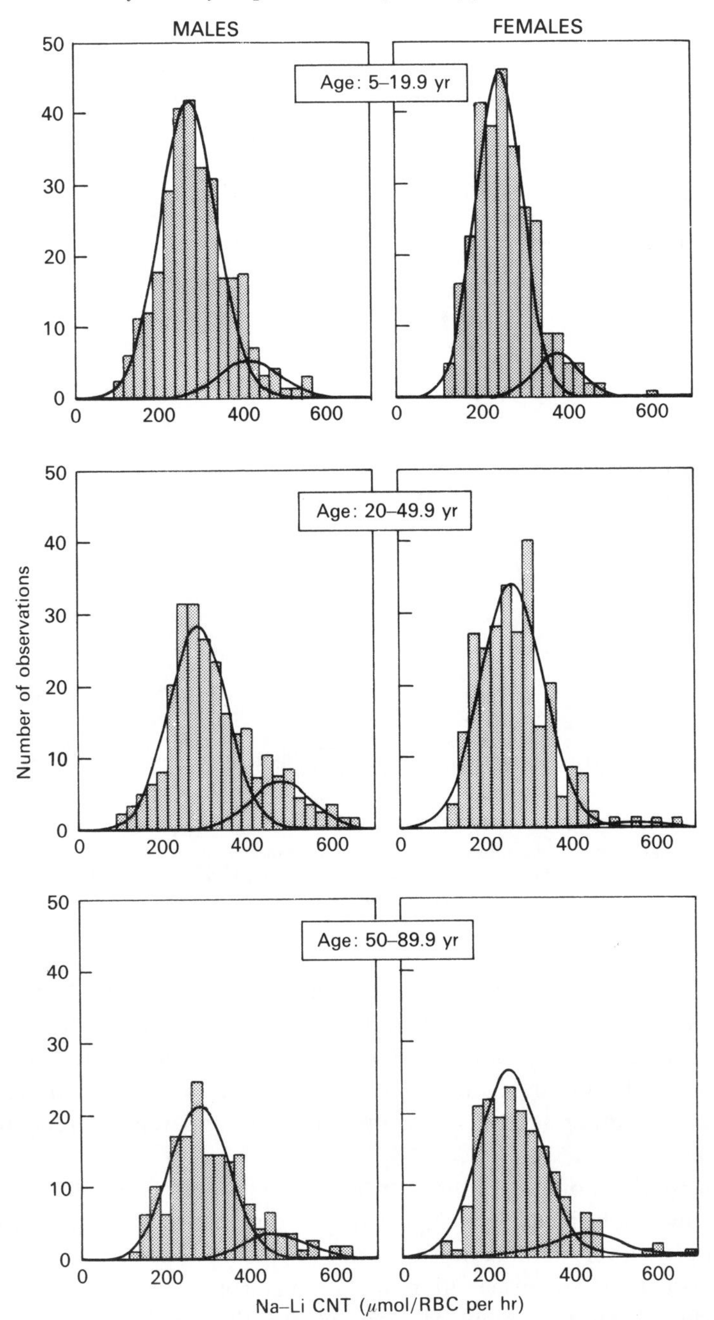

Figure 12.4. Age-specific distribution of Na–Li countertransport activity in a population. (From Turner *et al.*, 1989, by permission of the American Heart Association, Inc.)

Table 12.2. *Comparison between gender-specific and unstratified non-transmitted factor (NTF) and single major locus (SML) models for Na–Li countertransport (CNT)*

Parameters	NTF	NTF	SML	SML
Mean CNT values				
Males				
μ_{LL}	287		289	
μ_{LH}	(μ_{LL})		(μ_{LL})	
μ_{HH}	465		475	
Females				
μ_{LL}	293		295	
μ_{LH}	(μ_{LL})		(μ_{LL})	
μ_{HH}	418		424	
Whole sample		294		295
μ_{LL}		(μ_{LL})		(μ_{LL})
μ_{LH}		467		478
μ_{HH}				
p	0.752	0.752	0.718	0.772
h^2	0.234	0.234	0.174	0.183
σ	74	74	73	75
τ_1	(p)	(p)	(1.0)	(1.0)
τ_2	(p)	(p)	(0.5)	(0.5)
τ_3	(p)	(p)	(0.0)	(0.0)
$-\ln\mathcal{L}$	7430	7434	7429	7432
χ^2	—	7.6	—	6.7
d.f.	—	2	—	2

Notes: Items in parentheses were fixed by the model and not estimated. Mean Na–Li CNT values in μmol/l RBC per h (RBC, red blood cell).
Source: modified from Rebbeck *et al.*, 1991.

identical in value with the major locus genotypic values inferred for the SML families. This is a curious finding that raises the questions of what that NTF might be, or whether it may be that a much higher fraction of families actually are segregating at a SML, but that some ascertainment, diagnosis, or other artifacts are obscuring the apparent segregation probabilities.

In this study, a bimodal trait distribution seems correctly to imply an underlying single locus (with approximate dominance for one of the alleles). However, bimodality does not necessarily indicate a simple genetic situation. The distribution of plasma glucose is unimodal among non-diabetics, but as persons develop diabetes they move into a rather distinct upper mode in the glucose distribution (Stern, 1988). Unlike the CNT distribution, the glucose distribution changes with age (Rushforth *et al.*, 1971), so that mode membership cannot be assigned to specific genotypes – at least, the lower mode at young ages is a mix of susceptibles and non-susceptibles.

Many major genes may be identifiable without our having to worry about taking $G \times E$ interactions into account. But such interactions may be a sleeping tiger that will cause much difficulty for our effort to understand the entire causal spectrum. We have little information on this in humans, perhaps largely because most studies have involved chronic disease in industrialized nations where exposure to the relevant environments (e.g. high fat diets, automobile exhaust, electromagnetic radiation) may be nearly universal, and we have concentrated on strong effects detectable by the available samples. However, there is extensive and somewhat disturbing evidence from experimental systems, exemplified by the tomato crossing experiments discussed earlier, that subtle environmental differences can greatly affect QTL effects in a population.

When tomato plants from the same cross were grown in three different environments, not chosen for any specific characteristics relative to the plants, a substantial fraction of QTLs identified in one environment had no detectable phenotypic effects in other environments (Paterson *et al.*, 1991). Of 29 QTL × 3-environment combinations, using the same parental stocks, only 4 QTLs led to detectable phenotypic variation in all 3 environments, 10 in 2 environments, and 15, nearly half, in 1 environment only.

Risk factors for qualitative phenotypes

Modeling environmental risk factors

Regression methods can be used to model the effects of risk factors on the probability that an individual will ($\phi = 1$) or will not ($\phi = 0$) manifest a qualitative trait. One way to do this is by logistic regression (Chapter 3). When there is no $G \times E$ interaction one model is

$$\Omega_g(\phi) = \frac{\exp(\phi\{\alpha_G + \Sigma\beta_i x_i\})}{1 + \exp(\alpha_G + \Sigma\beta_i x_i)}, \qquad [12.4]$$

where G is the genotype of the individual. Here, the genotype-specific effects (the α_Gs) do not depend on the other risk factors (xs).

$G \times E$ interaction can be included as follows:

$$\Omega_g(\phi) = \frac{\exp(\phi\{\alpha_G + \Sigma\beta_{G,i} x_i\})}{1 + \exp(\alpha_G + \Sigma\beta_{G,i} x_i)}. \qquad [12.5]$$

This is algebraically straightforward but can involve a large number of parameters to be estimated. Regressions of this kind can be included as part of the penetrance function in the mixed model, or can be incorporated as part of the regressive genetic models {Bonney, 1986; Bonney *et al.*, 1989}.

The special case of age-related risk factors

For many chronic diseases, the effect of a risk factor seems mainly, or even only, to modify the age of onset; that is, exposure does not change the nature of the final phenotype, only its timing. For example, Figure 12.2 showed the way CR affects survivorship because of its affect on the risk of CHD.

How can we relate such effects to the hazard function? In the typical absence of biological guidelines as to how an exposure will affect the hazard, a general *proportional hazards* model can be used (Chapter 3). The hazard at age t for a given exposure level is expressed relative to some baseline hazard, $h_0(t)$:

$$h_g(t) = h_0(t)\,[\alpha_G + \exp(\Sigma\beta_{G,i}\,x_i)]. \qquad [12.6]$$

(The risk factors need not depend on genotype). The penetrance probability, conditional on genotype, for someone observed at age t is

Disease is not present: $S_G(t)$

Onset at age t: $S_G(t-1)\,h_G(t)$

Onset at an unspecified age prior to the current age t: $1 - S_G(t)$

(cf. [3.3]).

A modification of this method known as 'accelerated hazards' was used to determine the CR-specific hazard of CHD, when we computed the fitness effects of genes related to cholesterol levels in our illustrative example in Figure 9.10 (S. Weeks *et al.*, unpublished results).

Age of onset can be taken into account in other ways, depending on the problem to be solved [Abel *et al.*, 1989; Cupples *et al.*, 1989, 1991; Dawson *et al.*, 1990; Demenais and Bonney, 1989; Elston, 1973; Meyer and Eaves, 1988].

Precursor states for qualitative phenotypes

Many diseases are difficult to treat, so it is important to try to be able to identify *precursor phenotypes* that arise before the final disease so that prevention or early treatment can be applied. Precursor phenotypes are often useful in segregation and linkage analysis because their penetrance may be higher relative to underlying genotypes, to which they may be physiologically closer than the final disease, as noted above with regard to genes 'for' heart attacks. Thus, whereas we may gain more information about blood pressure genetics by looking at older individuals, for many other traits we want to look earlier.

Proliferative breast disease

Chapter 5 illustrated likelihood analysis in pedigrees with a study of PBD, which reported that PBD is controlled by a major locus (Skolnick *et al.*, 1990). PBD is itself a benign disease but Skolnick *et al.* asked whether PBD may be an important breast cancer precursor phenotype. To investigate this, they used a proportional hazards model like that given above. Each PBD genotype was assumed to modify the hazard of breast cancer by a constant proportion relative to the Utah population breast cancer hazard: $h_g(t) = k_g h_0(t)$. The authors estimated that the hazard in susceptible genotypes was elevated by a factor of about 8 to 11 for the genetic models shown in Table 5.3. PBD arises earlier and has a higher penetrance than breast cancer for alleles at this inferred locus.

Colon cancer

Likely precursors for colorectal cancer, one of the most common cancer types in Western nations, are *adenomatous polyps*, small growths that develop on the lumenal (interior) wall of the large bowel. A rare dominant allele for *familial adenomatous polyposis* (*FAP*), at a locus on chromosome 5, predisposes carriers to develop hundreds or even thousands of polyps by early adulthood (Chapter 15). The hazard of colorectal cancer in FAP allele carriers is increased by orders of magnitude over that in the general population; if not treated by prophylactic colectomy, 50% of individuals will experience malignant cancer of the colon by age 45 years, and nearly 100% by age 60, compared to a lifetime risk of about 0.5% in the general population (Lipkin *et al.*, 1980; Weiss and Chakraborty, 1984, 1990).

FAP is serious but is treatable and detectable, and the predisposing allele is highly penetrant. We have had no comparable phenotypic warning flag for the remaining vast majority of colorectal cancers. However, *isolated* colorectal polyps also constitute a risk factor for that cancer, and may be familial. A segregation analysis of a series of Utah families suggests that the fraction of colorectal cancers that arise from isolated, but familial, polyps may be quite high (Cannon-Albright *et al.*, 1988). Using the presence of proctoscopically detected polyps as the phenotype, the authors found a statistically significant LOD score for a

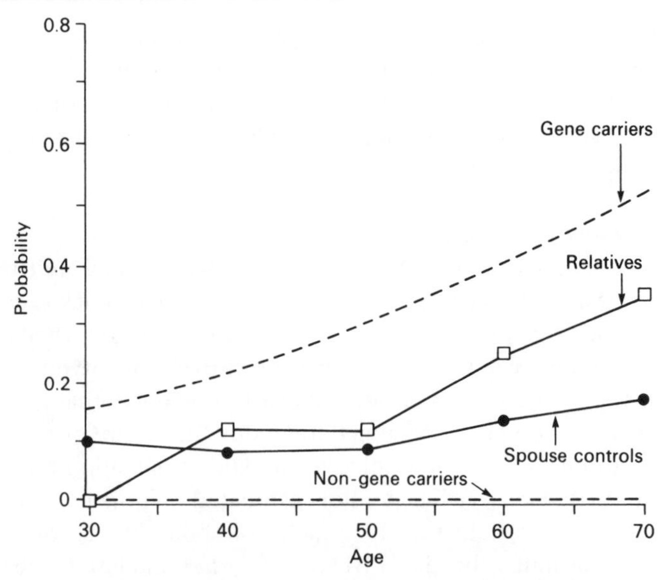

Figure 12.5. Age-specific probability of developing adenomatous colorectal polyps, based on segregation analysis model. (From Cannon-Albright *et al.*, 1988. Reprinted by permission of the *New England Journal of Medicine* **319**: 533–536.

major locus, with allele frequency 0.19, and estimated that 50–100% of colorectal cancers may arise from such inherited polyps. However, even in these families the penetrance of polyps in the susceptible genotypes is low, and of colorectal cancer even lower, so it is understandable that the disease does not seem to be obviously 'mendelian' – consistent with the common finding of weak familiality of risk of colorectal cancer.

Figure 12.5 shows the relevant hazard functions for polyps. The MLE values estimated from these data suggested that non-allele carriers may be essentially immune to such cancers. This is probably a statistical artifact; even if their risk really is low, I think it is unlikely to be zero, for reasons that are made clear in Chapter 15.

Other cancer examples

Gallbladder cancer (GBCA) is a generally rare disease but when it does arise it is usually fatal because it metastasizes to the liver before producing symptoms. The only major precursor state for GBCA is the presence of gallstones, usually formed from cholesterol crystals in cholesterol-saturated bile. When gallstones create symptoms they are removed, often with the gallbladder itself, and the person is no longer at risk for

GBCA. However, asymptomatic gallstones can remain present for years or even decades, and can eventually produce GBCA in some individuals (the proximate physiological mechanisms are unknown). In fact, some individuals predisposed to form gallstones, but who do not have symptoms, may face a risk of GBCA that is nearly as great as that of lung cancer in heavy smokers (Lowenfels *et al.*, 1985; Weiss *et al.*, 1984b).

GBCA strikes late in life and has until recent decades been poorly diagnosed, so that multiply affected families may be very rare, even though the disease is fully genetic and a high fraction of individuals are susceptible. This seems to be the case in Amerindians, where the risks of gallstones and GBCA are dramatically elevated (e.g., Weiss, 1985; Weiss *et al.*, 1984a,b). The proper high-penetrance phenotype to study in Amerindian families is the presence of saturated bile or of gallstones visualized by ultrasonography – not GBCA itself.

Although the common wisdom is that over 80% of all cancers are 'caused' by environmental risk factors (Doll and Peto, 1980), there are some reasons to think that such factors may select among a much higher prevalence of susceptible individuals than has been thought. For example, *ataxia telangiectasia* (AT) is a recessive disorder that leads to severe progressive loss of motor control (cerebellar ataxia) and other symptoms. Chromosomes in *AT* individuals, including carriers, are more susceptible to breakage by mutagens, specifically by ionizing radiation. *AT* carriers seem to be significantly more susceptible to cancer than is the general population (e.g., Morrell *et al.*, 1986; Swift *et al.*, 1987), and may be specifically vulnerable to the carcinogenic effects of exposure to ionizing radiation (see e.g., Ron *et al.*, 1988; Swift *et al.*, 1991). *AT* carriers alone comprise 1% of the population of the USA: how many other similar conditions might there be? (For some ideas about this, see Chapter 15).

Pleiotropy, multivariate phenotypes, and genetic syndromes

The genetics of correlated characters

A single locus may be pleiotropic, i.e. may have effects on several physiologically related quantitative characters, which will then be correlated in the population and in families. A model specifying genetic control of multivariate traits can follow the same forms given earlier for univariate traits. Phenotypic variation is assumed to be affected by major gene, polygenic, and environmental effects, and perhaps other risk factors. The model corresponds to that of [12.1] except that vectors and matrices are used rather than univariate terms. We have

$$\boldsymbol{\phi} = \boldsymbol{\mu}_g + \boldsymbol{PG} + \boldsymbol{E} + \boldsymbol{\beta}'x. \qquad [12.7]$$

(The notation $\boldsymbol{\beta}'$ indicates the transpose of the vector or matrix). For example, if there are two phenotypes and two risk factors (the numbers need not be the same), [12.7] will be

$$\begin{aligned}(\phi_1,\phi_2) &= (\mu_{g,1},\mu_{g,2}) + (PG_1,PG_2) + (E_1,E_2) \\ &\quad + \begin{bmatrix}\beta_{11}\,\beta_{21} \\ \beta_{12}\,\beta_{22}\end{bmatrix}\begin{bmatrix}x_1 \\ x_2\end{bmatrix} \\ &= (\mu_{g,1},\mu_{g,2}) + (PG_1,PG_2) \\ &\quad + (E_1,E_2) + (\beta_{11}x_1 + \beta_{21}x_2, \beta_{12}x_1 + \beta_{22}x_2). \qquad [12.8]\end{aligned}$$

From this, expressions for the variance conditional on major genotype and the measured environments can be written, also corresponding to the general quantitative genetic model given first in Chapter 6:

$$\boldsymbol{V} = \boldsymbol{G} + \boldsymbol{E} \qquad [12.9]$$

where each term is a matrix.The terms in $\boldsymbol{V}$ are, on the diagonals σ_i^2, the variances of the individual traits, and on the off-diagonals σ_{ij}, the covariances between the traits. The terms in $\boldsymbol{G}$ and $\boldsymbol{E}$ are the respective variances and covariances of the genotypic and environmental effects on the phenotype.

Using these models in segregation analysis is computationally demanding but is an extension of the general likelihood method for which specific modifications and approximations have been developed (e.g., Blangero and Konigsberg, 1991; Bonney and Elston, 1985a,b; Bonney *et al.*, 1988; Lalouel *et al.*, 1985; Lange and Boehnke, 1983).

Quantitative genetic syndromes

In the context of human diseases, a set of characters controlled by the same factors is known as a *syndrome* of conditions. In genetic syndromes the multivariate phenotypes can vary among family members, but their probability structure can be explained by a genetic model. Examples include the many classical developmental syndromes, (e.g., of craniofacial and neurological anomalies) that formed the core of pediatric genetics for decades. Syndromes of chronic disease are also known. Retinoblastoma carriers are at risk for retinal and bone cancers. The risks of breast and ovarian cancer are jointly elevated in some families, as is the risk of a variety of cancers in FAP families. The ultimate cancer example is the Li–Fraumeni syndrome, associated with mutations in a locus known as *p53*, which elevate the risk of many, perhaps most, tumor types. These are discussed in Chapter 15.

When single loci affect multiple quantitative traits, the vectors of trait means differ among genotypes, and the traits will be correlated among family members (perhaps with different correlations for each genotype).

Lipid system variation

The lipid system is a set of physiologically related traits, and variation at single loci can affect more than one of the traits. Lipoprotein particles transport both triglycerides (TGs) and cholesterol (C), in proportions determined in part by the different lipoproteins that form each type of particle. It is of interest, therefore, to see whether variation at lipoprotein loci might affect both C and TG.

Bivariate correlations between TG and C have been found by biometric analysis (correlations among family members) (Boehnke *et al.*, 1986), and by measured genotype approaches. As an example of the latter, the effects of apo-E genotypes on various lipid components were assessed on 563 normal blood donors from West Germany (Boerwinkle and Utermann, 1988). As shown in Table 12.3A, apo-E allelic variation accounted for 20% of the variation in apo-E, 12% in apo-B, and 4% in C concentrations.

In a separate study of 223 unrelated individuals from Nancy, France, apo-E alleles were shown to affect means and variances of β-lipoprotein, C, and TG levels (Boerwinkle *et al.*, 1987), as shown in Table 12.3B. The apo-E genotypes did not affect the correlation between β-lipoprotein levels and TG, or C, but did affect the TG–C correlations. The latter is shown in Figure 12.6, which gives the bivariate ellipses circumscribing 95% of the phenotype distributions for the three major apo-E genotypes (the relative proportion each mode comprises of the total population equals its corresponding genotype frequency, but cannot be seen in a contour plot). The TG mean values are about the same for the different genotypes (i.e., $\mu_{23} \approx \mu_{33} \approx \mu_{34}$), but the variances and correlations differ considerably. A major gene model could fail to find apo-E effects on TGs studied alone.

Reilly *et al.* (1991) provided extensive data from Rochester, Minnesota, on nine traits relative to apo-E genotypes, showing that there is substantial variation among the genotypes in their multivariate means and variances; that is, each individual is measured for nine variables, and generates a point in 9-dimension hyperspace. The clouds of points in that space differ among the genotypes.

Table 12.3A. *Effects of apo-E polymorphism on apo-E, apo-B, and total cholesterol levels in a West German blood donor sample*

Phenotype	Average effects ε_2	ε_3	ε_4	% of phenotypic variance explained
Apo-E	0.95	−0.04	−0.19	20.2
Apo-B	−9.46	−0.18	4.92	11.9
Total cholesterol	−14.20	−0.16	7.09	4.0

Note: Values are average effects in mg/dl.
Source: Boerwinkle and Utermann, 1988.

Table 12.3B. *Effects of apo-E polymorphism on plasma cholesterol, β-lipoprotein, and triglycerides levels from a Nancy, France, sample*

Phenotype	Average effects ε_2	ε_3	ε_4	% of phenotypic variance explained
Cholesterol	5.36	6.07	6.22	8.6
β-Lipoprotein	7.29	8.49	8.91	4.1
Triglycerides	1.15	1.01	1.11	~0

Source: Boerwinkle *et al.*, 1987.

A related approach has been taken to characterize the lipid effects of apo-B variation. Some hypo- and abetalipoproteinemias, deficits of apo-B and its lipoproteins, are characterized by a variety of symptoms including fat malabsorption. A large pedigree has been ascertained in which a mutation in or near the apo-B gene appears to produce the complex of phenotypes involved in *hypobetalipoproteinemia* (*HBLP*) (Leppert *et al.*, 1988). Apo-B RFLP variants were associated with correlated changes in total TG, total C, and LDL-C. A *discriminant function*, or linear combination of these values, was found that clearly separated the joint effects of the apo-E genotype classes. This is shown in Figure 12.7 and Table 12.4. This composite or artificial phenotype constitutes a statistical description of the 'syndrome' of apo-B effects.

Obesity, diabetes, and gallbladder disease in Amerindians

Another example of a major metabolic syndrome may be the susceptibility of Amerindians to develop gallbladder disease, NIDDM, and obesity, discussed earlier. We have called this a New World syndrome, to indicate the possibility of a common underlying genetic basis specific in

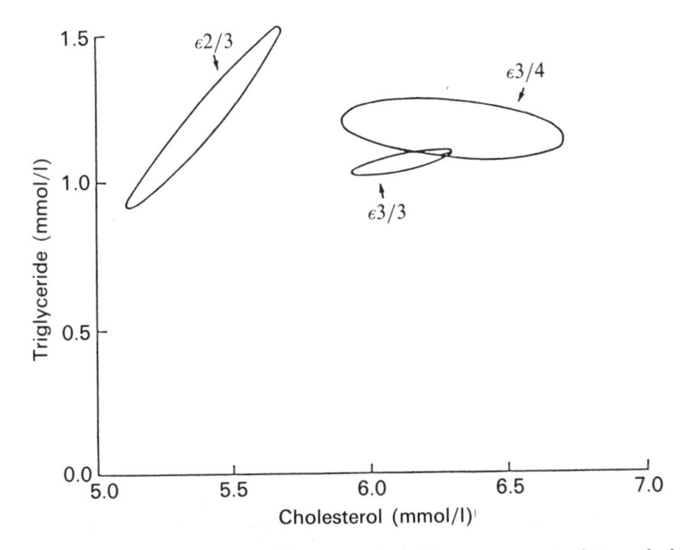

Figure 12.6. Bivariate effects of apo-E genotypes on plasma cholesterol and triglycerides. Each ellipse represents 95% of individuals with the indicated genotype. (From Boerwinkle *et al.*, 1987.)

Amerindians (Weiss *et al.*, 1984a). One reason why it has been difficult to find the genetic mechanism for this syndrome is that the allele frequency(ies) in Amerindian populations are so high that clear segregation of risk may not occur in families, because penetrance is variable and strongly affected by the environment, and because our understanding of precursor phenotypes is still inadequate. Potential precursor traits could include the saturation level of bile or silent gallstones (detected by ultrasonography), measures of obesity, and a quantitative phenotype that would indicate susceptibility to develop NIDDM (insulin response to a glucose challenge may prove to be such a trait).

This syndrome potentially illustrates many interesting aspects of genetic epidemiology. I now speculate about these, even though the appropriate studies have not yet been done. This syndrome exemplifies the extremes of phenotype amplification and $G \times E$ interaction. The pre-World War II absence of this syndrome in Amerindians shows that some environments exist in which basically *no* cases arise in any genotype; yet, in modern environments, Amerindians face a lifetime risk that approaches 100%, perhaps ten-fold higher than in North American Caucasians.

We can model this by modifying [12.7] to include genotype-specific ($\beta_{G,ij}$) as well as general (β_{ij}) regression effects for sets of measured

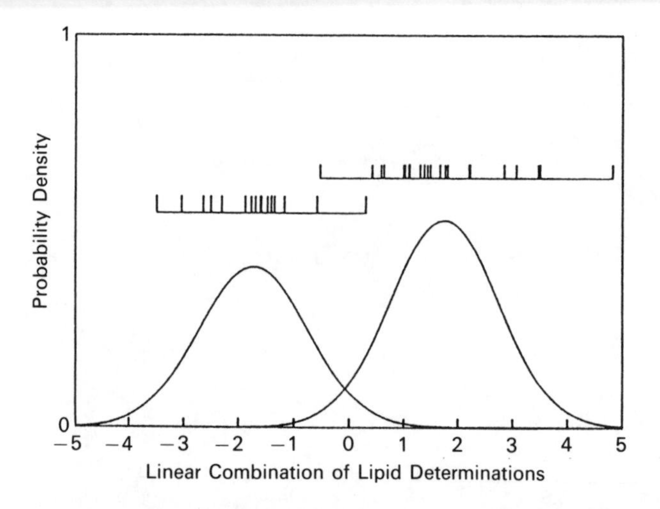

Figure 12.7. Discriminant function for hypobetalipoproteinemia. Discriminant function includes cholesterol, log-triglycerides, and LDL-C. Lower curves are distributions of the discriminant function expected in the population, based on the statistical model. Bars show values of this discriminant function among sampled individuals with (lower) and without (upper) the HBLP-associated RFLP. (From Leppert *et al.*, 1988.) Reproduced from the Journal of Clinical Investigation, **82**: 847–851, by copyright permission of the American Society for Clinical Investigation.)

Table 12.4. *Variables affecting genotypes for hypobetalipoproteinemia: individual v. joint effects*

Variable	Genotypic means		S.D.	LOD score
	Normal	Carrier		($\theta_{MLE} = 0.0$)
Variables tested independently				
Cholesterol	199	101	35	7.0
LDL cholesterol	133	38	37	5.0
Triglycerides	136	52	40	2.2
Apo-B	122	38	28	4.5
Variables tested as single discriminant function				
Cholesterol	190	96	32	7.6
Ln Triglycerides	5	4	0.5	
LDL cholesterol	118	37	35	

Note: Rounded to nearest whole unit.
Source: After Leppert *et al.*, 1988.

environmental risk factors, such as dietary consumption and energy expenditure (Weiss *et al.*, 1992). Here, I simplify by assuming that the same two risk factors are important both additively and interactively:

$$(\phi_1,\phi_2) = (PG_1,PG_2) + (E_1,E_2) + \left\{ \begin{bmatrix} \beta_{11}\,\beta_{21} \\ \beta_{12}\,\beta_{22} \end{bmatrix} + \begin{bmatrix} \beta_{g,11}\,\beta_{g,21} \\ \beta_{g,12}\,\beta_{g,22} \end{bmatrix} \right\} \begin{bmatrix} x_1 \\ x_2 \end{bmatrix} \qquad [12.10]$$

I have taken another liberty here, to make a point. In [12.10] I have eliminated the vector of major genotype means, $\boldsymbol{\mu}_g$ (cf. [12.8]). Written this way it is clearer that if we could identify all the factors that interact with the hypothesized major genotypes for a phenotype, among persons exposed to zero dose of the interacting risk factor(s), all major genotype terms drop out: there *is* no major gene effect! This may seem to stretch plausibility, for a trait that I think may be largely due to major locus variation, but in fact this *was* the situation for these conditions in Amerindians 50 years ago, and nicely illustrates the relativistic nature of major gene effects. See Chapter 6, note 3, regarding estimation.

This collection of conditions also illustrates other subleties. Several measured-genotype and RFLP studies of Amerindian NIDDM have identified small but significant effects on factors such as cholesterol, triglyceride and glucose levels, and obesity measures, that are associated with NIDDM. The loci include the apo-CIII region, LPL, insulin receptor, CETP, and others (e.g., Ahn *et al.*, 1991; R. Valdez *et al.*, unpublished results). These effects are too small to constitute the hypothesized major gene for Amerindian susceptibility to NIDDM or the other syndrome traits, but show that the level of relevant *polygenic* variation seems to be about the same as that in other populations: similar genetic associations have been found related to lipids, diabetes, and obesity in Caucasians.

Strategies using affected pairs of selected admixed matings were suggested in Chapter 10 as an approach to mapping genes responsible for this disease problem. Another related approach might be to take advantage of the possibility that the diseases may be clonal in the Americas (e.g., due to copies of the same original 'arctic' mutation(s)), and to use cladistic or other sequence-based strategies that search for regions of high sequence identity that are shared by affected individuals, because Amerindians are all descended from a limited number of original founding ancestors who expanded into the New World from northeast Asia. In this sense, Amerindians constitute a kind of super 'inbred' population in which the etiology may be much more homogeneous than in large, outbred populations.

Other allometric considerations

Considering allometric relationships is not unusual in biology but has not been thoroughly appreciated in epidemiology. It should be possible to use an understanding of the biological basis of a particular trait to develop more specific predictions of the allometric relationships associated with different genotypes. Many allometric relationships seem to have a rather generic form among vertebrates. Body size, M, alone is associated with the allometric scaling of many mammalian systems via a relationship such as $Y = aM^b$ {Calder, 1984}. This equation fits dimensional variables such as height or limb length, morphological variables such as skeletal shape, temporal variables such as lifespan and the length of childhood, and physiological variables such as basal metabolism. The relationship does not apply where it would be evolutionarily disadvantageous, for example red blood cell size and oxygen diffusion rate; it is efficient for large animals to have more, but not larger, red blood cells.

Now that extensive detailed genotyping has become routine it should be possible to look in more detail at the distribution of genotypic effects on quantitative traits. Theoretical considerations discussed in Chapter 9 lead to the prediction of a roughly exponential shape for such effects, within and between loci relative to a single trait, and of patterns of self-similarity among subsets of the population or of alleles controlling a trait.

Is it possible in any meaningful way to predict the 'shape' of the phenotypic change one would expect by substituting one allele for another? The answer probably depends at least in part on the measurement scale used to gauge the effects. A substitution may have linear (additive) effects on one scale, non-linear (e.g., multiplicative) on another. The expected effects must also depend on the biology of the locus concerned, knowledge of which we should bring to bear on phenotype amplification problems, that is, to increase the apparent penetrance of genotypes of interest.

If one allele has, say, higher kinetic activity per unit concentration of substrate, or higher binding affinity per molar concentration of ligand, how should that effect at the molecular level be reflected in phenotypes that we actually measure on individuals? For example, a cell-surface receptor whose effect is to mediate some enzymatic reaction (e.g., response to a growth hormone) may have a concentration-dependent effect per ligand 'received', but the amount of such reception may depend on the *area* of cell surface on which the receptors are displayed, and the effect might be proportional to the square of the dose change.

We know little about how allele substitutions can be expected to affect such traits as lipid levels, blood pressure, osmoregulation and kidney function in humans (see Dean *et al.*, 1988, for general principles). In the only attempt to look for this kind of thing in humans, of which I am aware, Sing *et al.* (1986) sought to account for Na–Li CNT genotypic variation by using a square-root transformation of Na–Li CNT phenotypes, but in this case, the transformation did not affect evidence for multimodality in the distribution.

Allometric allelic effects are especially important, and non-additive, when they are expressed as changes in the hazard function for some disease. Most human hazard functions, the age-specific risks of mortality from disease, are strongly non-linear (generally, exponential in nature). The proportional hazards models used to account for genotypic hazard effects are very non-linear with age. For example, in the PBD study discussed above, the high risk allele seemed to modify the hazard of breast cancer by a factor of about 10. This would alter the relative survivorship probabilities by a factor of e^9 even for a baseline hazard that did not change with age, making the absolute prevalence differences among genotypes very sensitive to the ages considered, as discussed above for hypertension in a French-Canadian sample.

Feedback relationships that affect genotypic allometry should be taken into account. Individuals with extreme phenotypes may be those in whom alterations in normal homeostasis obscure associations between important variables; in such individuals, the body may compensate for out-of-range physiological conditions, medical intervention, or dietary or other behavioral changes in ways that do not otherwise occur. For example, individuals with gallstones may experience severe pain shortly after eating fatty meals, and thus learn to avoid them; this may alter their lipid profiles. Symptomatic diabetes or hypertension lead to interventions that lower glucose and blood pressure, respectively. Thus, although there are statistical and sampling advantages to genetic studies on families ascertained through clinics treating patients with extreme phenotypes, parameter estimates in those families may not apply directly to the general population.

Relationships among risk factors, or between a risk factor and its outcome, may be affected by *thresholds* above or below which no further effects are possible. Obesity is a risk factor for NIDDM in Amerindians only up to a point (e.g. as discussed by Weiss *et al.* (1992)). After a certain level of obesity has been attained, the predictive power of obesity (for whatever underlying reason) is 'saturated'; that is, the allometric relationships are non-linear, sometimes very much so. Homeostatic patterns may also differ among individuals, if each person establishes a 'set point' based on his/her own environmental exposures early in life. Blood pressure is probably an example.

Conclusion

Phenotype amplification occurs as the effects of life experiences, including exposure to various environmental risk factors, accumulate. Some of these effects are very non-linear or non-additive. $G \times E$ interaction is an important aspect of chronic disease risk. If we can identify the major risk factors involved, and adjust for them – Presto! the major locus 'disappears'. The relativistic nature of our very definitions of major genes is shown clearly by the evanescent nature of QTL effects even under carefully controlled environmental and breeding conditions, as seen by the results in the tomato crosses. There are many environmental frames of reference that can make two of Einstein's trains, which from other frames of reference seem to be moving differently relative to each other, appear to run in parallel down adjacent tracks.

This chapter illustrated some of the possible approaches to understanding phenotype amplification and quantifying the factors involved. To date, however, it has been impractical to take more than one or two risk factors fully into account in genetic analyses in humans. We also have not dealt with alleles that have different effects at different ages, or whose penetrance functions are multimodal; for example, some apo-E genotypes can be associated with low or high cholesterol levels, depending upon genotypes at other loci, and other factors (Berg, 1987). It is difficult enough to find genes with high and dramatic penetrance. The situation can be intractable for genotypic effects that arise only late in life and/or after extensive phenotypic amplification by environmental risk factors. That is why it is so important to identify precursor phenotypes with a high penetrance and close relationship to underlying genotypes, or to characterize systematic allometric relationships they may improve prediction from genotype to phenotype. While it may be difficult to predict a specific phenotype from knowledge of a specific genotype, for complex traits such as lipid levels or blood pressure we can sometimes predict restricted phenotypic *ranges* for specific genotypes. I explore ways to do this in the Afterwords.

13 *Infectious disease: the response to biological challenge*

Which also we prove by the suddain jumps which the *Plague* hath made . . . which Effects must surely be rather attributed to change of the *Air*, than of the Constitution of Mens Bodies.

J. Graunt, *Bills of Mortality* (1662)

Offensive and defensive strategies against evolving threats

Infectious and parasitic diseases have been paramount among the threats to human health and survival for most of our (and our mammalian ancestors') evolutionary history. These diseases present three particular evolutionary problems. First is the essentially open-ended variety of pathogens to which we might be exposed. Secondly, while being able to attack nearly any kind of microorganism, we must be inhibited from attacking our own, very diverse, cells. Thirdly, we reproduce and can evolve genetic adaptations only slowly, but the pathogens that infect us typically reproduce in the millions, in days or even hours. These seem like evolutionary battles we could hardly win.

One way the immune system recognizes foreign pathogens is by their molecular structure; but in contrast to molecular recognition systems such as those between hormones and their receptors, which can be specifically programmed, the immune system has evolved to use a mechanism that can recognize an unlimited diversity of molecules {Roitt *et al.*, 1989}. That mechanism forces us to revise the standard assumption that the inherited genome is constant during the life of an individual.

Somatic recombination and intra-individual variability

The immune system is an intricate, highly evolved system for recognizing and destroying molecular intruders, or antigens (Ags), to which we can be exposed during our lives, in particular, Ags produced by or located on the surface of biological pathogens.

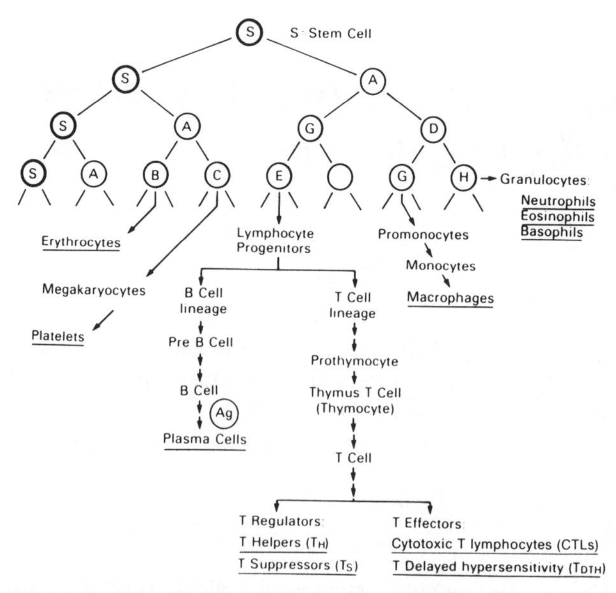

Figure 13.1. Schematic of hematopoietic cell differentiation. (From Eisen, 1990.)

The major cellular components of the immune system are the series of blood-borne cells produced by the process of *hematopoiesis*, illustrated in Figure 13.1. A pluripotent stem cell differentiates into several functionally distinct but interacting lineages that produce a mechanism for 'focused' molecular recognition and binding of Ags so that they, or the organisms that produce them, can be destroyed. This is accomplished in part by a systematic modification of the *host*'s inherited genome during development, to prepare the individual in advance against future infections.

The molecular defense rests primarily on the generation of an open-ended diversity of a class of molecular binding proteins known as *immunoglobulins* (Igs), produced and secreted by the class of white blood cells known as *B-lymphocytes*, and a comparably diverse set of proteins called *T-cell receptors* (TcRs) located on the surface of *T-lymphocytes*. Igs comprise about 20% of the proteins in blood plasma. Igs with known specificity to a given Ag are referred to as *antibodies* (Abs). Generally, B-cells are responsible for the *humoral* response, i.e., the release of Abs into the bloodstream to find and bind to foreign Ags, and T-cells are responsible for the *cell-mediated response* that depends on cooperation between the lymphocytes and HLA-coded proteins that reside on cell surfaces.

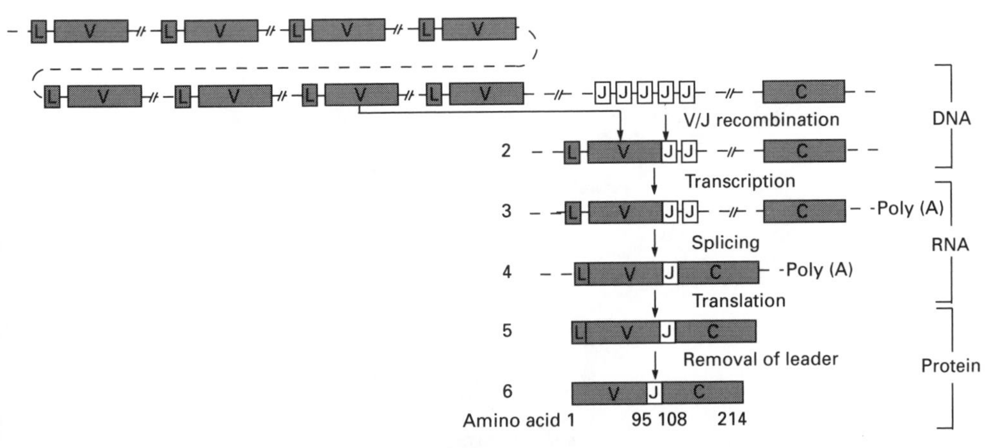

Figure 13.2. Schematic of immunoglobulin light chain gene region and somatic rearrangement. (From Suzuki *et al.*, 1989.)

Igs are complex hetero-tetramers composed of two copies of each of a 'heavy' (long) and a 'light' (short) polypeptide chain. Each chain is encoded by a *cluster* of contiguous genes, on chromosome 14 for the heavy chain, and alternative light-chain clusters, κ on chromosome 22, and λ on chromosome 2.

Each of these clusters is divided into sub-regions that contain sets of contiguous related genes; that is, members of multigene families. The chromosomal arrangement of the Ig light chain genes is shown in Figure 13.2; heavy chain regions are of similar structure with higher diversity of C regions (below). Each sub-region codes for a particular functional domain of the finished Ab molecule, as shown in Figure 13.3. The mature Ab contains two copies of a heavy chain polypeptide and two copies of either the κ or of the λ light chain product. A single Ab is thus the product of many Ig 'loci'.

In the final antibody the heavy and light chain products are composed of a variable (V) and constant (C) region gene product. The V and C chains are connected by a short joining (J) (and in the heavy chain also a D) region gene product (not shown in the figure). The V region is in the extracellular Ag recognition and binding part of the molecule.

Although the inherited Ig-coding regions contain many copies of duplicate V, D, J, and one or a few C genes, the final Ab uses *only one* of each of these, and only from one of the two homologous chromosomes, combined with the product of only one of the κ or λ light chains, from only one of the homologous chromosomes. Each lymphocyte expresses only *one* type of Ab molecule.

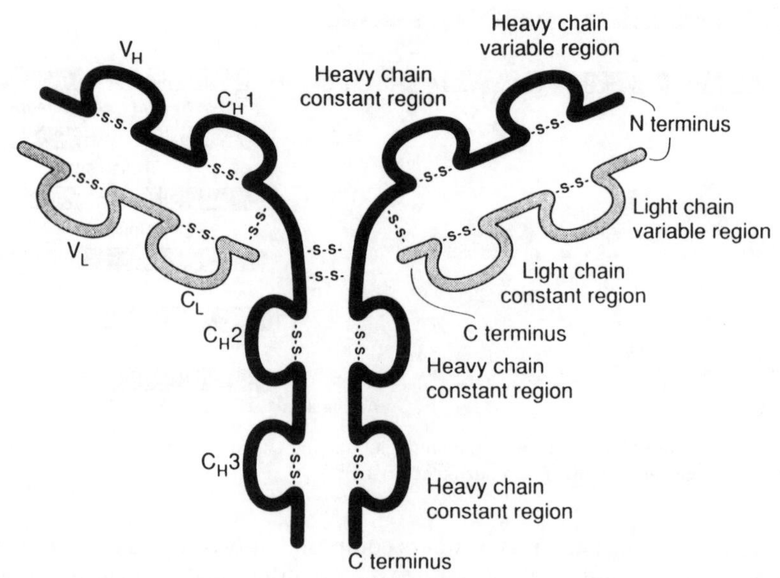

Figure 13.3. Schematic structure of mature antibody molecule. (From Weatherall, 1991.)

This complex structure results from *somatic rearrangement* of Ig genes, one of the most fascinating processes in all of biology. As illustrated schematically in Figure 13.2, early in the maturation of each type of lymphocyte, one each of the genes from the V region and J (and in the heavy chain assembly, D) region is randomly selected and the remaining, unselected, genes in these regions are excised from the chromosome, or made no longer functional in that lymphocyte and its descendent cells. The corresponding genes on the homologous chromosome set are inactivated by a poorly understood process of *allelic exclusion*. This rearrangement is faithfully inherited somatically in a clone of descendent lymphocytes, which can produce only a single Ab type, namely that for which the mRNA is coded by the chosen, remaining genes.

The Ig-producing gene clusters, the genes in the V, J, D, and C regions have arisen by gene duplication and their sequences have subsequently been modified by mutation. The V, J, and D regions, which are largely responsible for Ab specificity, are especially diverse. Table 13.1 shows the approximate number of copies of these genes in the heavy and light chain regions (the numbers vary within species). About $300 \times 6 \times 6 = 10\,800\ \lambda$, 1000 κ, and 108 000 heavy chain combinations can be generated by random selection of genes from these regions, or a total of about 10^{12}

Table 13.1. *Antibody diversity table*

Gene	Chromosome location	Number of copies of			
		V	J	D	C
Lambda (light)	2	<300	6	0	>6
Kappa (light)	22	≈260	4	0	1
Heavy chain	14	≈300	4	>10	9

Sources: Blackwell and Alt, 1989; Eisen, 1990; Lewin, 1990.

different Abs, not including diversity introduced by a high rate of somatic mutation, and sloppy joining during rearrangement.

Not all of the possible lymphocyte clones are actually produced, but the number is very large. Although each clone retains only the subset of the original genes in its stem cell precursor specific to the Ab that it produces, the germ line retains the full Ig gene regions, which are inherited by the next generation (with mutation, as in any other genes).

Rather than being randomly selected initially for Ab specificity, the heavy chain C region genes are used for *functional* differentiation, in the response to infection. During the course of an infection a pathogen is 'presented' to many lymphocytes as it circulates in the bloodstream (or, in the case of viruses, its coat protein remains on the surface of infected cells where it can be seen by B- or T-lymphocytes). Those lymphocytes whose Ab binds to the Ag are stimulated to proliferate and, in a process known as *class switching*, sequentially change the heavy chain C gene used in the production of their Abs, in their chromosomal order. Initially, the Ab is known as IgM and uses the C_μ gene, but Ag-binding stimulates the cell successively to express other C genes, deleting the upstream C genes in the process, and producing the IgD, IgE, and IgA molecules {Lewin, 1990; Roitt *et al.*, 1989}.

During the course of an infection not only do the lymphocytes modify to specialize the attack, but the response is *focused*. Initially, many different Abs may recognize various parts (*epitopes*) of the incoming pathogen, and all start an attack against it. But over time the complex, cooperative mechanisms of the immune system, involving B, T, and other cells such as macrophages, gradually select Abs that fit the Ag best; these proliferate more rapidly and eventually constitute the bulk of the mature immune response. Subsequently, the chosen clones persist, constituting an immunological memory of the attack.

The TcR is a structure like, and closely homologous to, the Ab molecule. Its coding regions are on chromosomes 14 and 7. The TcR functions to recognize a complex of an Ag plus a *host* HLA histocompatibility protein.

In this way, an enormous array of molecular defenses is established without the need to store millions of Ab genes in the genome – and without the need for people to evolve as fast as viruses! We thus stay approximately even in the evolutionary race with the endless diversity of pathogens to which we will be exposed during our lifetimes. However, since the specific response to a pathogen is generated somatically, no memory of specific responses (that is, of the specific V, J, D, and C combinations used) against a given pathogen is inherited. According to current understanding, therefore, the molecular recognition aspects of immune defense do not rely on natural selection for specific alleles at the V, C, or other regions. Evolution produced the system; individuals produce the diversity.

The Rusty Rule applies to the components of the immune system, adding variability, and sometimes pathology, to its functioning. For example, there are various *immunodeficiency diseases* that occur if the mechanisms for producing Abs do not function. Affected individuals can produce insufficient amounts of one or more of the Igs (IgM, IgD, IgA, etc.), or have impaired T-cell function, etc. (Blackwell and Alt, 1989; Roitt *et al.*, 1989). Essentially, there is a spectrum of severity and specificity of defect that is as broad as would be expected in a complex multiply interruptable system. (The immunodeficiency of AIDS is virally induced rather than inherited).

Empirical aspects of immune resistance and susceptibility

If gene rearrangement provides an individual with an *ad hoc* and somatic, rather than heritable, means of Ag recognition, then a newly exposed 'virgin' population should be as resistant to a given pathogen as an experienced population. The best opportunity to answer this question in humans has been the experience of Amerindians when Old World contagious diseases were introduced by European and African colonists. The latter two populations had long co-existed with numerous diseases that Amerindians, their most genetically distant human relatives, had never seen. Most current epidemic diseases probably arose with the settled populations and domestic livestock associated with the development of agriculture in the Old World, long after Amerindians had been physically isolated from the Old World (there were few indigenous contagious diseases in the New World).

Several Old World diseases swept through Amerindian populations upon early contact, killing a substantial fraction of their populations. For example, eyewitness accounts show that the densely populated Mayan coastal areas of the Yucatán peninsula in Mexico were greatly depopulated by smallpox, as part of a general decimation of much of native Mesoamerica (e.g., McNeill, 1976). This helped Hernán Cortés and a handful of Spaniards to overthrow the Aztecs in Mexico.

However, the evidence is largely that, while immunologically naive populations suffer somewhat more severe symptoms compared to experienced populations, they can typically mount effective immune responses. The apparent relative resistance of European and African populations seems largely due to chronic low-grade infection or to passive immunity from placental or lactational transfer of Abs. A major problem for Amerindians upon first-exposure was that the entire population was vulnerable, and nearly everyone became seriously ill at the same time (e.g., Nutels, 1968; Squire, 1992). Social structure quickly broke down, and there was no healthy person to provide even palliative help for affected individuals. Powerful supernatural forces were suspected, from which people expected to die. This appears largely to explain the virgin epidemics of measles observed in Yanomama Indian villages in the Orinoco River basin of Venezeula (Neel *et al.*, 1970) and the Tiriyo in Brazil (van Mazijk *et al.*, 1982; see also Black *et al.*, 1982); in the latter case, although the attack rate was 80%, medical care was available and mortality was very low.

In relatively controlled studies of the response to measles virus vaccine, previously unexposed Amerindians responded with a febrile (fever) response of about 0.4 deg.C greater than that of industrial populations, but without other complications (Black *et al.*, 1969, 1977). This difference does not appear to have been due to concurrent infections, poor nutritional state, or prior exposure to Ags that cross-react with measles to provide low-grade protection. Some measles outbreaks in Amerindians have involved a slightly higher mortality rate, even with adequate nursing care available, so the possibility of some inherent factors remains (Black *et al.*, 1982). There is also some evidence that measles Ab titer levels are affected by gene(s) linked to the HLA region on chromosome 6 (see below) (Black *et al.*, 1982).

Amerindians and Eskimos also seem to be more susceptible to infection by *Haemophilus influenzae*, bacteria that cause recurrent *otitis media* (middle ear infection) and its complications. The infection lasts longer and is more severe than in Caucasians. Apache children from the US Southwest exposed to a vaccine against the coat polysaccharide of *H.*

influenzae produce only one tenth the Ab concentrations produced by Caucasians of similar age (Siber *et al.*, 1990). Since lower reactions are not found for other vaccines, the difference is not due to generalized immune differences of environmental origin. Specific Ig constant regions may be associated with susceptibility to *H. influenzae* (Weatherall *et al.*, 1988).

Does natural selection affect the immunoglobulin genes?

There are some available data on the DNA sequence genetic diversity in the Ig regions. Comparison between germ line and mature Ig sequences (in rodents) shows a high rate of somatic mutation in the code for the extracellular region of the antibody that is involved in Ab recognition, accounting for a substantial fraction of the total expressed Ab diversity (Gojobori and Nei, 1986). Cladistic analysis of the heavy chain V genes shows that the genes are related by gene duplication, with subclades of genes more closely related to each other than to other genes in the cluster (or even more closely related to the corresponding gene in other species than to other V_H genes within species). Some of these genes are so different that they appear to have diverged millions of years ago, which suggests that gene duplication and conversion are rare in this system.

Nucleotide substitutions have accumulated at a rate about as high as that in non-functional genes so that, overall, purifying (allelic) selection seems to have been weak or absent for these genes (Gojobori and Nei, 1984). In the antigen recognizing, compared to structural, regions of the heavy chain cluster, non-synonymous substitutions have accumulated relatively rapidly compared to synonymous substitutions, suggesting that there is selection for diversity per se, as might be expected for such a system (Tanaka and Nei, 1989).

Of course, these overall tests do not preclude allelic selection for a few specific V_H alleles. It may be premature to dismiss totally the possibility of such selection at the Ig loci. A virulent infectious organism can exert stringent and immediate selection favoring individuals who are resistant. An Ag is an 'expert' at finding the few needles, in the haystack of Ab diversity produced by an individual, that can bind to it; it is possible that certain allelic variants in, for example, the V_H region have particularly effective binding affinity for the Ag. Individuals with such a variant could in principle be more efficient at recognizing the Ag and resisting the disease.

Thus, while this idea is contrary to current dogma in immunogenetics, it is worth searching for evidence of allelic selection in the Ig regions. A direct test might involve sequencing the Ig genes used by lymphocyte clones specific to a given Ag, to see whether certain alleles are used significantly more frequently than others or than their frequency in the general population. A less direct test would be to look for patterned levels of sequence diversity among the Ig genes (less diverse genes suggesting purifying selection), or for linkage disequilibrium or geographical distributions that are consistent with the biogeography of pathogens but not with known human population movements. Comparison between Amerindians and Europeans/Africans would be potentially most informative.

Other aspects of the immune system

Recognition and apprehension of an invading Ag or of a cell that has been infected by a pathogen is only the first step in an effective immune defense. Something must then be done with the prisoner. The immune system includes a series of other components designed to take care of this, including the generation of histaminic reactions, febrile response, and the destruction of infected cells or invading Ags.

The *complement* system involves a series of adjacent genes located in the HLA region. A sequential interaction among complement gene products known as the *complement cascade* lyses infected cells by opening holes in their membranes. Complement genes are used to induce histamine release from *mast cells* to destroy incoming foreign proteins, induce fever, help to confine an infection locally, and stimulate immune recognition. The complement system serves a diversity of rather non-specific reaction functions; it may have been an early mechanism for resisting infection that was refined by the subsequent evolution of the Ig system, much as we speculated about the globin gene mutations in Chapter 10.

These functions can all be interrupted by mutation; for example, complement deficiencies are known and lead to increased susceptibility to bacterial infection {e.g., Roitt *et al.*, 1989}.

Human leukocyte antigen: individual uniqueness and the friend or foe problem

The general nature of the HLA system and its parts

A second component of the immune system comprises the genes of the HLA region, or *major histocompatibility complex* (*MHC*), on chromosome 6. These genes interact with the components of the immune system already discussed {Bell *et al.*, 1989; Roitt *et al.*, 1989; Todd *et al.*, 1988a,b}. The HLA system was first studied intensively in the context of

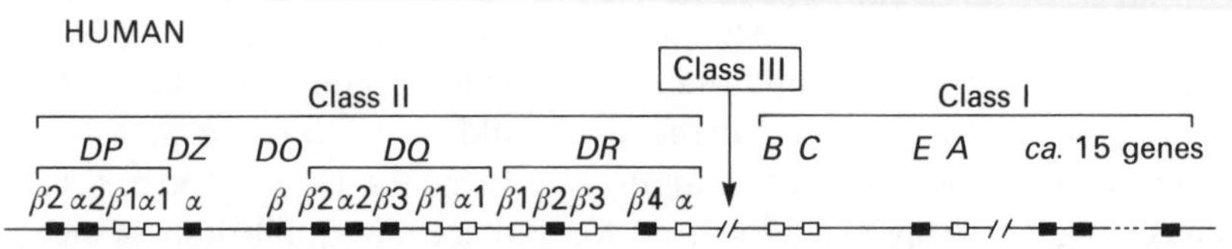

Figure 13.4. Organization of the HLA region on chromosome 6. (From Nei and Hughes, 1991.)

the problem of tissue transplant rejection. It is involved in many important general immune response functions that enable cytotoxic T lymphocytes to destroy infected cells, and is the major mechanism by which the body learns what is 'self', and not to be reacted to immunologically.

Figure 13.4 shows the map arrangement of the HLA region. The system consists of numerous loci coding for genes that are evolutionarily related as members of a local multigene family, part of a superfamily that includes the Ig genes and other more distantly related genes for cell-surface adhesion and cell-to-cell recognition molecules {Roitt *et al.*, 1989; Williams, 1985}. The genes, shown in Figure 13.5, share sequence-structural characteristics, with similar exons used for homologous purposes.

There are two sets of related HLA genes in this region. Class I, or *histocompatibility antigens*, include three (functional) loci, designated HLA-A, B, C, whose products are found on the surface of all nucleated cells, as self-Ags. Class I HLA proteins are heterodimers, composed of one copy of the HLA (A, B, or C) gene itself, attached to one copy of β_2 *microglobulin*, the locus for which is on chromosome 15. The different domains of the protein, like those of the Igs, are exon-specific (Figure 2.1, p. 21). Figure 13.6 shows a schematic of the molecular structure, with domains identified. In an infected cell, class I gene products bind to antigenic debris from the infecting pathogen, and this complex is recognized by TcR genes on cytotoxic T-cells, which then destroy the infected cell, as shown in Figure 13.7.

At the other end of the HLA region are the class II, or HLA-D, loci that include about six separate regions, at least three of which (denoted *DR*, *DQ*, and *DP*) are actively coding. Class II products are expressed on the surfaces of antigen presenting T lymphocytes; these bind to foreign Ags, and present the class II–Ag complex to helper (CD4) T-cells, which then stimulate a humoral response, i.e., trigger B cells that can recognize them to produce Ab. The class II proteins are dimers composed of one copy each of corresponding HLA-Dα and HLA-Dβ chains, coded by adjacent genes of similar structure within each of the class II regions (some

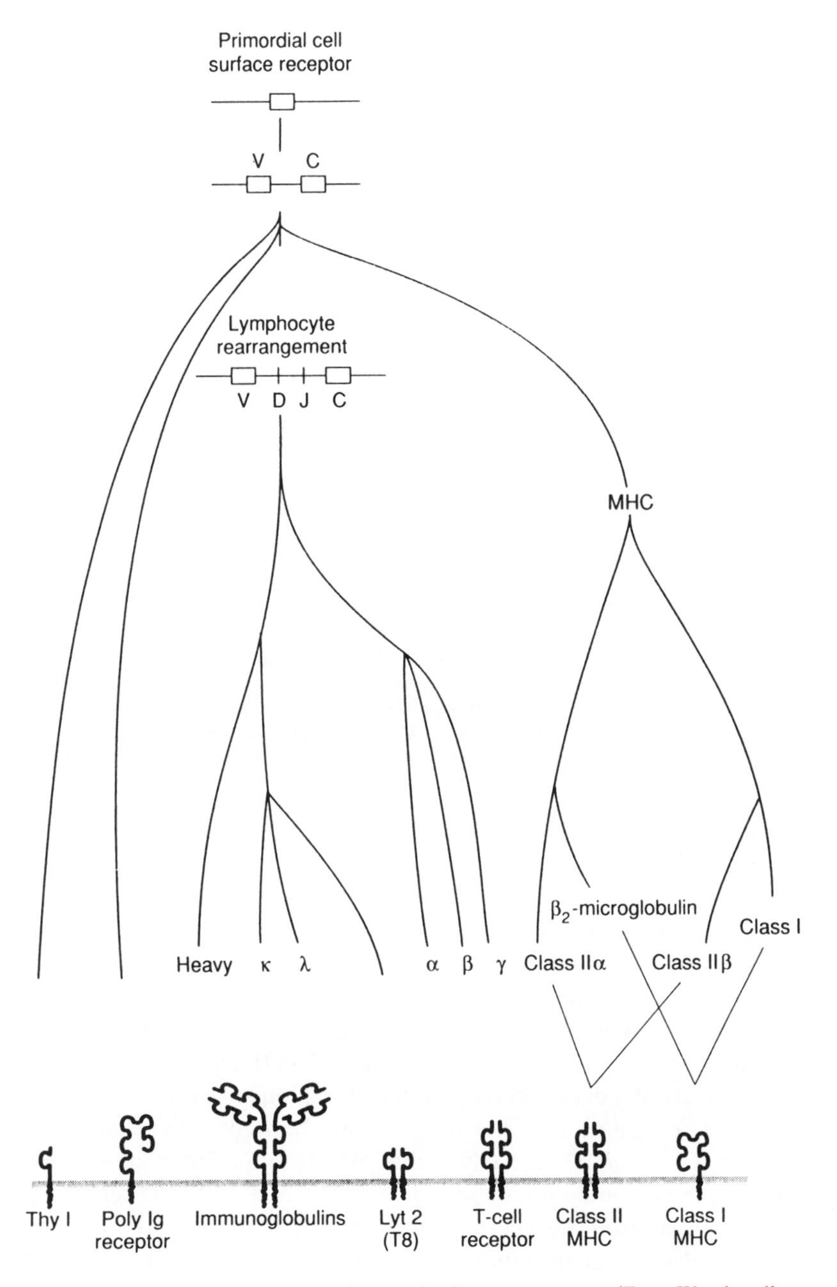

Figure 13.5. Immunoglobulin gene family sequence tree. (From Weatherall, 1991.)

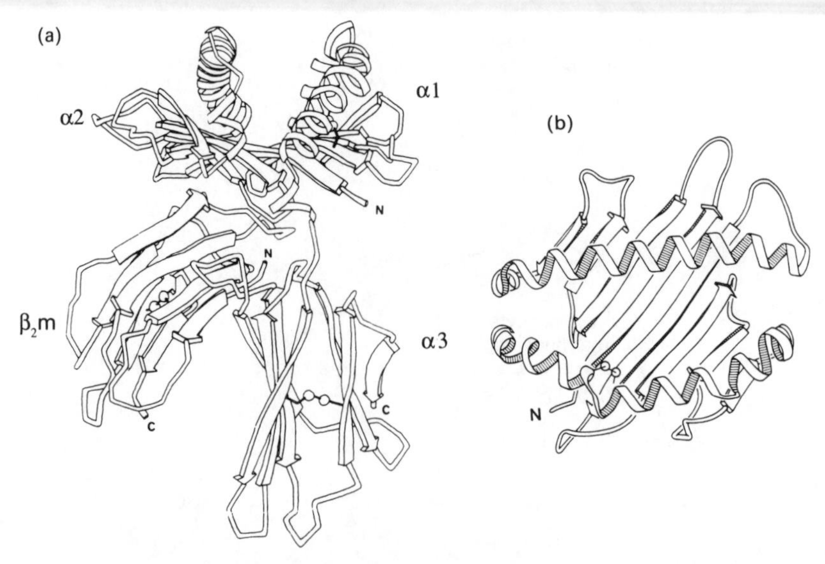

Figure 13.6. Schematic of the molecular structure of HLA molecules. The curled 'ribbons' represent helical polypeptide sections of the HLA protein that form the binding cleft, or pocket, that recognizes antigen. (a) Side view of the molecule, with the coding domains labeled (cf. Figure 2.1). (b) A view from above the cell surface. The upper and lower helical sections represent the binding cleft. The mutation related to IDDM is at the rightmost part where these upper and lower helices approach each other (From Bjorkman *et al.*, 1987. Reprinted by permission from *Nature* **329**: 506–512. Copyright © 1987 Macmillan Magazines Ltd.)

class II molecules are hetero-dimers of α- and β-chains from different D-region genes). Although the basic class II structure is α–β, the gene duplication history of this region has generated variability in the structure of each D 'locus', which differs in number, arrangement, and orientation of the α- and β-components. An Ag-recognition region is located in the external domains of both HLA class I and class II gene products.

Genes recently discovered in the class II region have class II characteristics but sequences that are intermediate between class I and II sequences (Cho *et al.*, 1991; Kelly *et al.*, 1991); the function of these genes is not yet known, but their sequence divergence, and conservation between human and mouse, suggest they may antedate the separation of the two classes of genes. Between the class I and class II gene regions is the cluster of genes for the complement system, sometimes referred to as HLA class III loci. Also, genes in the HLA region code for enzymes that help to process, transport and express the HLA substances themselves.

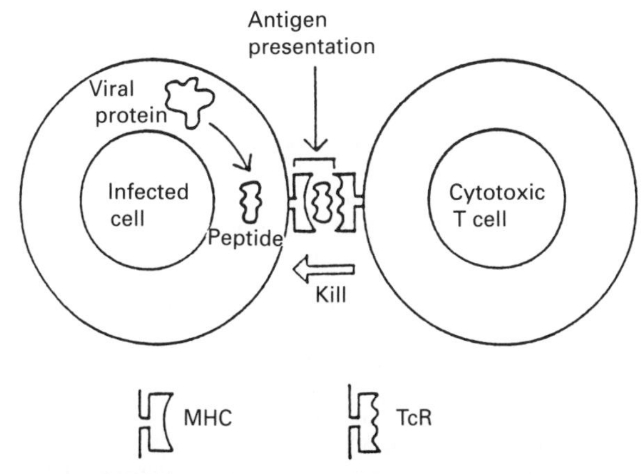

Figure 13.7. Schematic structure of the function of HLA molecules. Foreign viral-related protein enters the cell by an infectious process. It is degraded into small peptides which, bound to the HLA class I molecule, are transported to the cell surface and presented to cytotoxic T-cells, whose T-cell receptors recognize the HLA-protein complex. (From Nei and Hughes, 1991.)

The HLA-related diseases

In the last 20 years or so a great number of diseases have been associated statistically with the HLA system, as discussed in Chapter 3 (Table 3.5B). Two inflammatory diseases of joints and connective tissue, Reiter's syndrome and ankylosing spondylitis (AS) are strongly associated with class I alleles. So is IDDM in experimental animals. However, most major associations are with class II alleles; these diseases span the range of physiological systems and types of pathology. Many of these diseases involve *autoimmunity*, in which the affected individual produces Abs against his/her own tissues; examples are listed in Table 13.2.

For example, myasthenia gravis, a muscle weakness syndrome associated with neuromuscular transmitter deficiency, has been shown to involve anti-self Abs to a neurotransmitter receptor on muscle cells. Grave's disease, a form of hyperthyroidism, involves Abs against the receptor for thyroid-stimulating hormone. Systemic lupus erythematosus, a disseminated disease affecting the peripheral vascular system, kidney, and other organs, is caused by Abs against nuclear proteins found in many different kinds of cells (the cells most affected are those most accessible to the circulation).

A few diseases of purely infectious etiology are associated with HLA. The best example is leprosy, although the reasons for the association with HLA are not known (Abel and Demenais, 1988; Abel *et al.*, 1989; Serjeantson, 1983).

Table 13.2. *Some autoimmune disorders in humans*

Organ or tissue	Disease	Antigen
B cells of islets of Langerhans (pancreas)	Insulin-dependent ('juvenile') diabetes mellitus	?
Thyroid	Hashimoto's thyroiditis (hypothyroidism)	Thyroglobulin Thyroid cell surface and cytoplasm
	Thyrotoxicosis Graves' disease; hyperthyroidism)	Receptor for thyroid-stimulating hormone
Gastric mucosa	Pernicious anemia (vitamin B_{12} deficiency)	Intrinsic factor (I)
		Parietal cells
Adrenals	Addison's disease (adrenal insufficiency)	Adrenal cells
Skin	Pemphigus vulgaris	Epidermal cells
	Pemphigoid	Basement membrane between epidermis and dermis
Eye	Sympathetic ophthalmia	Uvea
Kidney glomeruli plus lung	Goodpasture's syndrome	Basement membrane
Red cells	Autoimmune hemolytic anemia	Red cell surface
Platelets	Idiopathic thrombocytopenic purpura	Platelet surface
Skeletal and heart muscle	Myasthenia gravis	Acetylcholine receptor
Brain	Allergic encephalitis	Brain tissue
Spermatozoa	Male infertility (rarely)	Sperm
Liver (biliary tract)	Primary biliary cirrhosis	Mitochondria (mainly)
Salivary and lacrimal glands	Sjögren's disease	Many: secretory ducts, mitochondria, nuclei, IgG
Synovial membranes, etc.	Rheumatoid arthritis	Fc domain of IgG; other
	Systemic lupus erythematosus	Many: DNA, DNA–protein, cardiolipin, IgG, microsomes, etc.

Source: Eisen, 1990.

Table 13.3. *Molecular mimicry: sequence similarities between microbial proteins and human host proteins*

Protein	Postition of starting (N-terminal) residue	Amino acid sequence
Human cytomegalovirus IE2	79	PDPLGRPDED
Human lymphocyte antigen DR	60	VTELGRPDAE
Poliovirus VP2	70	STTKESRGTT
Acetylcholine receptor	176	TVIKESRGTK
Papilloma virus E2	76	SLHLESLKDS
Insulin receptor	66	VYGLESLKDL
Rabies virus glycoprotein	147	TKESLVIIS
Insulin receptor	764	NKESLVISE
Klebsiella pheumoniae nitrogenase	186	SRQTDREDE
Human lymphocyte antigen B27	70	KAQTDREDL
Adenovirus 12 E1B	384	LRRGMFRPSC
A-gliadin	206	LGQGSFRPSQ
HIV p24	160	GVETTTPS
Human IgG constant region	466	GVETTTPS
Measles virus P3	13	LECIRALK
Corticotropin	18	LECIRACK
Measles virus P3	31	EISDNLGQE
Myelin basic protein	61	EISFKLGQE

Source: Eisen, 1990.

The reasons for the HLA associations may involve altered forms of self-antigens. Lymphocytes with anti-self Abs are normally quiescent, and can react against a self-Ag only if a cognate T-helper cell recognizes a different part of the same self-Ag. The current view is that such T-cells are purged in the thymus gland. However, somatic mutations or chemical alterations that modify the self-Ag could trigger an anti-self reaction if the self-Ag contains a new epitope (recognition site) that no longer looks like 'self'.

Another mechanism that may be common in HLA-associated diseases is *molecular mimicry* {Eisen, 1990}. A number of human proteins contain amino acid sequences that resemble those of bacteria or viruses (Table 13.3). An exogenous agent that mimics self (even to some slight extent) may trigger an antibody response that cross-reacts with cells presenting similar Ags, causing autoimmune cell destruction.

We do not as yet have a clear picture of the etiology of any of these diseases. None is caused solely by the presence of a given HLA allele or haplotype, and all appear to require other factors. MZ twin concordance, for example, is often only about 50% or so. Within affected sibships high risk haplotypes may be as common in unaffected as in affected siblings.

HLA-B27 and ankylosing spondylitis

One of the strongest HLA-related disease associations is the worldwide association between AS and the *HLA-B27* allele. AS is an inflammatory disease that produces vertebral adhesion and pain, ulcerative colitis, psoriasis, and a general increase in mortality from many causes. The primary pathology of spondylitis is still poorly understood as is the reason for the HLA association. Molecular mimicry involving a pathogenic microorganism is suspected, or an attack on the B27 molecule in association with some pathogen, but no good candidate has been established for which the evidence is consistent (Bell *et al.*, 1989).

HLA and type I diabetes (*IDDM*)

IDDM, type I diabetes, is due to the gradual disappearance of insulin production. The disease typically arises in childhood. One characteristic of IDDM is the presence of anti-self Abs against the β-islet cells that produce insulin in the pancreas. These cells are systematically destroyed as the disease progresses. The autoimmunity, and various regional and seasonal patterns of IDDM, suggest an infectious etiology. For example, carriers of anti-β-cell antibodies can rapidly develop IDDM after viral infections, presumably because such infections stimulate the immune system generally, increasing the aggressiveness of the existing anti-self activity.

IDDM is clearly familial and has strong associations with alleles in both class I (HLA-B) and class II (HLA-D) loci (e.g., Field, 1988; Thomson, 1988; Thomson *et al.*, 1989). But no single locus or allele is sufficient to explain the occurrence of the disease in families. The epidemiology regarding the HLA-D region is complex and apparently heterogeneous (Thomson, 1988; Thomson *et al.*, 1989). Black–white admixture data have suggested dominant effects of the alleles at the *DR* locus. Generally, the *DR3* allele appears to act in a recessive way relative to the *DR4* allele, *DR4* is dominant or co-dominant in the absence of *DR3*, and *DR3*/*DR4* heterozygotes are at especially high risk. Sequence subsets of these serologically determined alleles are probably predisposing and/or various *DR* alleles vary slightly in their protective or predisposing effects.

Mutations substituting other amino acids for aspartic acid at codon 57 of the HLA-DQβ gene product are associated with IDDM; that is, Asp57 is dominantly protective. The mutation is found in a high percentage of IDDM patients, and though also found in normal humans, the relative risk in diabetics is over 100 (Morel *et al.*, 1988). The same mutation has been shown experimentally to produce diabetes in rodents (the DQβ gene is highly conserved between humans and rodents (Hughes and Nei, 1989)). Position 57 is in the binding region of the DR gene product; in the homologous class I structure shown in Figure 13.6(b), this is at the rightmost end of the lower ribbon helix (Bell *et al.*, 1989; Todd *et al.*, 1988a,b).

Absence of class I alleles has been shown in one diabetes-prone mouse strain to be associated with IDDM, if either β_2 microglobulin is deficient or there is a mutation in a putative HLA peptide transporter that helps the translated class I protein to be presented properly on the cell surface (Faustman *et al.*, 1991). However, loci on several chromosome regions other than HLA are involved in IDDM in both humans and mice (Cornall *et al.*, 1991; Garchon *et al.*, 1991; Julier *et al.*, 1991; Todd *et al.*, 1991). The HLA-DR4 associations depend on alleles segregating at other loci. Some of these are good candidate genes (e.g., the insulin region on chromosome 11). The loci identified to date in mice are related to severity and the course of the disease (Avner, 1991). This pattern is consistent with the general evolutionary pattern for complex traits.

Other HLA-associated diseases are genetically less well understood, but also seem to have complex etiology. The evidence for environmental and/or other-locus involvement includes differences in risk among populations. For example, rheumatoid arthritis (RA) is an autoimmune disorder of the joints whose frequency is elevated in some Amerindian populations (see e.g., Atkins *et al.*, 1988; Del Puente *et al.*, 1989). Multiple sclerosis (MS) is a disease characterized by progressive, episodic degeneration of the myelin sheath of motor nerves; risk of MS is associated with spending one's childhood in temperate climates, and there have also been well-documented epidemics, suggesting that viruses or other biological pathogens may be able to trigger the disease (e.g., Waksman, 1989).

Indirect evidence in the HLA system

Variation in the HLA system appears to be under strong control by natural selection {Hedrick *et al.*, 1987; Hughes and Nei, 1988, 1989, 1990; Hughes *et al.*, 1990; Nei and Hughes, 1991}. Compared to other human loci and considering a plausible range of values for critical evolutionary parameters such as N_e and μ, the diversity of the classes I and II loci is considerably greater than would be expected if the loci were evolving neutrally. The details of nucleotide substitution support this inference.

In the Ag recognition sites of both classes I and II genes, heterozygosity and amino acid diversity are very high, similar to the pattern noted above for V_H genes, and suggesting selection for variability per se. The Ag recognition site can be further divided into sub-regions. The most functionally critical region appears to be a 'binding cleft', lined by charged amino acids, into which the Ag fits and in which changes in the charge structure of the amino acids would affect binding specificity; human and mouse HLA class I haplotypes show a preference for charge-changing substitutions in the binding cleft (Figure 13.6(a)), but not in functionally less constrained regions of the Ag recognition site where heterozygosity is less (Hughes *et al.*, 1990).

There is also indirect evidence from linkage disequilibrium for specific haplotype selection in the HLA region. The mean disequilibrium between alleles at the HLA-A and B loci is greater than the amount that would be caused by drift alone. Some pairs of alleles are in strong positive disequilibrium, for example the *A1B8* haplotype, while other *A1B−* or *A − B8* haplotypes are in negative disequilibrium. This would be expected if the *A1B8* haplotype has specifically been selected for in the past (Hedrick *et al.*, 1987).

Defensive strategies by non-immunological mechanisms

The immune system is an offensive system: it actively seeks and detects incoming pathogens, or cells infected by them, The Ab-producing system uses somatic rearrangement to generate diversity. However, there are numerous *ad hoc* defensive genetic mechanisms against infection. One is to deprive the pathogen of the site(s) it uses to enter a target tissue or cell. The defenses involve normally inherited genes whose primary function has nothing to do with protection against infection.

The best-known examples of such passive defense are the genetic mechanisms for malarial resistance. Sickle cell hemoglobin changes the shape of the red blood cell so that the *P. falciparum* parasite is less able to enter the cell, where it must spend part of its life cycle. The mutation-producing deficiency in the enzyme G6PDH deprives the malarial parasite of chemical nutrients it needs while in the red cell. And the 'negative' alleles of the Duffy blood group system prevent the parasite that causes *vivax* malaria from binding to red cells.

Chapter 3 showed associations between ABO blood types and various non-infectious diseases but ABO is also associated with infectious diseases, including smallpox, cholera and other enteric diseases, bubonic plague, syphilis and others. The strength of the evidence varies. The reasons may have to do with molecular recognition and binding. Along with several other unlinked loci, called H, Lewis, and Secretor, the ABO system codes for enzymes that produce oligosaccharide molecules found attached to various epithelial tissues including the lining of the bladder, urinary tract, and intestinal tract. The structure of these molecules, and whether they are secreted into tissue fluids, vary and are determined by the genotypes at these loci (e.g., Levitan, 1988).

The ABO system is antigenically related to Ags produced by the bacterium *Escherichia coli*, which colonizes the digestive system, and with which we have a necessary symbiotic relationship. The mammalian ABO polymorphism may be used for protection against infection by *E. coli* and related types of bacterium, some of which can be seriously pathogenic. For example, ABO similarities between host and bacteria that may impair host defensive strategies have been associated with more severe infection (Vogel and Motulsky, 1986). These bacteria can establish pathogenic infections outside the gut, for example in the urinary tract, if they can bind to the epithelial surfaces there. This requires that bacterial-surface proteins called adhesins bind with oligosaccharide receptor sites on the epithelial surface. There is some evidence that in persons with the non-secretor genotype, the blood group oligosaccharides that are normally expressed on the uroepithelium, where they obscure less prominent adhesin receptors, are absent, leading to persistent or chronic urinary tract infection (e.g., Sheinfeld *et al.*, 1989).

Undoubtedly many more non-immune defense mechanisms remain to be discovered. They will explain variation in susceptibility and severity of infectious disease. Pathogens that evade detection and destruction by the immune system must live in some target tissue to do their work. The nature of the host's biochemical, metabolic, and physical environment will vary. Genetic defenses that attack vital parts of the pathogen's life cycle or physiology may be difficult for the pathogen to circumvent, and could evolve in the host by normal allelic selection.

Conclusion

Two of the most interesting aspects of immune resistance are the evolution of somatic rearrangement as a way of adapting to rapidly evolving pathogens, and the high level of heterozygosity in the HLA system used to identify self from non-self. The critical element of both of these is the evolution of *diversity* per se. Gene duplication has been important in both systems.

The non-immune mechanisms for defense against infection involve traditional allelic selection. As discussed in earlier chapters, a diversity of mechanisms is always expected when there is phenotypic selection for complex traits (here, resistance to infection). Attacking pathogens use every form of subterfuge to defeat detection and destruction {Goodenough, 1991; Roitt *et al.*, 1989}. In response, we mount defenses or attacks at all vulnerable points in the biology of the pathogen.

The evolution of the immune system is a fascinating subject, far too intricate to go into here, although we have seen that some elements, such as general febrile responses, appear in organisms without complex immunoglobulins and appear to be earlier defense mechanisms. Geneticists are beginning to work out the mechanisms that produce somatic rearrangement of the Ig genes, including the identification of the recombinase enzymes that are involved.

To my knowledge, no other human systems have been shown to involve a similar rearrangement system. However, a mechanism so currently highly developed and coordinated must have had numerous ancestral uses, must involve many duplicate genes with related function (but not part of the immune system per se), and so on. I anticipate that we will shortly learn of other systems that use homologous mechanisms. Whether these will involve recognition of environmental agents (e.g., odors, pheromones) or will have to do with chromosome management or other functions is an open question. But it would be very surprising if homologous versions of such an elaborate mechanism were not used in some other ways.

14 *Variation within the inherited genotype*

A beautiful theory slain by an ugly fact.

After T. H. Huxley 'Biogenesis and abiogenesis' (1892)

An exception disproves the rule.

A. Conan Doyle, *The Sign of the Four* (1890)

Some of our basic ideas about genes and their transmission have been established for nearly a century. One of these is Mendel's Law of Segregation, from which we derived the transmission probabilities (τs) given in Chapter 5. However, as noted earlier in this book, some investigators estimate these probabilities from the general likelihood in families rather than specifying them, and the estimates are sometimes significantly different from the mendelian values of 0, $\frac{1}{2}$ and 1. A second basic principle of classical genetics is that genes are faithfully transmitted during mitosis, so that all of our somatic cells have the original inherited genotype. The process of somatic rearrangement in immunoglobulin and TcR genes shows that this postulate, also, is not always correct.

This chapter shows various other ways in which violations of these classical assumptions have important effects on the pattern of genotypes in families and the relationships between genotype and phenotype. The term non-traditional inheritance is often used for such phenomena {Hall, 1990a,b; Holliday, 1987, 1989; Solter, 1988}, but the mechanisms have always been there whether part of our 'traditions' or not!

Mechanisms for non-mendelian inheritance

A few mechanisms can modify the probability that a given allele will be transmitted from parent to offspring.

Meiotic drive and gametic selection

Natural selection usually affects diploid organisms one genotype at a time. However, part of our natural life cycle is spent in a haploid condition, namely the time between the formation of a gamete and the production of a fertilized egg to form the zygote. Gametes are subject to natural selection just as zygotes are. For example, egg cells and their precursors must successfully manage their routine housekeeping physiological functions for decades, and spermatocytes must express genes for cell division and differentiation thousands of times, before fertilization. Perhaps selection in these gametes is responsible for the pervasive and subtle ways in which the DNA sequence differs from random nucleotide usage at millions of nucleotides and tens of thousands of genes (Chapter 9) {Hartl and Clark, 1989; Li and Graur, 1990; Nei, 1987}.

When selection affects gametes, the offspring will not reflect a 1:1 proportion of parental alleles. This *segregation distortion* due to *gametic selection*, also known as *meiotic drive*, may differ for maternal and paternal parents because egg and sperm cells have different environments and requirements. There is no dominance in haploid gametes, so gametic selection may be more directly allelic than selection in diploids, and might more efficiently affect the frequency of alleles. However, there are few documented examples of meiotic drive in humans, none well established.

Insulin-dependent diabetes mellitus

A possible example of meiotic drive is IDDM, which is transmitted more often to the offspring of diabetic women than those of diabetic men. Meiotic drive here may involve the HLA class II genes (Field, 1988; Julier *et al.*, 1991; Vanheim *et al.*, 1986; Warram *et al.*, 1984). HLA-DR4/− or DR3/− heterozygote parents who are unaffected by IDDM transmit the HLA alleles to non-diabetic children with the expected mendelian frequencies, but diabetic mothers transmit their DR4 to offspring with higher probability than do diabetic fathers. However, genetic imprinting (see below) may be the explanation rather than meiotic drive.

Other mechanisms can produce non-traditional inheritance patterns. There may be selection associated with gestation in mammals, because the mother must be host to a fetus that is half non-self. The fetus is only incompletely isolated by the placental barrier, but maternal–fetal reactions, such as those associated with ABO and Rh antigens, may adversely affect fetuses with certain genotypes {Vogel and Motulsky, 1986}. There is an idea that maternally HLA-compatible fetuses may be selected against; this could help to explain the level of HLA polymorphism, but

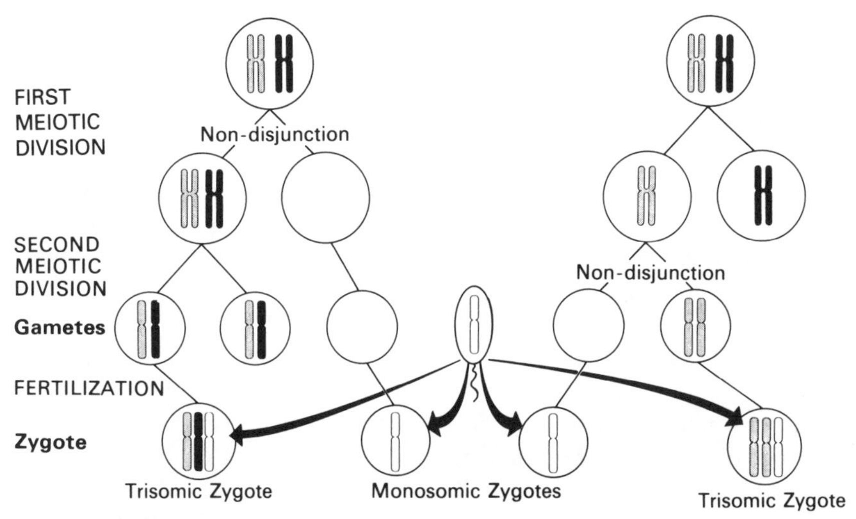

Figure 14.1. Chromosomal consequences for the offsprong of meiotic non-disjunctions. (From Gelehrter and Collins, 1990.)

the evidence is ambiguous (Nei and Hughes, 1991). This would be zygotic rather than gametic selection, but would appear in non-mendelian segregation proportions in surviving offspring.

Chromosomal non-disjunction

Meiosis involves two cell divisions, as shown in Figure 14.1. If there is *non-disjunction* in either of these divisions, i.e., if the chromosomes fail to separate properly, the descendent cells will inherit an anomalous chromosomal count (*aneuploidy*). Non-disjunction in the first meiotic division produces gametes either with no copies of a given chromosome or with a complete set of both parental chromosomes. If the latter are successfully fertilized, they will be true triploids. Non-disjunction in the second meiotic division produces gametes with no parental chromosome or with a duplicate copy of one of the parental chromosomes and none from the other. Zygotes in this situation will be haploid or triploid, but in the latter case always with two doses of one allele from the non-disjunction parent. The best-known example of non-disjunction effects is DS, which results from triplication of parts of chromosome 21, usually from the mother (Antonarakis *et al.*, 1991).

Many aneuploidies are aborted during development because anomalous gene doses are incompatible with some of the delicately timed events in differentiation. However, some aneuploid fetuses do survive. Sex chromosome aneuploidies are well known. Turner's syndrome is an X– genotype, Klinefelter's XXY, and there are others whose symptoms reflect the karyotype and are explicable in terms of gene dosage during development (see e.g., Gelehrter and Collins, 1990; Levitan, 1988; Vogel and Motulsky, 1986).

Some cases of non-disjunction lead to phenotypes that mimic known single-locus conditions, rather than gross anomalies, providing another route to non-mendelian segregation proportions. An example is *uniparental disomy*, duplicate copies of the same chromosome from a given parent. Many instances may be benign and unnoticed, but if that parent happens to be heterozygous for a recessive trait, an offspring who inherits two copies of the same mutant chromosome will be affected, but (unless the rare non-disjunction occurs again) none of his/her siblings will be affected (assuming the other parent to be homozygous normal). This could lead us to infer wrongly that the other parent is a carrier heterozygote. Maternal disomy explains a fraction of CF patients (Hall, 1990a,b), and a substantial fraction, perhaps 20% or more, of cases of Prader-Willi syndrome (see below; Hall, 1990a,b; Nichols *et al.*, 1989), and others.

The translocation of an intact gene to another chromosome can lead to offspring with one, two, or three doses of the gene. For example, if a small piece of chromosome 21 containing the *DS* gene is translocated, and if the receiving chromosome and the remaining normal 21 happen to segregate to a given gamete, the fertilized zygote will be triploid for the *DS* gene, and will have the DS disorder.

Phenotypic variation within the same genotype

Somatic mosaicism

Somatic mutations of all types can occur at any time during the life-history of an individual (Hall, 1988). Such mutations produce *somatic mosaicism* in the affected tissue, because not all of its cells share the same genotype. *Germ line mosaicism* occurs if the developing gonadal cells experience mutation; then, a fraction of the gametes produced will carry the mutant allele. The fraction will depend on when during gonadal development the mutation occurred; if early, a higher fraction of the final tissue will be affected. Generally, if a fraction p of the gametocytes in an Aa heterozygote are $A\alpha$ mutants, the segregation proportion of the gametes will be

A: $\frac{1}{2}$

a: $(1 - p)/2$

α: $p/2$.

For example, if the mutation arose at an embryological stage when there were only two gonadal progenitor cells, 25% of the gametes would be mutants. The affected parent can produce three different gametes at the locus. Such mutations probably occur at some point in the development of the gonads of most individuals. However, they are likely to be rare and hence to affect only a small fraction of the total number of gamete producing cells. Hence $p \approx 0$, and it is unlikely that more than one offspring of that person would inherit the mutation.

Somatic mosaicism arising from mutations occurring during or after organogenesis results in a patch of mutant genotypes in all tissues that contain cells descended from the orginal mutant cell. The phenotype of the individual, or of a particular organ, can be comparably patchy, or *variegated*. The earlier in organogenesis that such mutations occur, the more impact it can have on the final tissue.

An important type of anomalous event is known as *mitotic recombination*. Occasionally during mitosis, a somatic cell distributes two copies of one of its parental chromosomes (or chromosome pieces) to a daughter cell rather than one copy of each chromosome to each daughter cell. For any locus at which the individual is a *constitutive* (inherited) heterozygote, the resulting daughter cells will be homozygotes (e.g., an *Aa* cell produces *AA* and *aa*, rather than two *Aa* mitotic daughter cells). The phenotypic effect in cells in which the genes are expressed will depend on the degree of dominance, and the earlier the event occurs the greater the resulting mosaicism. In the millions of cells and cell divisions that occur during our lifetimes, huge variation is likely to be generated by somatic recombination of this kind. Most of the time the overall phenotypic effects will go unnoticed because only a few individual cells among the sea of normal cells will have deviant phenotypes. But we see in the next chapter that cancer is a very significant counter-example.

We do not yet know how much of the substantial variation in clinical phenotypes observed among individuals with the same constitutive genotype might be due to somatic mutation and variegation. However, it seems likely that considerable phenotype variance is somatically induced. Such variation will appear as individual-specific environmental variance (σ^2_{E}) in segregation analysis with standard genetic models.

An example of this kind of variability is seen in the disease neurofibromatosis 1 (NF1), caused by a mutation at a locus on chromosome 17, characterized by the development of numerous discolorations and growths on the skin, among other problems (e.g., Gelehrter and Collins, 1990). The location, number, timing of appearance and severity of these lesions is highly variable among individuals with the same NF1 mutation, suggesting that somatic mutations are responsible.

Imprinting: inherited genetic variation that is not sequence based

Not all genes are expressed all the time in all cells. How is the regulatory timing of gene expression controlled? The answers are very incompletely known at this stage {Lewin, 1990}, but one important component is DNA methylation, described in Chapter 1. Methylation patterns are inherited (though with some error) via methylation enzymes active during DNA replication. The process of methylation is itself controlled by genes, but the pattern of methylation is inherited as a non-coding characteristic of a gene region, and for that reason is known as *epigenetic* inheritance. The means by which specific methylation sites are recognized, or how the timing of their methylation and demethylation is regulated, are poorly known {Hall, 1990b; Monk, 1990; Solter, 1988}, but a number of human phenotypes seem to be affected by DNA methylation patterns.

Two basically different kinds of marking take place. First, programmed sites all along the genome are marked at the start of each generation. Because of the importance of this marking for development, and the fact that the marking pattern has memory in descendent cells, this process is known as *genomic imprinting*. Secondly, different individual loci appear to be marked, and unmarked, variously in different tissues and times in the life cycle, as the demands of their tissue-specific expression dictate.

It is interesting to consider some evolutionary aspects of imprinting as suggested by the experimental evidence to date. Perhaps one of the first uses of genomic imprinting in higher vertebrates was for the inactivation of one of the two X-chromosomes in females. Experimental evidence suggests that across the mammalian genome the level of gene expression has become relatively fixed and difficult to alter. Serious dose problems can arise between males and females in physiological pathways that involve X-linked genes. Shutting down one X-chromosome in females keeps the dose of X-linked genes hemizygous in both sexes (though actually a few genes remains expressed on the 'inactive' X (Davies, 1991) and a few are active on the Y chromosome). X-inactivation is the most thoroughly understood imprinting system, and has been shown to involve methylation as an important mechanism (Grant and Chapman, 1988).

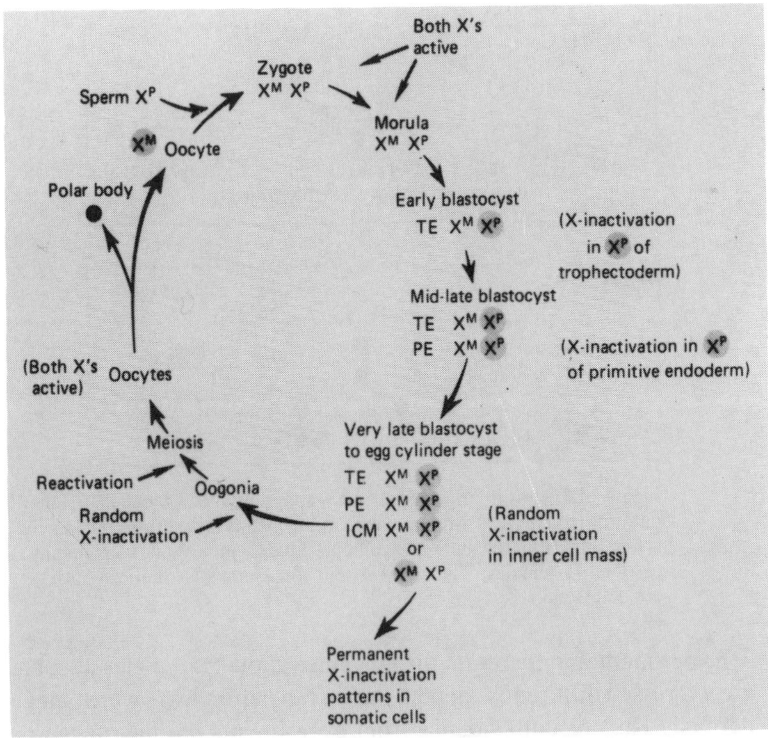

Figure 14.2. Schematic of the cycle of X chromosome inactivation and reactivation in the human life cycle. X^m is maternal and X^p paternal chromosome. Embryonic tissue layers: ICM, cell mass of the embryo; TE, trophectoderm; PE, primitive ectoderm., Chromosomes within gray circles are inactivated. (From *Patten's Foundations of Embryology*, 5th edn, by B. M. Carlson. Copyright 1988 McGraw-Hill, Inc. Reproduced by permission of McGraw-Hill, Inc.)

The imprinting 'life cycle' of the human X-chromosome is illustrated schematically in Figure 14.2

Another potentially important evolutionary use of genomic imprinting in a sexually reproducing species is to prevent parthenogenesis, requiring both sexes to contribute to the genome. Imprinting may also have evolved to support placentation. Paternal genes appear to be preferentially expressed in extra-embryonic tissues of the blastocyst, or early embryo (e.g., trophoblast layer of the developing embryo), with maternal genes preferentially expressed early in the development of the embryo itself (inner cell mass, or ICM, of the embryo) (e.g., Hall, 1990b).

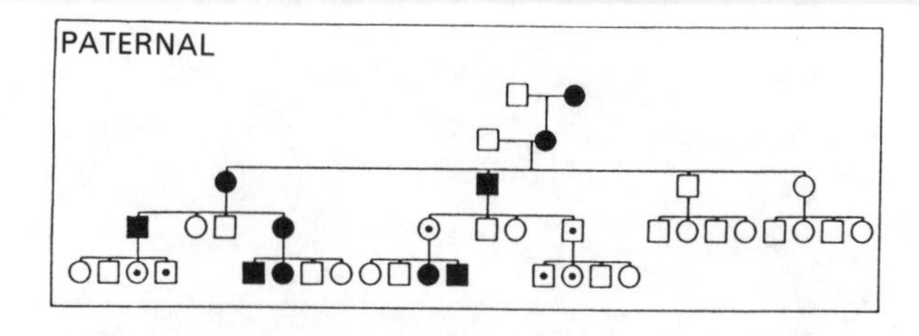

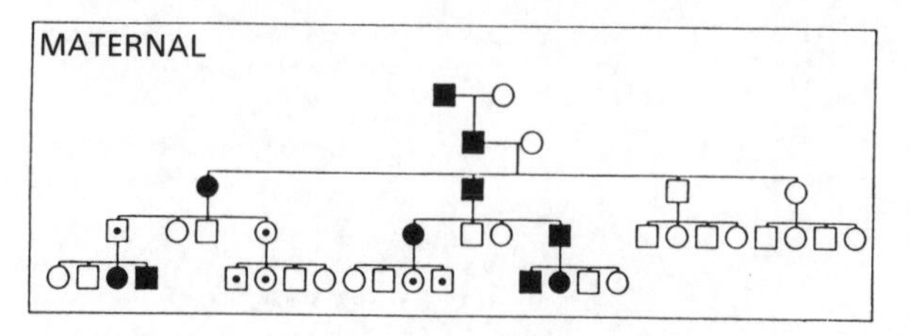

Figure 14.3. How imprinting generates apparently 'non-mendelian' inheritance patterns in pedigrees. Maternal imprinting means no expression of a gene if it is inherited from the mother. Symbols with dots indicate non-expressing carriers; ○, females; □, males. (From Hall, 1990a. © American Society of Human Genetics.)

Experimental uniparental disomic mice show that alleles at some loci are expressed differently, depending on the parent from which they were derived. The current evidence from experiments on mice suggests that such *parental origin* effect may apply only to a modest fraction of all genes, but the effect can be dramatic when it does occur, and what we know about evolution (Chapter 9) tells us that lesser effects must also exist.

Parental imprinting constitutes a source of deviation of phenotypic (but not genotypic) inheritance patterns from their mendelian expectations in humans {Hall, 1990a,b; Holliday, 1987, 1989; Solter, 1988}. Paternal gametes are highly methylated, but maternal gametes are not. Thus, the same allele may be differently expressed depending on whether inherited from the mother or the father. This can lead to apparently 'non-mendelian' transmission in families. Figure 14.3 illustrates this with idealized pedigrees for maternal and paternal imprinting.

Prader-Willi syndrome

Prader-Willi syndrome (PWS) is a leading cause of 'syndromal' obesity in humans, and involves other symptoms including small stature and mental deficiency. The *PW* locus is on the long arm of chromosome 15. Maternally disomic cases of PWS show that the cause is the absence of a functioning paternal chromosome 15, rather than a maternal double dose effect or a deleterious mutation (Butler, 1990). Most 'normal' cases of PWS are due to a deletion of the relevant region of the paternal 15. Remarkably, there is another disorder, known as Angelman syndrome, that involves a form of hyperactivity that produces symptoms 'opposite' to those of PWS, caused by a deletion of a very closely linked (or even the same) region, on the *maternal* chromosome 15 (Magenis *et al.*, 1987).

Cancers

Two cancers that have both heritable and sporadic forms show parental imprinting effects. Wilms' tumor, a kidney tumor, can arise as a result of mutations in several different genes, with correspondingly variable phenotypic manifestations. Sporadic cases of Wilms' tumor often involve mutations in paternal chromosome 11; in familial Wilms' tumor, which is usually due to genes on another chromosome, it is also the father who transmits the initial mutation.

Perhaps also involving the Wilms' tumor region of the DNA is a disease of fetal overgrowth called *Beckwith-Wiedemann syndrome* (*BWS*). BWS involves some phenotypic overlap with Wilms' and includes cancer predisposition. The syndrome is caused by uniparental disomy involving an insulin-like growth factor gene (Henry *et al.*, 1991).

Retinoblastoma (*RB*) is a congenital cancer of the retina caused by mutations at the *RB* locus on the long arm of chromosome 13. About 85% of cases of RB are sporadic, the remaining 15% heritable. The same *RB* mutations also cause dramatically elevated risk of osteosarcomas (bone cancers) in individuals with heritable RB or when they occur somatically in osteoblasts. Although sporadic cases of RB itself arise equally by somatic mutations on either parental chromosome, new familial cases seem preferentially to arise from paternal germ line mutation (see e.g., Dryja *et al.*, 1989; Zhu *et al.*, 1989). However, sporadic osteosarcomas (i.e., cases not in RB families) are caused preferentially by mutations in the *RB* locus on the maternal chromosome (Toguchida *et al.*, 1989). Individuals who carry the *RB* gene but who are fortunate not to experience a retinal tumor, are preferentially the children of affected *RB*-gene bearing mothers rather than fathers (Scheffer *et al.*, 1989). This shows that imprinting at the *RB* locus can have tissue-specific effects.

Huntington's disease

Were the effects of imprinting missed by human geneticists in a century of classical segregation analysis in nuclear families? The answer is unclear. For one thing, dramatic cases of imprinting effects are rare. Also, the pattern is at best statistical, and sometimes silent, at the phenotype level, and can be confirmed only by direct DNA analysis. However, subtle imprinting effects must occur, and it may be that the statistical noise and phenotypic uncertainty that affect many traits include a component that is due to imprinting.

The severity of cases of HD depends in an interesting way on the parental origin of the *HD* gene (Hall, 1990b). The disease-causing mutation(s) is(are) completely dominant; that is, severity is as great in heterozygotes as in homozygotes. The age of onset is quite variable, ranging from less than 5 years to old age (mean, 38), but the onset age as well as the severity depend to some extent on the parent from whom the mutant allele is inherited. Juvenile onset is characteristically associated with paternal inheritance. A small fraction of heterozygous cases inherited from the father are more severe than cases inherited from the mother. However, there are exceptions to these statements, so that simple imprinting is not a complete explanation.

In fact the HD gene has recently been identified and a striking new mechanism discovered. The gene contains repeats of the triplet sequence $(CAD)_n$. The copy number varies rapidly by mutation and the higher it is, beyond about 40 copies, the worse and earlier the disease occurs. As n increases 'anticipation' may occur; the disease becoming progressively worse in successive generations. Other genetic diseases are now known to be affected by a similar mechanism; the repeats may be in the gene as in HD or near it. The fragile-X syndrome is a case of the latter.

Fragile-X syndrome

The fragile-X syndrome is an X-linked cause of mental retardation, common especially in males. It is the most common inherited form of mental retardation, with an incidence as high as 1 per 1500 males in Caucasian populations. The syndrome is associated with cytologically visible breakages in a region of the long arm of the X-chromosome. These breakages can be induced by folate deprivation that impairs the thymidine synthesis that is essential to DNA replication.

The biology of fragile X is not well understood and has seemed puzzlingly complex. Penetrance is incomplete and variable both within and between families. Normal transmitting males (NTMs) have a detectable fragile-X chromosome but are phenotypically unaffected and can transmit the gene to unaffected daughters who in turn bear seriously affected sons and daughters. It appears that for the fragile X to be expressed the gene must first be passed through a female (the mother), potentially implicating the sex-specific X-imprinting apparatus.

Table 14.1. *Diagnostic results of probing fragile-X chromosomes with two probes*

	Length pattern (probe StA22)		Methylation pattern (probe StB12.3)		
Fragile-X phenotype	Normal	Abnormal	Absent	Medium/high	Total
Males					
Normal	59	0	50	0	0
Non-transmitting	9	9	4	0	0
Affected	2	44	1	5	27
Females					
Normal	41	0			
Positive carriers	1	43			
Negative carriers	0	17			
Untested carriers[a]	0	17			
Unknown	21	7			

Notes: Methylation patterns: relative intensity of methylated to unmethylated bands in region as detected by StB12.3 when region is cut by *Ban*1. Length pattern: pattern detected by the probe is abnormal if the 1.15 kb band is absent or if abnormal size fragments were present.
[a]Carriers not tested cytogenetically. See source for details.
Source: Oberle *et al.*, 1991.

There is a CG island in the fragile-X region, such as is typically found near expressed genes. The fragile-X syndrome occurs in individuals with a methylated C in this island, perhaps thereby inactivating some nearby gene (Heitz *et al.*, 1991; Oberle *et al.*, 1991; Vincent *et al.*, 1991; Yu *et al.*, 1991). If this is correct, some other locus, responsible for the methylation enzyme(s), is responsible for fragile X (Laird, 1987); at least, the syndrome does not appear to be due to allelic variation at the X-linked site itself.

Recently, a $(CGG)_n$ segment in this region has been identified that is highly variable in length (Oberle *et al.*, 1991; Yu *et al.*, 1991). In phenotypically normal individuals with the fragile-X karyotype, but not in karyotypically normal individuals, there is an insert of about 150–500 bp in this region. The insert is much longer (i.e., variably 1–4 kb) among phenotypically affected individuals. These results are shown in Table 14.1 and Figure 14.4

The *small* insert may serve as a kind of 'premutation' (Oberle *et al.*, 1991) that causes chromosomal instability. NTMs carry the insert but do not express the trait, nor do their daughters who inherit the insert, the fragile-X karyotype but not the disease syndrome, and the region is

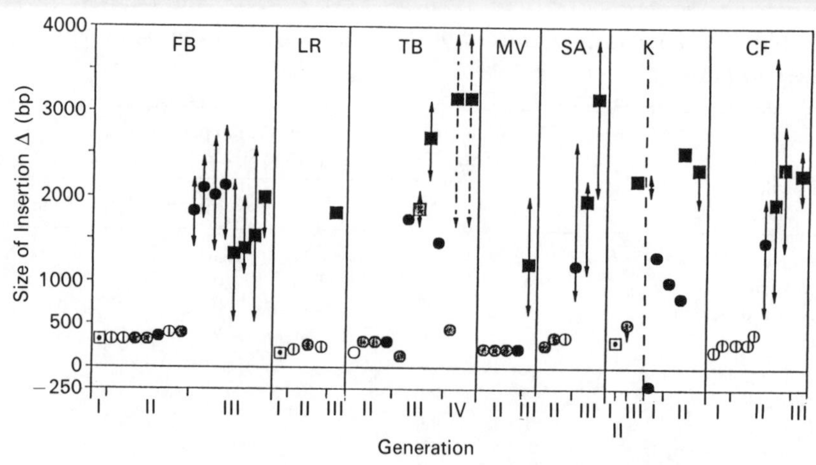

Figure 14.4. Fragment length changes in phenotypically affected fragile-X individuals. Each vertical panel with letters represents a separate family. Within each family, the roman numbers on the horizontal axis indicate generations. The ordinate represents size differences between the normal insert and the abnormal fragments. Each individual is represented by a symbol, defining his/her cytogenetic status. Only carriers of fragile-X mutations are shown. Squares with dots, NTMs; open circles, cytogenetically untested carrier females; divided circles, females without fragile-X detected; gray symbols, <4% fragile-X chromosomes; dark symbols, ≥4% fragile-X. Males with ≥4% fragile-X were retarded but not all females. Arrows approximate the extent of size heterogeneity in an individual when present. Dashed arrows represent faint smear. The first six families have a NTM in generation 1. See source and text for details. (From Oberle *et al.*, 1991, *Science* **252**: 1097-1102. Copyright 1991 by the AAAS.)

generally unmethylated in these carriers. However, in a high fraction of the daughters a much larger (1–4 kb) insert occurs, which becomes methylated, and in the offspring (the NTMs' grandchildren) the disease occurs. It is not clear whether both these steps are needed, and there are exceptions to the inheritance pattern.

Imprinting prospectus: how important?

Current knowledge of the effect of imprinting on non-mendelian inheritance patterns for human traits is largely speculative and based on only a few isolated examples that presently are curiosities as much as important new biological facts. There are suggestions, however, that the phenomenon may have more widespread effects. More than 20 chromosome regions that are affected by imprinting in the mouse have homologous regions in humans known to contain genes that may be relevant to tumors, digestive and circulatory diseases, and growth problems (these are tabled by Hall, 1990b).

Conclusion

Some of the complexities in segregation proportions may be explicable in terms of 'non-traditional' forms of inheritance, and phenotypic variability among individuals with the same constitutive genotype may be due to somatic genetic rather than heritable factors.

Imprinting could also provide a false impression of selection against heterozygotes (Chakraborty, 1989). Suppose, for example, that at a particular diallelic locus a fraction x of all maternal gametes are inactivated and not expressed, but that all paternally inherited alleles are expressed. Assuming random mating, the frequency of *apparent AA* homozygotes will be equal to $p^2 + p(1 - p)x$; the second term represents inactivated *as* in *Aa* individuals that are phenotypically transparent. There will be an apparent deficit of heterozygotes that could wrongly be inferred to reflect selection.

Other strange potentially non-mendelian events have been observed, for example occasional 'showers' of 'rogue' cells, with seriously aberrant karyotypes that can effect up to 1% of all circulating lymphocytes (Awa and Neel, 1986). No known exposure to risk factors are responsible, and the pattern may be due to infectious episodes. There may thus be many 'strange' causes of somatic or even germinal mutational change.

Modifications of the genome are a natural part of the life history, and our statistical models may need to be changed to take them into account. At present the mechanisms are too poorly understood for us to know just how to do that, or how important the phenomena may be.

15 *Cancer and aging: a microcosm of evolution during life*

I wish I had the voice of Homer
To sing of rectal carcinoma,
Which kills a lot more chaps, in fact,
Than were bumped off when Troy was sacked.

J. B. S. Haldane, 'Cancer's a Funny Thing' (from Clark, 1968)

Contrary to the verse by the great population geneticist J. B. S. Haldane, written just after being treated for the cancer that was to take his life, his was among the first generations in which a lot of chaps were bumped off by cancer. Previously most people died from other causes, such as infectious diseases. Although cancers are genetic diseases, selection has prevented most cancers from being directly heritable, but cancer demonstrates the important effects that somatic mutations can have. Cancer reflects evolution in a microcosm, occurring among our own cells during our own lifetime rather than among the individuals in a population over evolutionary time.

Cancer age patterns reflect the time required for somatic mutations to produce their complex cellular phenotypes. This leads naturally to a general consideration of the age patterns of chronic disease and, consequently, of aging in general. These are the subjects of this chapter.

Somatic mutation: the genotype changes with age

An individual begins life as a single cell, but an adult organism is the product of a very large number of subsequent cell divisions that occur as the zygote grows, develops, and renews its tissues. Somatic mutations occur in these somatic cells, and are inherited by their mitotic descendents during the life of the individual. The growing amount of genetic variation, across the genome, that accumulates among the set of *stem cells* (dividing, tissue-renewing cells) in any tissue causes a corresponding change in the distribution of phenotypes among the cells. The rate at which such variation accumulates will depend on the number of cell generations, number of dividing cells, mutation rate, number of loci contributing to the cellular phenotype, and the distribution of allelic effects on, and the intensity of selection against, the resulting cellular phenotypes.

These are all the elements that generate phenotypic diversity in evolution (except for recombination). The mutational changes in any given cell generally are swamped by the normal behavior of the millions of other cells in the same tissue, and do not affect the overall phenotype of the individual. But there are important circumstances in which the phenotype may change. In particular, somatic mutations may occasionally confer a growth advantage that causes a stem cell and its descendants to proliferate without observing the normal tissue-context growth-inhibition signals. This amplifies the effects of mutations on a single cell and makes them detectable, indeed, life threatening. In addition to cancers, diseases that can result from somatic mutations include autoimmune disorders involving the proliferation of lymphocytes with anti-self Abs and atherosclerotic plaques that form around proliferations of mutant endothelial cells on the interior arterial walls, perhaps triggering clotting reactions that lead to myocardial infarction (heart attacks).

Cancer: somatic mutation, somatic evolution

A cancer is a clonal growth of cells that fail to obey the normal rules for *contact inhibition*, i.e., context-specific tissue structure. The cells are said to be *transformed*. Their normal genetic program for the particular type of cell has been changed by the inactivation of necessary genes, by the 'ectopic' expression of such gene(s) (expression at incorrect times or places), and/or by functional mutations in the coded protein. As the mutant clone grows, it competes rather than cooperates with the surrounding normal cells. Additional mutations produce competing subclones whose evolving characteristics can be very dangerous to the host. One of these is *metastasis*, the ability of cells to slough off into the bloodstream to colonize distant sites, making cancers life-threatening diseases.

Somatic mutations are rare, but there are millions of cells at risk. The probability that at least one cell will be transformed depends on the mutation rate, cell number and division rate, and number of changes that must occur. Environmental risk factors that affect these are risk factors for cancer. Exogenous mutagenic substances such as chemicals and ionizing radiation are almost always carcinogens, as are endogenous mutagens such as metabolic intermediates and by-products. Mutagens are known as cancer *initiators*, and produce dose–dependent relationships between exposure and risk. Environmental factors that stimulate cell differentiation and tissue growth are known as cancer *promoters*, because dividing cells are vulnerable to transforming mutations. Promotors include tissue irritants and growth hormones.

Cancer risk factors affect only the specific cells exposed to them, and the disease usually arises in naturally mitotic (dividing) cells, at ages when the tissue is more rapidly dividing. Thus, cancer age- and sex-patterns reflect the natural life-history stages of the tissues involved. For example, there is a peak in risk of bone cancers in growing bone plates (e.g., at the ends of the long bones) during the adolescent growth spurt. The risk of cancer in some female reproductive tissues (e.g., ovary, endometrium) stabilizes at around the age of menopause, at which time cellular proliferation, such as that associated with menstruation, lactation, and the like, slows. Since *terminally differentiated* (non-mitotic) cells seem rarely able to be mutated to grow, neuronal and muscle-cell cancers are rare in adults.

The genes involved in cancer

The genes involved in the machinery of cell division, if ectopically or aberrantly expressed, can contribute to uncontrolled cell growth and cancer {Gelehrter and Collins, 1990; Lewin, 1990}. Figure 15.1 provides a schematic map of some of these pathways. The cellular processes involved in carcinogenesis are similar among tissue types, and the genes involved are members of a number of multigene families used in cell division {Bishop, 1987, 1991}. These include hormone and growth factor genes and their receptors {Aaronson, 1991}, genes for intracellular message transduction (Hanks *et al.*, 1988), and nuclear DNA-binding proteins such as Ras {Astrin and Costanzi, 1990; Herrlich and Ponta, 1989; Klein, 1988}.

Oncogenes (transforming genes) are genes whose product is involved in cell proliferation; such genes can act in a cellularly *dominant* way in that if their product is present the effect occurs. Expression out of normal context and/or of mutant versions of oncogenes can contribute to transformation. These genes were originally found in cancer-causing viruses, but the same genes exist in the *host*'s genome, where they are expressed under host control during appropriate times. The viruses had apparently picked up the genes through their own genetic life cycle in infected hosts. The endogenous versions of these genes are called *proto-oncogenes*; some of the 20 or more that are known are shown in Figure 15.1. An

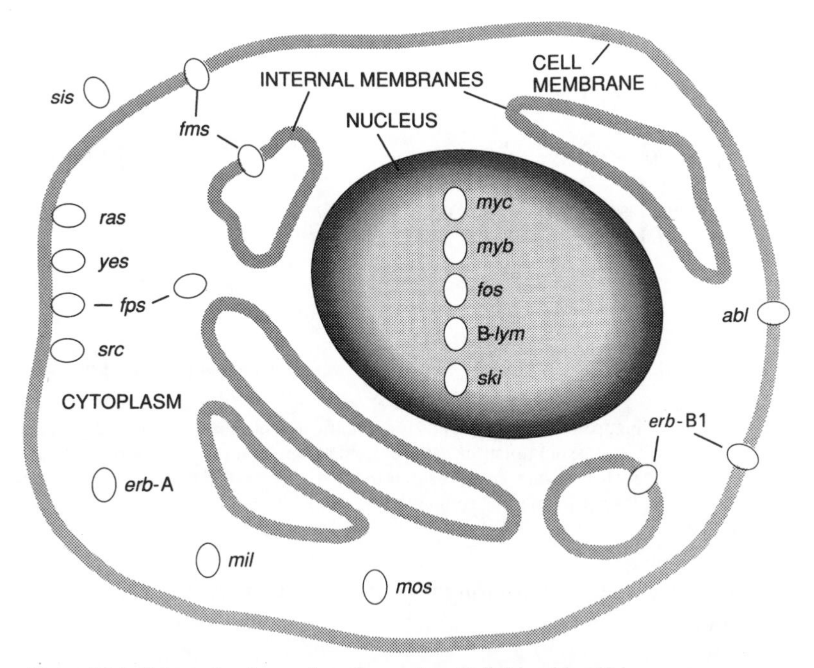

Figure 15.1. Schematic of generic pathways in cell division identifying genes commonly mutated or aberrently expressed in cancer. For example, *sis* encodes a secreted growth factor; *erb*B1 encodes a membrane-bound growth factor receptor; *ras* gene product transmits signals from receptors to the nucleus; *fos* and *myc* gene products are in the nucleus and are related to DNA binding and gene expression. (From Gelehrter and Collins, 1990.)

example is the *ras* gene family, whose members are located on several different chromosomes and are often mutated in human cancers.

A second class of cancer genes normally acts to suppress cell growth, for example by binding to growth factors. At the level of the cell, mutations in these genes act *recessively* to transform cells; the effect occurs only if both copies of the gene are dysfunctional. These are called *tumor suppressors* {Hollstein *et al.*, 1991; Scrable *et al.*, 1990; Weinberg, 1991}. Examples are the *p53* and *RB* gene products.

It is unlikely that both copies of a given gene will undergo inactivating somatic mutation in any given cell, but if there are enough cells and enough time this can happen to at least one cell. Point mutation, aberrations of normal mitosis, perhaps even imprinting can be responsible, as shown in Figure 15.2. Mitotic recombination (Chapter 14) seems to make this more likely, and a heterozygote for normal and dysfunctional suppressor genes can produce a daughter cell that is a dysfunctional

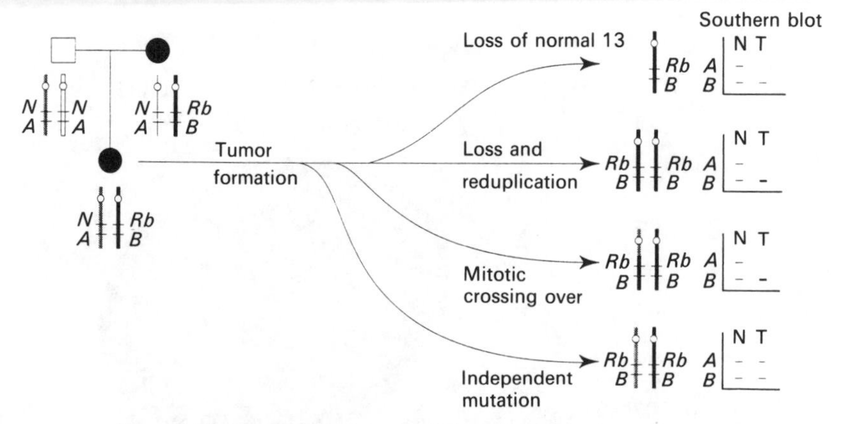

Figure 15.2. Mechanisms for somatic loss of recessive locus (e.g., tumor suppressor) function. *Rb*, retinoblastoma mutation; *N*, normal allele; *A* and *B*, alleles at a marker locus that make detection of *RB* possible. *N*, normal; *T*, transformed. (From Gelehrter and Collins, 1990.)

homozygote, i.e., no longer protected by at least one good allele at the locus.

Inherited major genes for cancer

Some loci seem to act as 'major genes' in regard to cancer in that mutations in those genes have major transforming effects. Their action may suffice to transform in experimental cell systems, but seems not to be sufficient in human cancers; that is, cancer is a *multistage process*, and several events are usually required before a normal cell is transformed. Dominantly acting oncogene mutations may be incompatible with embryogenesis, and such mutations seem rarely to be heritable. However, mutations at suppressor loci *are* heritable, because the fetus can develop normally if it inherits at least one normal allele at the locus. Nonetheless, such persons undergo millions of mitoses during life and if even one of these involves an inactivating mutation in the remaining allele, that cell will have a 'recessive' genotype, and the individual can develop cancer. Since the probability that this will occur in at least one cell can be quite high, cellularly recessive suppressor-locus mutations have high penetrance at the level of the individual and segregate as *dominant* alleles in families.

Cancer suppressors were first discovered in RB (Knudson, 1971). Sporadic RB occurs only when both copies of the *RB* locus on chromosome 13 have been inactivated by mutation. The total number of retinal cells is only a few million, retinal cells are neurons that do not divide after the retina is formed, and the probability of double mutation during the brief period of fetal development is small. Sporadic RB is correspondingly rare, and usually involves only a single tumor in one eye. By contrast, fetuses who inherit one mutant *RB* allele need await only a second event, which occurs with sufficient probability that most manifest two or more independent tumors, involving both eyes.

RB seems to be an exception that requires homozygosity only at a single suppressor locus. In other tumors, mutations at suppressor loci act in a manner contributory to a more complex carcinogenic process. With age, the cells in a given tissue accumulate mutations in several relevant loci and need await only some transforming combination of subsequent mutations before a tumor arises. For example mutations at a locus on chromosome 5 are associated with FAP (Chapter 12), a trait that appears first to involve the growth of numerous adenomatous (gland-like) polyps, on the inner lining of the large bowel, that are precursors to colorectal cancers. Benign cysts of the breast, certain types of moles and warts (papillomas), and numerous other conditions seem to be cancer precursors that may reflect partial genetic transformation, and are important to identify.

The QTL map of cancer

Suppressor loci were detected, in part, by the discovery of somatic *loss of variation* (*LOV*) mutations, in which tumor cells contained only one of two alleles present at the locus in the constitutive (inherited) genotype. Mitotic recombination is one mechanism that can produce LOV changes, as was found first in association with the *RB* and other pediatric tumor loci. However, by using a battery of highly polymorphic VNTR or RFLP markers, LOV mutations are found in most if not all tumor types that have been tested.

Somatic LOV is found at some specific loci for a variety of tumors. For example, LOV at the *RB* locus occurs in some breast, bone, and lung cancers. A locus called *p53*, on chromosome 17, codes for a suppressor protein and experiences mutation in a high fraction of a wide variety of tumors including connective tissue (Mulligan *et al.*, 1990), esophagus (Hollstein *et al.*, 1990), colorectum (Baker *et al.*, 1989), leukemia (Ahuja *et al.*, 1989), brain, breast, lung (Nigro *et al.*, 1989), and bladder (Sidransky *et al.*, 1991), as well as in families with a syndrome of different tumors in different family members (Srivastava *et al.*, 1990).

If so many tumors involve *p53* mutations, would it not be expected that a variety of other somatic mutations are also occurring as we age? Indeed, once the genome was searched systematically this is exactly what was found, including many different LOVs along with other types of mutation (see e.g., Hollstein *et al.*, 1991).

Figure 15.3 shows the distribution of LOV mutations among individuals with colorectal cancer (Vogelstein *et al.*, 1989). Similar genome-wide patterns of mutation have been found for cancers of the kidney (Morita *et al.*, 1991), bladder (Perucca *et al.*, 1990), breast (see e.g., Devilee *et al.*, 1989), lung (Weston *et al.*, 1989), glial (dividing) cells in brain tumors (Bigner *et al.*, 1988), melanoma (Fountain *et al.*, 1990), and others, and this seems to be generic among cancers.

Each tumor type seems to involve a small number of specific loci that are mutated in many cases. For example, colorectal cancers frequently involve mutations in *p53* on chromosome 17p, *FAP* and a nearby locus region on 5, and a locus on 18q. Breast cancers show locus *p53* changes, but also frequently involve other mutations on 17p, 17q, 18 and a number of other genes or gene combinations (see e.g., Cropp *et al.*, 1990; Davidoff *et al.*, 1991; Hall *et al.*, 1990a,b). Changes on 17p may be heritable in the form of susceptibility to early breast cancer (Hall *et al.*, 1990b); whether there is a connection with the possible PBD-predisposing locus is unknown, but initial tests did not find linkage between the latter and the putative chromosome 17 region (Skolnick *et al.*, 1990).

These tumor-characteristic loci are involved in a high fraction of cases of a given tumor type, and in that sense act as major genes. However, such mutations are not always found, and numerous other mutations that vary among cases of the same tumor type also occur, suggesting that transformation requires additional events. These may include chromosomal rearrangements and mutations in oncogenes such as *ras*.

For a given tumor in a given individual, the set of somatic mutations that have arisen are known as the tumor *allelotype*; since it is likely that not all mutations that have occurred are detectable by current methods, the allelotype is only a partial representation of the full tumor genotype. Currently, we know little about heritable cancer polygenotypes (e.g., Easton and Peto, 1990), but we do know that the allelotype varies among individual tumors even of the same organ {Scrable *et al.*, 1990; Solomon, 1990; Weiss, 1990b, 1990d, 1992}.

Genotype to phenotype in cancer

Although cancers have traditionally been thought of as qualitative (presence/absence) traits, their genetic complexity has all the characteristics that typify complex traits controlled by multiple loci. As with other traditionally classified qualitative diseases such as PKU and the hemoglobinopathies, when we look more closely what we see is a distribution of

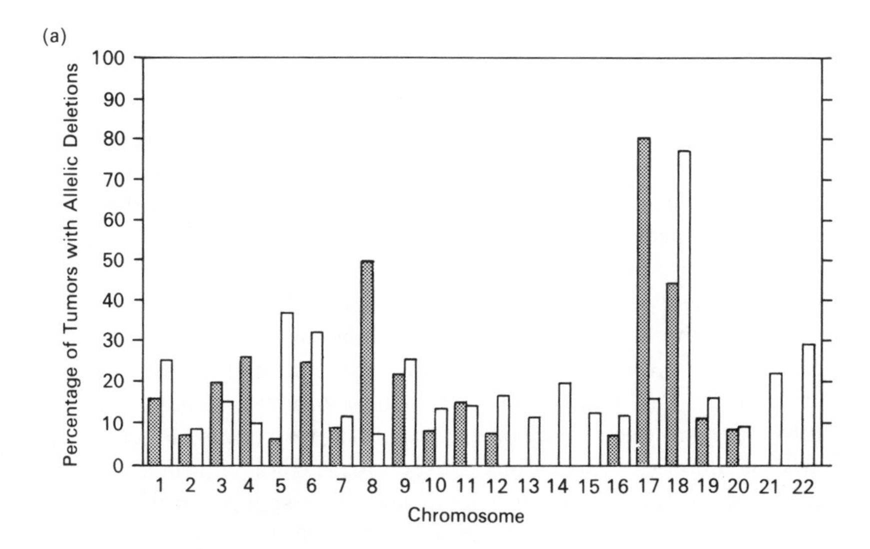

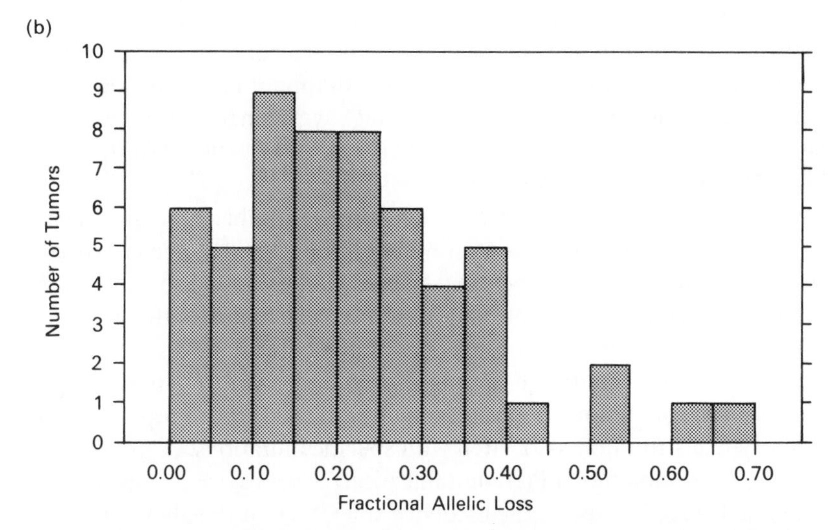

Figure 15.3. Colorectal cancer allelotypes. (a) Each chromosome arm (hatched = short (p) arm; open = long (q) arm), showing percentage of tumors that had lost RFLP heterozygosity between tumor and normal adjacent colonic mucosa. (b) The relative frequency with which tumors had lost heterozygosity at specific fractions of informative sites. Fractional allelic loss is the fraction of chromosome arms on which at least one RFLP heterozygosity was lost. (From Vogelstein *et al.*, 1989, *Science* **244**: 207–211. Copyright 1989 by the AAAS.)

phenotypes. Cancers of the same organ differ among patients in their cell type, ability to metastasize, response to therapy, growth rate, and so on. Different tumour subtypes may be associated with specific environmental risk factors, onset age, or sex. Tumors continue to evolve new phenotypic characteristics after they are detected and/or in response to therapy.

As might be expected, some associations between tumor allelotype and tumor phenotype have been documented. For example, colorectal tumor phenotypes can be staged; class 1 adenomas are small tubular adenomas with low grade dysplasia (tissue structure anomalies); class 2 adenomas are more proliferative growths that still do not show transformation to carcinoma; class 3 adenomas are growths with some areas of invasiveness (penetrate the basement membrane of the colorectal epithelium), and finally there are adenocarcinomas, true invasive cancers. Figure 15.4 shows the empirical associations between these colorectal tumor stages and mutations at *ras* proto-oncogenes and three suppressor genes, the *FAP* gene on chromosome 5, another gene commonly mutated in colorectal cancer on 18, and the *p53* gene on 17 (Vogelstein *et al.*, 1988, 1989). In addition to specific locus associations, there is a quantitative phenotype association with the *number* of mutations detected. The allelotypes of kidney and other cancers that have been tested appear to be asssociated with severity in similar ways (Ahuja *et al.*, 1989; Morita *et al.*, 1991; Nigro *et al.*, 1989; Perucca *et al.*, 1990).

Although these results are heterogeneous, variable, and statistical, there seems to be no fixed order or number of events that needs to occur in a given tumor. A major caveat that must be added is that so far the data compare the genotype of thousands of tumor cells to that of thousands of adjacent normal cells; we do not yet have good data on the distribution of the same types of mutation among *normal* cells as a function of age. Ascertaining via tumor, therefore, may be a biased way to represent the phenotype distribution associated with specific mutations. A proper test would require cloning (or PCR amplification) of genes from many pairs of individual cells. However, the evidence suggests that the allelotypes are in fact largely causal; for example, the colorectal cancer allelotype distribution does not vary with age (Vogelstein *et al.*, 1989), and some of the tumor-associated mutations have been shown experimentally to be involved in cell function related to cancer.

How much inherited variation is there in cancer genes?

There are only a few tumors that seem to be truly familial, in the usual sense that alleles at a major gene that dramatically elevates risk are inherited. These include RB, Wilms' tumor of the kidney, colon cancer associated with FAP, several endocrine tumor syndromes, and a scattering of others that usually constitute only a tiny fraction of all tumors of a given organ. Almost all of these instances involve recessive suppressor loci.

Current opinion is thus that there is little in the way of less-dramatic heritable cancer susceptibility. In many instances, the loci for which somatic changes have been detected differ from those responsible for clear-cut heritable tumor susceptibility (see e.g., Hall *et al.*, 1990a). However, many factors conspire to make weaker cancer familiality difficult to detect. These include late onset age, competing causes of death, small sibship sizes, and the fact that effective diagnosis has only recently been widely available (Weiss, 1985). What we know about complex traits suggests that a spectrum of susceptibility variation must exist, even if truncated by selection against genotypes that cannot survive embryogenesis.

This speculation is consistent with the complex, and variable, allelotypes, with the fact that early onset tumors are more likely to be familial, and that there is an overall slight statistical excess risk among family members, both of same-site and sometimes all-site cancer risk (Skolnick *et al.*, 1981; Weiss *et al.*, 1986). There are also suggestions that a high fraction of colorectal (Cannon-Albright *et al.*, 1988) and breast (Skolnick *et al.*, 1990) cancers develop from heritable precursors. Heritable *p53* mutations appear to be associated with certain familial cancer syndromes, in which different family members experience cancer of different types, an example of which is the Li-Fraumeni syndrome (Malkin *et al.*, 1990; Srivastava *et al.*, 1990). Indeed, recent reports suggest that very subtle increases in cancer risks are due to heritable mutations at this locus (Malkin *et al.*, 1992; Toguchida *et al.*, 1992). Susceptible genotypes at the *AT* locus, that are common in the population, were described in Chapter 12.

When we have become able to screen the genome for *inherited* cancer-related 'allelotypes', we may discover a distribution of susceptibility that is associated with age of onset, cell type, and other tumor characteristics. Clear cases such as FAP are probably the extreme tails of a susceptibility distribution.

I think the picture just given is consistent with the way evolution produces traits, as I have tried to describe in this book. Cancer can be viewed as a quasi-quantitative trait that is affected by a few major loci and many contributing polygenic loci, with high levels of allelic equivalence and many genetic pathways to disease. Nonetheless, it is only fair to acknowledge that a number of investigators think that cancer is much

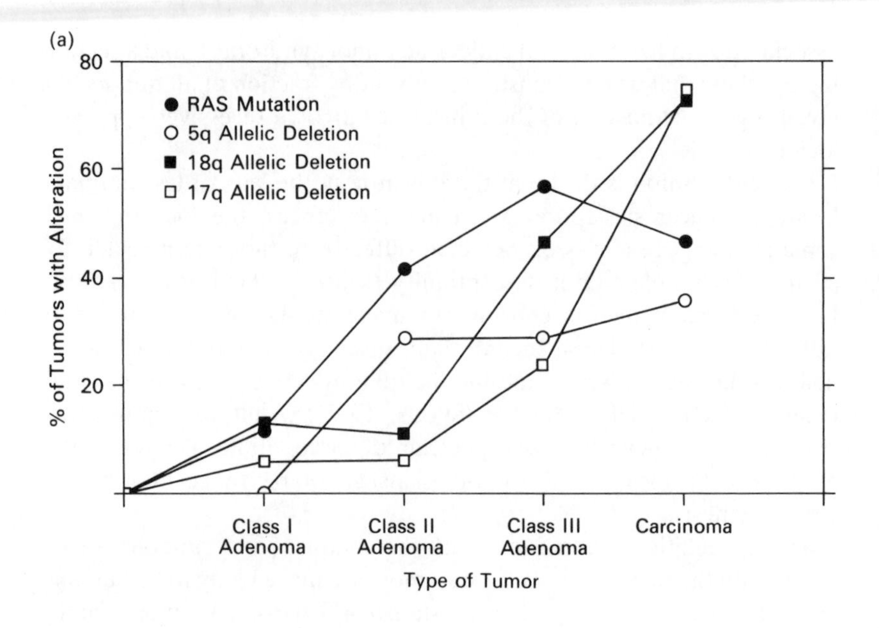

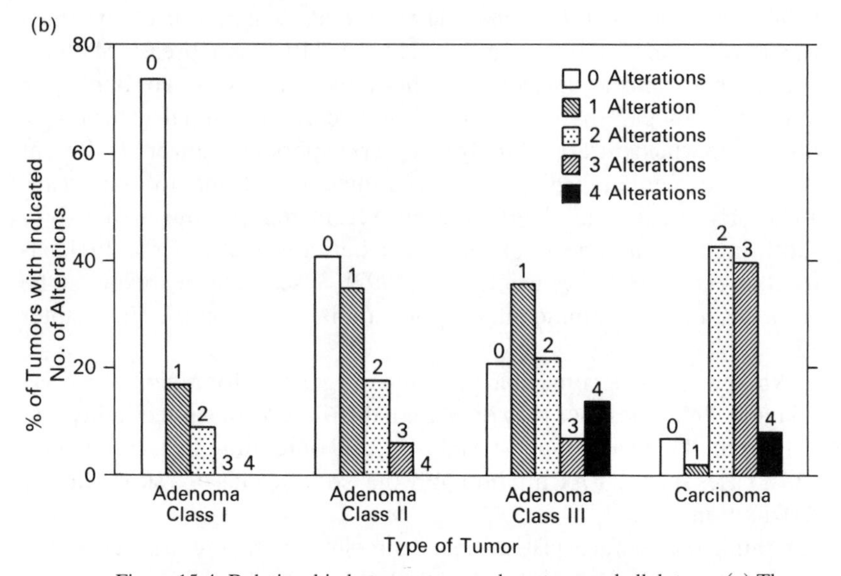

Figure 15.4. Relationship between tumor phenotype and allelotype. (a) The relative fraction of tumors with a selection of given somatic mutations (relative to adjacent normal tissue) and tumor cellular phenotype. (b) The relative number of these alterations and phenotype. (c) The phenotypic stages associated with specific mutations. (From Vogelstein *et al.*, 1988. Reprinted by permission of the *New England Journal of Medicine* **319**: 525–532 (1988).)

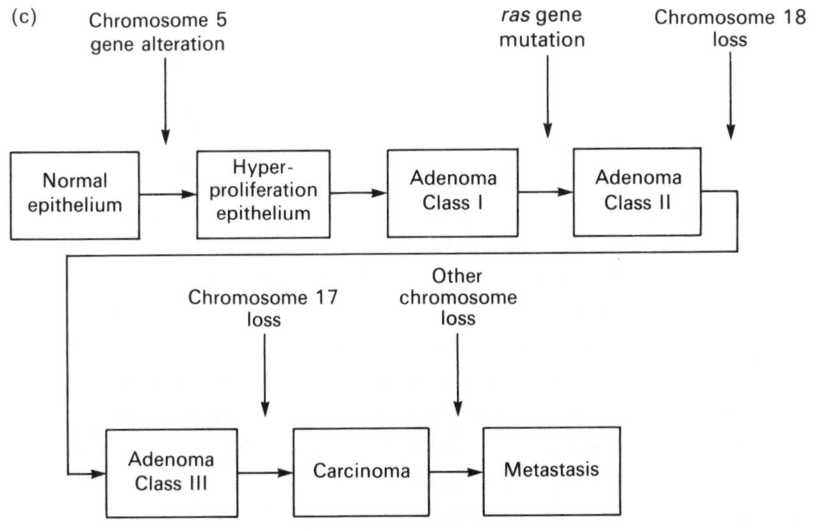

Figure 15.4. *Continued.*

more like a qualitative trait, caused by only a few important transforming events. In particular the consistency of cancer hazard patterns with two- or six-hit multistage hazard models has been interpreted in this way (e.g., see Moolgavkar, 1992).

Evolution in vivo

Cancer may provide an opportunity to study many of the aspects of the evolution of complex phenotypes. For example, the colorectum has about 10^7 stem cells, undergoing cell division approximately 100 times each year. Roughly speaking, in a given lifetime of 75 years, an individual will generate 7.5×10^{10} or more stem cell divisions. This is comparable in scale to the number of individuals in the history of the human species (Weiss, 1984). Except for recombination, the evolution of the distribution of stem cell genotypes may thus approximate the long-term evolution of an entire species. Studies of somatic mutation will be relevant to a number of other traits as well, such as hair color patterns, hair graying, age-spotting, and other aspects of senescence.

The calibration of the human lifespan and the age-pattern of disease

Many diseases strike with increasing probability with age. These are known as *chronic* (degenerative, or senescent) diseases, and constitute the overwhelming bulk of causes of death in industrial societies (and much of the topic of this book). It seems natural to expect that useful information about the pathogenic (and genetic) processes that are involved in such diseases is contained in the age of onset patterns.

Indeed, the hazard functions for a wide variety of chronic diseases are mathematically similar; that is, the hazards have similar *shape* when plotted against age. In addition to this, there are clear associations between overall aging rates and natural lifespans among various biological characteristics of mammalian species and their typical lifespans, as shown schematically in Figure 15.5. These facts, reviewed encyclopedically by Finch {1990}, have led many students of the biology of aging to suggest that senescent causes of death are calibrated by some common genetic mechanism rather than each tissue degenerating in its own way.

This raises the basic question of the evolution of genetic mechanisms to calibrate the length of the human lifespan itself. What could this process have been, and how do we find it? If the chronic diseases that affect different tissues involve different genes with tissue-specific expression, as

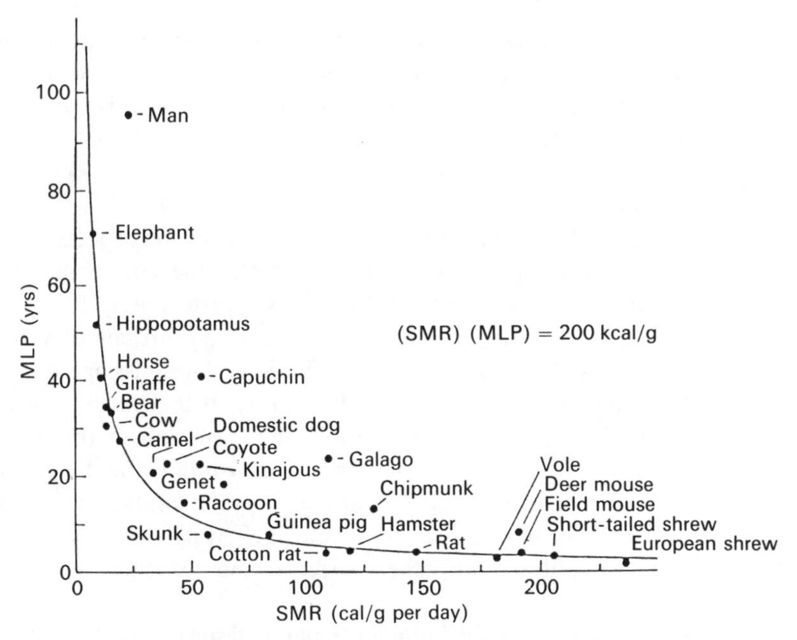

Figure 15.5. Relationship between estimated maximum lifespan potential (MLP) for members of various species and the standardized metabolic rate (SMR), showing the idea that the lifespan is programmed in some simple way. (From Cutler, 1978.) 1 cal = 4.184J.

we have seen throughout this book, how can some other factor control the rate at which this disparate pattern of damage occurs?

Cancer genotypes as a model for aging?

Because carcinogenesis is a relatively pure stochastic process, cancers have often been taken to be the archetype of senescent disease and as models of the genetics and biology of aging itself. For example, there are strong similarities in the shape of hazard functions for different adult cancers, for different tissues, even though the absolute values of the hazards differ by orders of magnitude {Chakraborty and Weiss, 1989; Cook *et al.*, 1969; Moolgavkar, 1992; Peto, 1978; Weiss, 1990b,d, 1992; Weiss and Chakraborty, 1984}.

Given what we know about evolution by gene duplication and the relationship between the structure of the genome and its control of complex phenotypes (e.g., Chapter 2), it is reasonable to speculate that cancers are in fact *homologous* diseases. That is, the mechanisms for the production of terminally differentiated cells from precursor stem cells may have arisen once and been re-used in the evolution of diverse tissue (e.g., Hall and Watt, 1989). It is no surprise that tumors are associated with mutation in genes that are members of multigene families, and that the same or homologous genes are involved in tumors of different organs.

If different tissues have similar mechanisms of differentiation, there will be similar ways to disrupt those processes, which could explain why cancers have similarly shaped hazard functions (the absolute differences would be due to the number of cells in a given tissue, their turnover rates and level of exposure to mutations, and so on). Indeed, a variety of stochastic multistage epidemiological models have been shown to fit closely to cancer hazard functions.

However, the inference that similar shape implies similar etiology can be misleading. Many processes can lead to similar hazards. We can see this first by considering the case of cancers. Think of the bed of stem cells of a given organ. These cells begin life with the same (inherited) genotype. Referring back to the evolution of quantitative phenotypes as discussed in Chapter 9, we can assume a Laplace distribution of allelic effects of somatic mutations as they accumulate with age, leading to a roughly normal distribution of phenotypes among the stem cells, on some cancer-relevant scale related to their growth behavior.

As shown schematically in Figure 15.6, the phenotype distribution of these cells will spread (develop increasing variance) with age, as mutations introduce genotypic, and hence phenotypic variance. There may, of course, be some small amount of truncation selection removing cells

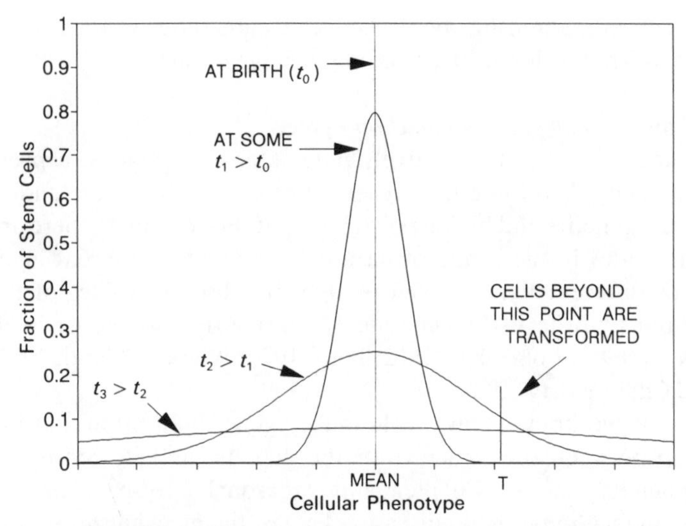

Figure 15.6. Schematic of mutational process by which a distribution of genotypic effects among stem cells in a tissue is generated with age, and relationship between that distribution and cancer risk. t, time; T, threshold.

that develop lethal mutations. If the prevalence of cellular phenotype ϕ at age t is $\Pr(\phi|t)$, and if we can assume, approximately, that cancer arises in a cell whose phenotype exceeds some threshold, i.e., $\phi \geq T$, then the probability that a given *cell* will be transformed at age t, measured in cell-cycle generations, $r(t)$, is

$$r(t) = \int_T^\infty \Pr(\phi|t)\mathrm{d}\phi, \qquad [15.1]$$

where the integral is over all 'transformed' phenotypes. This is like the liability threshold models discussed briefly in Chapter 6 for quantitative phenotypes. The probability that an *individual* experiences cancer at age t is proportional to the probability that at least one of his/her N cells is transformed at that age:

$$h_t \propto 1 - [1 - r(t)]^N. \qquad [15.2]$$

(Since t here is in cell generations, a discrete measure for time, I use slightly modified hazard-like notation). The absolute value of $r(t)$ is very small, and the terms in [15.2] that involve powers of $r(t)$ higher than 1 will be much smaller than $r(t)$ itself. Thus, we can approximate the age-specific hazard from [15.2] by

$$h_t \propto 1 - Nr(t). \qquad [15.3]$$

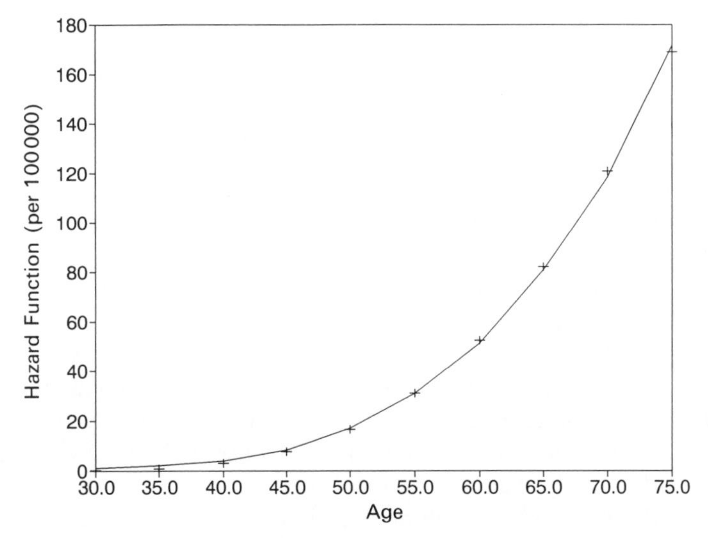

Figure 15.7. Fit of quantitative somatic mutation model (line) to US 1970 colorectal cancer hazard function (points).

Essentially, the hazard is approximately a linear function of the size of the pathological tail of the distribution of genotypic effects among the stem cells represented in the figure by the area to the right of T at each time, t_0, t_1, etc.). A simple first-order simulation of this mathematically complicated process shows that the increasing tail area with age can represent the shape of observed cancer hazard functions reasonably closely, including the similar shape but absolute difference of the hazards for different tumors (Weiss, 1992; and unpublished results). A close-fitting example, for colorectal cancer, is given in Figure 15.7.

This may seem to confirm the idea that cancers are homologous diseases or are calibrated by some common mechanism, but in fact the model just sketched out is rather general. I assumed an accumulating distribution of allelic effects of somatic mutations, but many processes can generate an increasingly dispersed susceptibility with age that is approximately normally distributed. For example, the distribution of person-years of exposure to levels of serum cholesterol, or of blood pressure, may have roughly such shapes. If the risk of disease is proportional to the tail of the distribution of such exposures, the hazard will have a shape that resembles – for purely generic reasons – the hazards associated with cancers.

Many degenerative processes can be characterized in this way. Genotype-specific *differences* in the hazard among individuals may suggest that those genotypes are related to the disease, but I think the information contained in the *shape* of the hazard may be rather generic and devoid of information about disease-specific pathogenic processes. The comparable shape of hazards for diverse diseases probably does *not* indicate the presence of common underlying aging mechanisms. Of course, some aspects of the hazard are informative, for example, changes in shape associated with stages of growth or maturation, or after exposure to risk factors.

I do not think that the lifespan is programmed by a specific active mechanism, except by general metabolic characteristics that may passively affect the relative speed with which various otherwise independent processes occur. Different causes of death are essentially dependent on the environmental and genetic factors that affect them specifically. This is in line with what we know about evolution (e.g., Rose, 1991; Weiss, 1989; S. Weeks *et al.*, unpublished results).

Aging, by this view, is the result of many independent processes but is not a unitary phenomenon. It seems to me doubtful that a highly tuned senescence process could evolve to determine the human lifespan in any unitary way. The causal spectrum for every trait we have looked at is complex and subtle, at least at its 'edges'. The more diffusely defined a trait, the more complex and variable that spectrum seems to be. Aging is a very diffuse kind of 'phenotype'. The overall human mortality hazard function is a complex composite of a distribution of the individual susceptibility, or 'frailty', of members of the population (Manton and Stallard, 1988; Vaupel *et al.*, 1979). Each member of the population has a distribution of relative susceptibilites to different causes of death due to his/her genotype at the loci relevant to those causes, and due to different environmental exposures. The job of genetics is to elucidate the nature of the mix (Weiss, 1990c).

Conclusion

He says he feels like I'm his own grandma.
Time's turnt me nearbout brown enough to be.
Look at these speckeldy marble-cake hands.
I could be a mixed marriage all in myself.

A. Gurganus 'Oldest Living Confederate Widow Tells All' (1984)

Carcinogenesis reflects processes very similar to those that generate genetic variation among individuals over time. The evolving tissue mosaic created by somatic mutation may simulate many evolutionary mechanisms in microcosm. Our methods are only now able to begin to investigate whether somatic mutation may play a greater role in chronic diseases (in addition to cancer) than was previously thought. Accumulating damage may stochastically erase imprinting information in an increasing fraction of cells in a given tissue, may reduce the number of cells with effective message receptors on their surface, and so on, and thus in many ways the proportion of functioning cells in any tissue can diminish with age.

An analysis of cancer at the cellular level shows that the disease encompasses a distribution of phenotypes, and suggests that there may be much more variation in inherited cancer susceptibility than has been thought, even if much of that variation is subtle and only modestly affects the age-specific hazard.

Afterwords: towards a unified general model

Perhaps some day, an encyclopedic biochemist will accomplish the feat of including in a single map all the metabolic reactions in the biosphere . . . I will venture a prediction: it will indeed be a single map on which any point can be reached from any other point.

De Duve (1991)

I chose Vesalius' classic figure of the human circulation for the Frontispiece of this book, because although it itself is not about genetics it is an appropriate metaphor for the genetic architecture that is typical of so many traits. The circulation is a redundant, anastamotic system that provides many vascular pathways to deliver blood to the same tissue, in a hierarchically structured way that has a few major arteries grading into many smaller arterioles and capillaries.

The spectrum of causation and the gradation of phenotypes

Biologists, perhaps like all people, seem to be divided into those who think qualitatively and those who think quantitatively, but the distinction is one of degrees rather than kind. In many instances phenotypes can be divided into categories, such as the presence or absence of a disease, for which associations with different genotypes can be documented generally by case-control studies or more precisely from mendelian patterns in families. For practical purposes we can say that some genotypes 'cause' a particular phenotype.

For traits that are measured on a continuous quantitative scale, we typically think of genetic causation as involving a *set* of loci. The distribution of allelic effects at these loci appears to be roughly 'exponential' in shape: a few alleles have strong effects, while many other have individually small effects that we may be able to treat satisfactorily as an aggregate.

However, in either case, genotype–phenotype associations in families are essentially statistical in nature, and when we look closely the distinction between qualitative and quantitative traits becomes less clear. Either PKU is present or it is absent . . . or is it? At both the clinical level and the level of allelic effects there is a gradation from mild to severe, variation that is essentially continuous and can be related to levels of the PAH enzyme activity.

There is a similar story to be told for essentially every disease that has been studied in enough detail to provide the evidence. We have seen that variation at even a single locus can produce a quantitative pattern of variation, mimicking polygenic inheritance, an interesting twist on Edwards' (1969) comment that multifactorial diseases can 'simulate mendelism'! In fact, the apparent complexity of a trait partly depends on how precisely we define it. We have difficulty with vague terms such as mental retardation or other neuropsychiatric traits (Kidd, 1992), or hypertension. More precisely defined enzymatic traits such as PKU, TSD, or Na–Li CNT are more tractable, because they are dominated by a single locus. Nonetheless, even these traits have a 'full' spectrum of genotypic effects at these loci.

Evolution and diversity

We can explain the genetic architecture of biological traits in terms of a few basic principles of evolution. There are many ways to interrupt or modify the physiological pathways that affect biological systems. Mutations arise in all of these, are usually unique in the population in which they arise, often have similar effect on darwinian fitness, and generate a high level of variation 'noise' that is tolerated by natural selection. There is a steady flow of variation through every population, new mutations generating new variation, much of which is quickly lost to genetic drift.

This leads to the Rusty Rule, that whatever can go wrong will go wrong. Chronic diseases such as cancer and heart attacks usually strike late in life, when natural selection acts weakly if at all, and a particularly large amount of variation may affect the risk of such traits. Despite our personal interest in such diseases, there is probably little connection between their severity and the selective intensity against them, except to maintain a mutation–selection balance for alleles that cause very early cases (or in some way affect fertility). For the majority of alleles, which evolve nearly neutrally, population genetics theory provides a powerful means to predict many aspects of the causal spectrum, such as the distributions of allele frequencies and of allelic effects.

However, all statements in science (except this one itself), have exceptions. We typically expect a trait to be controlled by a small number of rare alleles with large effect, and many alleles with small effect. But we know of circumstances in which some genotypes seem to have strong effects and also to be common. This can be artifactual, as when we pool mutations together that cause similar phenotypes, as we saw with the thalassemias and LDL receptor mutations. Because there are many mutational ways to inactivate a gene, the aggregate of such severe effects is much greater than is predicted by an exponential distribution, but in fact at the DNA level the mutations are individually rare.

There are of course true examples of multiple alleles with large effect at high frequency. Usually, strong natural selection is involved. The division of the population into males and females is the most dramatic example. A specific globin mutation can be preserved by selection due to malaria, and can spread to many different populations. Selection can specifically favor genetic diversity per se, as in the immune system.

Even in these situations the Rusty Rule and weak phenotypic selection imply that for any trait there will be substantial genetic variability – and etiological heterogeneity – both within and between populations. What are we to do? How can we *find* this large array of alleles? Clearly, as mass genotyping technology becomes available, along with a more complete polymorphic map, we will be able to document an increasing fraction of the alleles that have consistently strong effects. The landscape, however, is subtle and complex. The inevitable rush of enthusiasm to screen samples, families, or populations for causal alleles for every type of trait will produce many irreproducible results and excessive claims. I think we will be forced to accept that we cannot understand a trait well by enumerating all of its individual 'causes', which will be quixotically ephemeral and environmentally plastic. Instead, we need to identify deeper structures that can reduce the dimensionality of variation and explain it in a simpler way.

Evolution and structure

Classical population genetics provided a set of rules or evolutionary laws for the distribution of genetic variation. These rules are at the core of our genetic models for the relationship between genotype and phenotype, for example those used in segregation and linkage analysis. But many aspects of genetic variation were not predicted by this theory. For example, non-traditional inheritance such as genomic imprinting, somatic mutation, and somatic mosaicism are not a part of the old order, nor is the use of DNA sequences for regulatory purposes.

The old laws, which were based on the idea of recurrent mutations among a limited set of possible alleles, did not deal explicitly with the kind of dependence we see among alleles – at the level of their sequence *and at the level of their effects* – that occurs because they have a cladistic (tree-like) structure due to the uniqueness of each mutation. The classic picture of genetic architecture was of a more homogeneous structure, which could be generated and molded in parallel in different populations, at both genotypic and phenotypic levels. We now know, however, that a biological trait is more like a Bonsai tree (see Chapter 9), with a full branching structure at the DNA sequence level but constrained phenotypic variation, than it is like the normal distribution of infinitessimal homogeneous components of classical polygenic models.

Gene duplication is another fundamental but unpredicted aspect of evolution, whose importance we have seen in the examples of lipids, collagen, globins, immunoglobulins, and oncogenes. Related genes have related function and related sequence structure. Gene duplication provides a mechanism for the evolution of new function and of dominance, both of which are old problems in biology. The evolution of dominance of existing variants, using redundancy and complexity to buffer them against the effects of new mutations, reduces 'genetic load' (or the amount of over-reproduction needed for a population to rid itself of deleterious mutations) and increases the tolerance for mutational 'noise'. Gene duplication is also a crude way to provide new genetic material for the adaptations of the future – Goldschmidt's 'hopeful monsters' in modern guise (for discussion, see Goldschmidt, 1940).

In Chapter 9 I suggested that genetic variation has what I called Stochastic Self-Similarity, a kind of quasi-fractal characteristic implicitly documented in subsequent chapters. The causal spectrum can be arranged hierarchically; descendent branches on the causal genetic tree have internal characteristics (numbers of alleles, distributions of allelic effects, cladistic sequence topology, etc.) similar to those of the higher structures that contain and spawned them. This property may simplify our characterization of the *structure* of the causal spectrum for a given trait, even if chance events determine which, and how many, *specific* alleles are involved and how common they are.

In this sense, most people would agree that we understand the circulation of the blood even though the system is complex, highly anastamotic, hierarchically nested, and there is a point at which its vascular components become too small and numerous to be named individually. It is not a failure of understanding if we do not name all the capillaries – so long as we know the points at which to stop a hemorrhage into a given area.

Improved understanding by more directed questions

The genetic basis of complex traits such as cholesterol levels is complex enough that the prediction from *specific* genotype to *specific* phenotype is rather weak, except for a few rare variants. We might organize our question, however, in a way that provides some sense of our practical powers of prediction. The following presents Adam Connor's and my attempt to do this, without having to go beyond traditional notions of additive allelic effects (Connor, 1992; and K. Weiss, unpublished results).

We can usually specify *selection sets*, $\mathbf{F} = \{\phi_i\}$, of phenotypes and $\mathbf{G} = \{g_i\}$, of genotypes that are of interest to us. The phenotype selection set might include an enumerated collection of qualitative phenotypes such as 'affected' or a range of quantitative phenotypes, such as 'cholesterol >200 mg/dl'. Similarly, the genotypic selection set can include a list of specific genotypes, or a range of genotypic effects.

We can define two measures of the relationship between these two sets in a given population, based on the penetrance functions. First, is the *predictance* of genotypes in **G** for phenotypes in **F**:

$$\left.\begin{aligned}\text{Pred}\,(\mathbf{F}|\mathbf{G}) &= \Pr(\phi\in\mathbf{F}|g\in\mathbf{G}) \\ &= \Sigma_{g\in\mathbf{G}}\,\Sigma_{\phi\in\mathbf{F}}\,\Omega_g(\phi) \qquad \text{for discrete phenotypes, or} \\ &= \textstyle\int_{\mathbf{G}}\int_{\mathbf{F}}\,\Omega_G(\phi)\,\mathrm{d}\phi\,\mathrm{d}G \quad \text{for continuous phenotypes.}\end{aligned}\right\} \quad [\text{A.1}]$$

The terms represent sums over all phenotypes in **F** and all genotypes in **G**. The predictance quantifies how sampling on selected genotypes 'captures' (identifies, or accounts for) important phenotypes. Even if we cannot predict the *specific* phenotype accurately, we may be able to predict membership in a useful range.

For example, suppose we are interested in identifying persons with cholesterol above 300 mg/dl, or cholesterol ratios (CR = total chol/HDL chol) above about 6.0, known to lead to elevated risk of heart disease. Based on published cholesterol values, the predictance of FH genotypes (*LDLR* mutations) for such phenotypes is about 0.84: most FH heterozygotes have elevated cholesterol. The comparable predictance for a person with an average genotype (CR ≈ 4.0) is only 0.05, about 15-fold less than that of FH genotypes. As another example, the predictance of colorectal cancer phenotypes by FAP genotypes is nearly 100%.

The second measure goes the other way. The *detectance* of phenotypes in **F** for genotypes in **G**, is the fraction of individuals with phenotype in **F** who have genotypes in **G**:

$$\left.\begin{aligned} \text{Det}\,(\mathbf{G}|\mathbf{F}) &= \Pr(g\in\mathbf{G}|\phi\in\mathbf{F}) && \\ &= \Sigma_{\mathbf{F}}\,\Sigma_{\mathbf{G}}\,P_g|\phi && \text{for discrete phenotypes, or} \\ &= \textstyle\int_{\mathbf{F}}\int_{\mathbf{G}} P_G|\phi\,\mathrm{d}G\,\mathrm{d}\phi && \text{for continuous phenotypes.} \end{aligned}\right\} \qquad [\text{A.2}]$$

If we ascertain persons with a given set of phenotypes, how well will we identify interesting genotypes?

The distribution of cholesterol among *LDLR* mutations is not understood with precision although, as with PKU and other traits, the effect depends on the specific mutation involved (Goldstein and Brown, 1988). To attempt to understand the effects of such genotypes we might sample individuals with, say, CR > 8.0. However, the detectance of FH mutations by such a sample would only be between about 0.07 and 0.26 (Connor, 1992); we would have to go further out on the CR scale to raise this. Indeed the genetic basis of FH was identified by concentrating on families with very extreme cholesterol values. In the colorectal cancer example, up to 80% of colorectal tumors manifest some type of somatic mutation at the *p53* locus (e.g., reviewed by Hollstein *et al.* (1991)).

These values illustrate an interesting point. We often tend to confuse predictance with detectance; that is, to assume that because the detectance is high (diseased persons often have mutations at a specific locus) most persons with the mutations have the disease. In fact, about 20% of persons with *LDLR* mutations have cholesterol well within the normal range and would never appear at a clinic (Roy *et al.*, 1991). Similarly, it may be a mistake to assume that only a few loci explain cancers of a given type, such as the association of *FAP* and *p53* with colorectal cancer. The detectance of such somatic mutations is high, but we do not yet know what the predictance is, because we have not yet been able to screen random colorectal stem cells for these mutations.

These measures provide set-to-set mappings between genotypes and phenotypes, and we may wish to know how efficient the mapping is. To be general we can define a function $H(g)$ as a measure of the impact of genotype g (we could also consider the reverse, the mapping of phenotype to genotype). H could be the predictance, but we might want some other measure. We can define the *Gain*, $\Psi(H(g),b)$, as the sum of the values of H, over a selection set of genotypes $\mathbf{G} = \{g|b\}$, chosen by some criterion b, weighted by the relative proportion of each such genotype in **G**:

$$\Psi(H(g),b) = \Sigma_{g\in G}\Pr(g)H(g)/\Sigma_{g\in G}\Pr(g). \qquad [A.3]$$

In particular, if we can *order* the genotypic effects of a set of identified alleles, we could define $\mathbf{G} = \{\mu_g \geq b\}$ as all genotypic values exceeding a cutoff point b. The Gain sum (or integral, for continuous traits) is the expected or average value of H of the set of genotypes defined by criterion b, and thus provides a measure of how H changes as we change **G**.

Of particular interest is how this information changes as we change the definition of **G** in some quantitative way. This is expressed by the *Gain function*:

$$\psi(H(g),b) = \Delta\Psi(H(g),b)/\Delta b \qquad [A.4]$$

where Δ indicates change (for a quantitative scale we would write the Gain function as $\partial\Psi(H(g),b)/\partial b$). This makes most sense for genotypic effects that are on a continuous scale. The Gain function then measures the rate of gain in phenotypic information (H) per unit change in the intensity of our genotypic screening, as defined by the cutoff point, b, of genotypes that are identified.

Example: the Gain for genes that affect cholesterol

Gain concepts could be applied to situations such as PKU screening: what is the cost-effectiveness of screening newborns for genotypes whose known mean PAH levels are below some cutoff point? However, as an example here we can look at the effects of the CR, using the Framingham data as discussed earlier (p. 174). Here, we define genotypes by their genotypic values on the CR scale, expressed in terms of standard deviations from the mean. What we really care about is not just CR but its impact on CHD. A useful measure of that impact is $H(g) = LE(g)$, the mean number of years of life expectancy lost due to CHD, among persons with a particular genotypic value.

In Chapter 9 we computed survivorship functions specific to each CR level in Framingham; those survivorships can be used to predict the life expectancy at each age t, for a given CR level, relative to the life expectancy (LE) of persons of the same age with the mean cholesterol value (i.e., relative to $LE(g = 0\sigma)$)[1]:

$$\begin{aligned} LE(g,t) &= LE(0,t) - 68.145 - 1.02t - 3.596g - 0.016tg \\ &\quad + 0.003t^2 + 0.0007\,t^2g \\ &= [3.596 - 0.016t + 0.0007t^2]g \qquad [A.5] \end{aligned}$$

(S. Weeks *et al.*, unpublished results).

Assuming, as in Chapter 9, a Laplace distribution of additive allelic effects, with parameter λ, translated (convoluted) into diploid genotypic effects, Connor (1992) worked out the gain function to be

$$\psi[LE(g),b \geq 0] = (3.596 - 0.016t + 0.0007t^2)\frac{(1 + b/\lambda)^2(3 + b/\lambda)}{(2 + b/\lambda)^2} \quad \text{[A.6]}$$

for $b \geq 0$ (with a comparable expression for $b < 0$). The first factor is nearly linearly dependent on the age at which one samples an individual, and the second factor is a 'kernel' that depends on the relative dispersion of allelic effects (λ) and the point (b) that defines what effects we are considering.

The kernel reflects the rate at which the average explained years of life changes per unit change in the cutoff, b. With the structure of this example, the amount of information (mean lost life expectancy in the ascertained selection set) initially rises faster than the cutoff value itself, but eventually the increase in information becomes linear with increase in b; that is, each unit in b yields only one more unit of information. Essentially, at first, as our criterion b changes, we rapidly ignore small-effect genotypes disproportionately and hence gain rapidly in information. Other examples will behave differently, although this pattern probably applies, at least qualitatively, to a broad range of phenotypes.

Figure A.1 does not consider the frequency of the effects in the population as a whole. To do this, we can weight the Gain by multiplying [A.3] by $\Pr(\text{CR} \geq b)$ (that is, cancel the denominator); this weighted Gain function now measures how genotypes in **G** contribute to the *population* mean of $H(g)$. A weighted Gain function is plotted in Figure A.2 for the effects of CR on life expectancy[2]. Explained life expectancy increases roughly linearly with genotypic effect [A.5], but the frequency of the genotype drops off roughly exponentially. As our cutoff b moves away from the mean (in the positive direction), the Gain initially drops off rapidly because, although genotypes with near zero mean effect have individually small effects, they are very common. In the tail (high b cutoff), the genotypic effects are strong but their frequency in the population is very small.

Identifying major disease genotypes in a specialty clinic may tell us about those genes, and may lead us to a better understanding of the physiology of the trait (as it did for the FH/LDL receptor story). However, this tells us little about the broader public health issues involved in heart disease. For that, we should perhaps screen for the more common, if less dramatic, genetic effects. Of course, whether persons

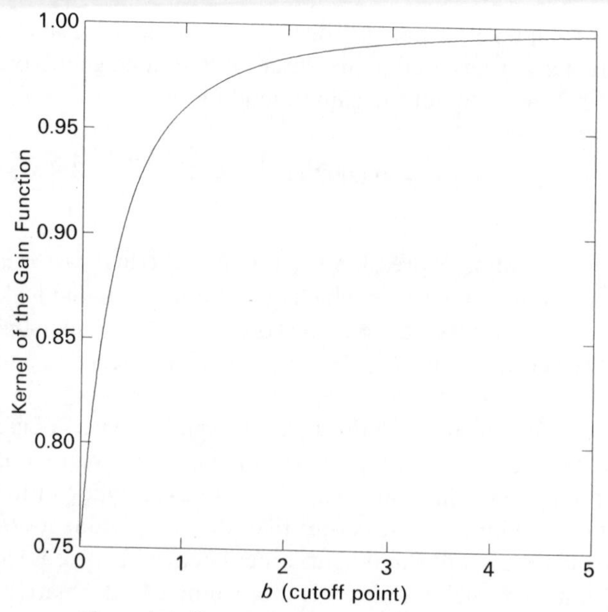

Figure A.1. Kernel of gain for modeled cholesterol ratio ($\lambda = 3$). (From Connor, 1992.)

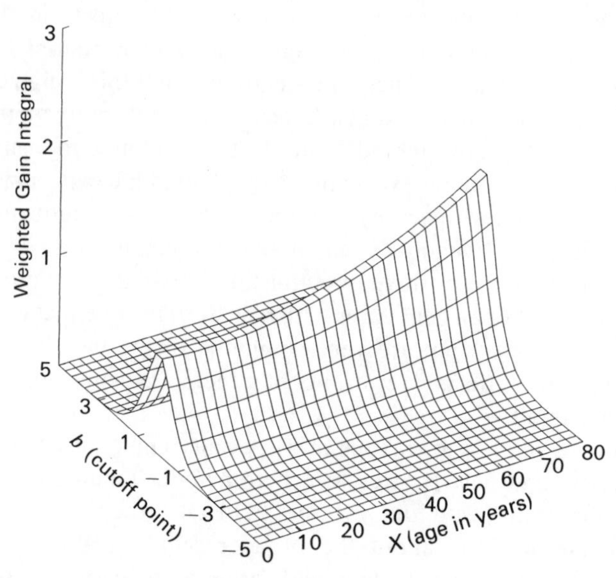

Figure A.2. Weighted gain integral for modeled cholesterol ratio ($\lambda = 3$). (From Connor, 1992.)

with commoner multilocus genotypes, whose risk is only slightly higher than average, would observe preventive measures is a separate question.

This modeling has considered only the additive effects of alleles at a single locus. However, the concept applies equally to individuals genotyped at multiple loci, and as our understanding of the genetic architecture of disease-related traits increases, and we can order genetic effects cladistically, we may have powerful predictive power.

Here I have described selection sets and Gain functions (H) in relation to biomedical objectives – risk factors for disease. Evolution uses the very specific criterion of *fitness*; that is, effects on the Net Maternity Function. A similar analysis would show that the unweighted Gain in fitness, relative to the effects of drift, was large only in the extreme tail of the distribution.

Conclusion

De Duve's (1991) quote at the top of the Afterwords is appropriate, I think. The glamor of finding 'the' gene for a given disease overshadows the greater effort to understand its full causal spectrum. Clearly there are a few rare alleles at major genes that affect a trait. These may be of great biomedical importance, but they are of little population importance. It is from the many more genotypes with individually small effects that the principles of evolution, which mold the pattern of variation, are to be understood. Biomedical research is concerned mostly with abnormal variation, but most variation, in most traits, in most people, for most of their lives is within the *normal* range, and deserves more attention. Diseases are just part of the natural phenotype distribution that we happen to choose to study.

Classical biometric analysis treated the polygenes controlling quantitative traits as a homogeneous aggregate whose individual effects need not be identified. That approach had great success in controlled breeding situations, but is poor when dealing with individuals. We know that the causal effects are not homogeneous, and there are times when we would like to understand the internal structure – especially if we want to identify its specific effects for screening, counseling, or therapeutic purposes.

Unfortunately, the complexity generated by evolution makes the *enumeration* of all but the strongest effects very difficult. We may not be able to identify all the allelic variants that affect a trait – and these are always changing in any case, as people die and new mutations arise. Enumeration is never an ultimate goal in science, and is a rapidly obsolescing way to think about the relationship between genotype and phenotype. A list cannot tell us about the 'emergent' phenomena that comprise a substantial component of the genetic causal spectrum that much of this book has been about.

The ultimate goal in science is *synthesis*, and what we really need to understand is the *order* in the genetic architecture of biological systems. We have recently been confronted with a flood of new DNA sequence data, and other genetic details, much of it seemingly chaotic and specific to each situation. But order can be found in complexity if we know what to look for. I have tried to suggest that it is in the context of evolution that we are being led to such a synthesis.

Evolution places strong constraints on variation: mendelian segregation, gene duplication, the use and re-use of structures at the genetic as well as phenotypic level, and the nearly faithful replication of DNA modified occasionally, in an essentially directional way, in that mutations generally create reconstructable clades of haplotypes unique to each population. These effects are generic at the gene level, for all phenotypes, a *sine qua non* for the development of a general theory. As De Duve predicted, everything is connected through the evolutionary tree of gene duplication and mutation. We could hardly ask for a more structured subject to study.

Notes

1. Since in this example H increases approximately linearly with CR the net effect of a genotype, $LE(g)$, can be equated to the effect at its genotypic value, $LE(\mu_g)$. For more complex situations, we would have to compute the weighted average of H over the entire penetrance function for the genotype.
2. In this example, the plot is symmetrical because the effects of CR on life expectancy are linear.

References

Aaronson, S. A. (1991). Growth factors and cancer. *Science* **254**: 1146–1153.

Abel, L. and Demenais, F. (1988). Detection of major genes for susceptibility to leprosy and its subtypes in a Caribbean island: Desirade Island. *American Journal of Human Genetics* **42**: 256–266.

Abel, L., Mallet, A., Demenais, F. and Bonney, G. E. (1989). Modeling the age-of-onset function in segregation analysis: a causal scheme for leprosy. *Genetic Epidemiology* **6**: 501–516.

Ahn, Y. I., Valdez, R., Reddy, A. P., Cole, S. A., Weiss, K. M. and Ferrell, R. E. (1991). DNA polymorphisms of the apolipoprotein AI/CIII/AIV gene cluster influence plasma cholesterol and triglyceride levels in the Mayans of the Yucatan Peninsula, Mexico. *Human Heredity* **41**: 281–289.

Ahuja, H., Bar-Eli, M., Advani, S. H., Benchmol, S. and Cline M. J. (1989). Alterations in the p53 gene and the clonal evolution of the blast crisis of chronic myelocytic leukemia. *Proceedings of the National Academy of Sciences, USA* **86**: 6783–6787.

Amos, C. I., Dawson, D. V. and Elston, R. C. (1990). The probabilistic determination of identity-by-descent sharing for pairs of relatives from pedigrees. *American Journal of Human Genetics* **47**: 842–853.

Amos, C. I. and Elston, R. C. (1989). Robust methods for the detection of genetic linkage for quantitative data from pedigrees. *Genetic Epidemiology* **6**: 349–360.

Amos, C. I., Elston, R. C., Wilson, A. F. and Bailey-Wilson, J. E. (1989). A more powerful robust sib-pair test of linkage for quantitative traits. *Genetic Epidemiology* **6**: 435–449.

Anderson, K. M., Wilson, P. W. F., Odell, P. M. and Kannel, W. B. (1991). An updated coronary risk profile: a statement for health professionals. *Circulation* **83**: 356–362.

Annest, J. L., Sing, C. F., Biron, P. and Mongeau, J. G. (1979). Familial aggregation of blood pressure and weight in adoptive families. 2. Estimation of the relative contribution of genetic and common environmental factors to blood pressure correlations between family members. *American Journal of Epidemiology* **110**: 492–503.

Antonarakis, S. E. and the Down Syndrome Collaborative Group (1991). Parental origin of the extra chromosome in trisomy 21 as indicated by analysis of DNA polymorphisms. *New England Journal of Medicine* **324**: 872–876.

Antonarakis, S. E., Kazazian, H. H. and Orkin, S. H. (1985). DNA polymorphism and molecular pathology of the human globin gene clusters. *Human Genetics* **69**: 1–14.

Antonarakis, S. E., Orkin, S. H., Kazazian, H. H., Gof, S. C., Boehm, C. D., Waber, P. G. *et al.* (1982). Evidence for multiple origin of the $ß^E$-globin gene in Southeast Asia. *Proceedings of the National Academy of Sciences, USA* **79**: 6608–6611.

Armitage, P. and Berry, G. (1987). *Statistical Methods in Medical Research*, 2nd edn, Oxford, Blackwell Scientific.

Astrin, S. M. and Costanzi, C. (1990). The molecular genetics of colon cancer. *Seminars in Oncology* **16**: 138–147.

Atkins, C., Reuffel, L., Roddy, J., Platts, M., Robinson, H. and Ward, R. H. (1988). Rheumatic disease in the Nuu-Chah-Nulth native Indians of the Pacific Northwest. *Journal of Rheumatology* **15**: 684–690.

Avigad, S., Kleiman, S., Weinstein, M., Cohen, B., Schwartz, G., Woo, S. and Shiloh, Y. (1991). Compound heterozygosity in nonphenylketonuria hyperphenylalanemia: the contribution of mutations for classical phenylketonuria. *American Journal of Human Genetics* **49**: 393–399.

Aviv, A. and Gardner, J. (1989). Racial differences in ion regulation and their possible links to hypertension in blacks. *Hypertension* **14**: 584–589.

Avner, P. (1991). Sweet mice, sugar daddies. *Nature* **351**: 519–520.

Awa, A. A. and Neel, J. V. (1986). Cytogenetic "rogue" cells: what is their frequency, origin, and evolutionary significance? *Proceedings of the National Academy of Sciences, USA* **83**: 1021–1025.

Baker, S. J., Fearon, E. R., Nigro, J. M., Hamilton, S. R., Preisinger, A. C., Jessup, J. M. *et al.* (1989). Chromosome 17 deletions and p53 mutations in colorectal carcinomas. *Science* **244**: 217–221.

Barton, N. H. and Turelli, M. (1987). Adaptive landscapes, genetic distance and the evolution of quantitative characters. *Genetical Research* **49**: 157–173.

Barton, N. H. and Turelli, M. (1989). Evolutionary quantitative genetics: how little do we know? *Annual Review of Genetics* **23**: 337–370.

Beaty, T. H. (1980). Discriminating among single locus models using small pedigrees. *American Journal of Medical Genetics* **6**: 229–240.

Bell, J. I., Todd, J. A. and McDevitt H. O. (1989). The molecular basis of HLA–disease association. In Harris, H. and Hirschhorn, H., editors, *Advances in Human Genetics*, vol. 18, New York, Plenum, pp. 1–41.

Berg, K. (1983). Genetics of coronary heart disease. *Progress in Medical Genetics*, n.s. **5**: 35–90.

Berg, K. (1987). Genetics of coronary heart disease and its risk factors. In Bock, G. and Collins, G., editors, *Molecular Approaches to Human Polygenic Disease*, Ciba Foundation symposium No. 130, Chichester. J. Wiley, pp. 23–28.

Bigner, S. H., Bark, J., Burger, P. C., Mahaley, M. S., Bullard, D. E., Muhlbaier, L. H. and Bigner, D. D. (1988). Specific chromosomal abnormalities in malignant human gliomas. *Cancer Research* **88**: 405–411.

Bishop, J. M. (1987). The molecular genetics of cancer. *Science* **235**: 305–311.

Bishop, J. M. (1991). Molecular themes in oncogenesis. *Cell* **64**: 235–248.

Bishop, D. T. and Williamson, J. A. (1990). The power of identity-by-state methods for linkage analysis. *American Journal of Human Genetics* **46**: 254–265.

Bjorkman, P. J., Saper, M. A., Samraoui, B., Bennett, W. S., Strominer, J. L. and Wiley, D. C. (1987). Structure of the human class 1 histocompatibility antigen, HLA-A2. *Nature* **329**: 506–512.

Black, F. L., Hierhozler, W. J., Lian-Chen, J. F., Berman, L. L., Gabbay, Y. and Pinheiro, F. de P. (1982). Genetic correlates of enhanced measles susceptibility in Amazon Indians. *Medical Anthropology*, Winter 1982: 37–46.

Black, F. L., Pinheiro, F. de P., Hierholzer, W. J. and Lee, R. V. (1977). Epidemiology of infectious disease: the example of measles. In Ciba Foundation Symposium No. 49, *Health and Disease in Tribal Societies*, Amsterdam, Elsevier/North Holland, pp. 115–135.

Black, F. L., Woodall, J. P. and Pinheiro, F. de P. (1969). Measles vaccine reactions in a virgin population. *American Journal of Epidemiology* **89**: 168–175.

Blackwell, T. K., and Alt, F. W. (1989). Mechanism and developmental program of immunoglobulin gene rearrangement in mammals. *Annual Review of Genetics* **23**: 605–636.

Blangero, J. and Konigsberg, L. W. (1991). Multivariate segregation analysis using the mixed model. *Genetic Epidemiology* **8**: 299–316.

Boehnke, M. (1986). Estimating the power of a proposed linkage study: a practical computer simulation approach. *American Journal of Human Genetics* **39**: 513–527.

Boehnke, M., Lange, K. and Cox, D. R. (1991). Statistical methods for multipoint radiation hybrid mapping. *American Journal of Human Genetics* **49**: 1174–1188.

Boehnke, M. and Moll, P. P. (1989). Identifying pedigrees segregating at a major locus for a quantitative trait: an efficient strategy for linkage analysis. *American Journal of Human Genetics* **44**: 216–224.

Boehnke, M., Moll, P. P., Lange, K., Weidman, W. H. and Kottke, B. A. (1986). Univariate and bivariate analyses of cholesterol and triglyceride levels in pedigrees. *American Journal of Medical Genetics* **23**: 775–792.

Boehnke, M., Young, M. R. and Moll, P. P. (1988). Comparison of sequential and fixed-structure sampling of pedigrees in complex segregation analysis of a quantitative trait. *American Journal of Human Genetics* **43**: 336–343.

Boerwinkle, E., Chakraborty, R. and Sing, C. F. (1986a). The use of measured genotype information in the analysis of quantitative phenotypes in man. 1. Models and analytical methods. *Annals of Human Genetics* **50**: 181–194.

Boerwinkle, E. and Sing, C. F. (1987). The use of measured genotype information in the analysis of quantitative phenotypes in man. III. Simultaneous estimation of the frequencies and effects of the *apolipoprotein E* polymorphism and residual polygenic effects on cholesterol, betalipoprotein and triglyceride levels. *Annals of Human Genetics* **51**: 211–226.

Boerwinkle, E., Turner, S. T. and Sing, C. F. (1984). The role of the genetics of sodium lithium countertransport in the determination of blood pressure variability in the population at large. In Brewer, G. J., editor, *The Red Cell: Sixth Ann Arbor Conference*: New York, Alan R. Liss, pp. 479–507.

Boerwinkle, E., Turner, S. T., Weinshilboum, R., Johnson, M., Richelson, E. and Sing, C. F. (1986b). Analysis of the distribution of erythrocyte sodium lithium countertransport in a sample representative of the general population. *Genetic Epidemiology* **3**: 365–378.

Boerwinkle, E. and Utermann, G. (1988). Simultaneous effects of the apolipoprotein E polymorphism on apolipoprotein E, apolipoprotein B, and cholesterol metabolism. *American Journal of Human Genetics* **42**: 104–112.

Boerwinkle, E., Visvikis, S., Welsh, D., Steinmetz, J., Hanash, S. M. and Sing, C. F. (1987). The use of measured genotype information in the analysis of quantitative phenotypes in man. II. The role of the apolipoprotein E polymorphism in determining levels, variability, and covariability of cholesterol, betalipoprotein, and triglycerides in a sample of unrelated individuals. *American Journal of Medical Genetics* **27**: 567–582.

Bonney, G. E. (1986). Regressive logistic models for familial disease and other binary traits. *Biometrics* **42**: 611–625.

Bonney, G. E. (1988). On the statistical determination of major gene mechanisms in continuous human traits: regression models. *American Journal of Medical Genetics* **18**: 731–749.

Bonney, G. E., Dunstan, G. M. and Wilson, J. (1989). The use of regressive logistic models for ordered and unordered polychotomous traits: application to affective disorders. *Genetic Epidemiology* **6**: 211–215.

Bonney, G. E. and Elston, R. C. (1985a). Integrals of multinormal mixtures. *Applied Mathematics and Computation* **16**: 93–104.

Bonney, G. E. and Elston, R. C. (1985b). Likelihood models for multivariate traits in human genetics. *Biometrical Journal* **5**: 553–563.

Bonney, G. E., Lathrop, G. M. and Lalouel, J.-M. (1988). Combined linkage and segregation analysis using regressive models. *American Journal of Human Genetics* **43**: 029-037.

Breslow, N. E. and Day, N. E. (1980). *Statistical Methods in Cancer Research*, vol. 1 *The Analysis of Case-Control Studies*, Lyon, International Agency for Research on Cancer.

Breslow, N. E. and Day, N. E. (1987). *Statistical Methods in Cancer Research*, vol. 2 *The Design and Analysis of Cohort Studies*, Lyon, International Agency for Research on Cancer.

Bulmer, M. G. (1989). Maintenance of genetic variability by mutation–selection balance: a child's guide through the jungle. *Genome* **31**: 761–767.

Butler, M. G. (1990). Parder-Willi syndrome: current understanding of cause and diagnosis. *American Journal of Medical Genetics* **35**: 319–332.

Byers, P. H. (1990). Brittle bones – fragile molecules: disorders of collagen gene structure and expression. *Trends in Genetics* **6**: 293–300.

Byers, P. H. (1991). Osteogenesis imperfecta: translation of mutation to phenotype. *Journal of Medical Genetics* **28**: 433–442.

Calder, W. A. (1984). *Size, Function, and Life History*, Cambridge, Massachusetts, Harvard University Press.

Canessa, M. L., Adragna, N., Solomon, H. S. and Tosteson, D. C. (1979). Increased sodium-lithium countertransport in red cells of patients with essential hypertension. *New England Journal of Medicine* **302**: 772–776.

Cannings, C. and Thompson, E. A. (1977). Ascertainment in the sequential sampling of pedigree. *Clinical Genetics* **12**: 208–212.

Cannings, C., Thompson, E. A. and Skolnick, M. H. (1978). Probability functions on complex pedigrees. *Advances in Applied Probability* **10**: 26–61.

Cannon-Albright, L. A., Skolnick, M. H., Bishop, T., Lee, R. G. and Burt, R. W. (1988). Common inheritance of susceptibility to colonic adenomatous polyps and associated colorectal cancers. *New England Journal of Medicine* **319**: 533–536.

Carey, G. and Williamson, J. (1991). Linkage analysis of quantitative traits: increased power by using selected samples. *American Journal of Human Genetics* **49**: 786–796.

Carlson, B. M. (1988). *Patten's Foundations of Embryology*, 5th edn, New York, McGraw-Hill.

Cavalli-Sforza, L. L. and Bodmer, W. F. (1971). *Genetics of Human Populations*. San Francisco, W. Freeman.

Chakraborty, R. (1986). Gene admixture in human populations: models and predictions. *Yearbook of Physical Anthropology* **29**: 1–43.

Chakraborty, R. (1989). Can molecular imprinting explain heterozygote deficiency and hybrid vigor? *Genetics* **122**: 713–717.

Chakraborty, R., Ferrell, R. E., Stern, M. P., Haffner, S. M., Hazuda, H. P. and Rosenthal, M. (1986). Relationship of prevalence of non-insulin-dependent diabetes mellitus to Amerindian admixture in the Mexican Americans of San Antonio, Texas. *Genetic Epidemiology* **3**: 435–454.

Chakraborty, R., Kamboh, M. I. and Ferrell, R. E. (1991). 'Unique' alleles in admixed populations: a strategy for determining 'hereditary' population differences of disease frequency. *Ethnicity and Disease* **1**: 245–256.

Chakraborty, R., Smouse, P. E. and Neel, J. V. (1988). Population amalgamation and genetic variation: observations on artificially agglomerated tribal populations of Central and South America. *American Journal of Human Genetics* **43**: 709–725.

Chakraborty, R. and Weiss, K. M. (1986). Frequencies of complex diseases in hybrid populations. *American Journal of Physical Anthropology* **70**: 489–503.

Chakraborty, R. and Weiss, K. M. (1988). Admixture as a tool for finding linked genes and detecting that difference from allelic association between loci. *Proceedings of the National Academy of Sciences, USA* **85**: 9119–9123.

Chakraborty, R. and Weiss, K. M. (1989). Age-specific risks for cancer as determined by multi-stage models of carcinogenesis. In Krishnan, T., editor, *Medical Statistics*, Bombay, Himalayan Publishing House, pp. 64–91.

Chakraborty, R., Weiss, K. M. and Ward, R. H. (1982). Evaluation of relative risks from the correlation between relatives: a theoretical approach. *Medical Anthropology*, Summer 1980: 397–414.

Chakravarti, A., Badner, J. A. and Li, C. C. (1987). Tests of linkage and heterogeneity in mendelian diseases using identity by descent scores. *Genetic Epidemiology* **4**: 255–266.

Chakravarti, A. and Chakraborty, R. (1978). Elevated frequency of Tay-Sachs disease among Ashkenazic Jews unlikely by genetic drift alone. *American Journal of Human Genetics* **30**: 256–261.

Cho, S., Attaya, M. and Monaco, J. J. (1991). New class II-like genes in the murine MHC. *Nature* **353**: 573–576.

Clark, A. G. (1991). Causes and consequences of variation in energy storage in *Drosophila melanogaster*. *Genetics* **123**: 131–144.

Clark, R. W. (1968). *The Life and Work of J. B. S. Haldane*, Oxford, Oxford University Press.

Cochran, W. G. (1976). *Sampling Techniques*, 3rd edn, New York, J. Wiley.

Conneally, P. M. and Rivas, M. L. (1980). Linkage analysis in man. In Harris, H. and Hirschhorn, K., editors, *Advances in Human Genetics*, vol. 10, New York, Plenum Press, pp. 209–266.

Connor, A. J. (1992). The effect of extreme genotypes upon functions of the phenotype. Ph.D. dissertation, Pennsylvania State University.

Cook, P., Doll, R. and Fellingham, S. A. (1969). A mathematical model for the age distribution of cancer in man. *International Journal of Cancer* **4**: 93–112.

Cooper, D. N. and Youssoufian, H. (1988). The CpG dinucleotide and human genetic disease. *Human Genetics* **78**: 151–155.

Cornall, R. J., Prins, J.-B., Todd, J. A., Pressey, A., DeLarato, N. H., Wicker, L. S. and Peterson, L. B. (1991). Type I diabetes in mice is linked to the interleukin-1 receptor and *Lsh/Ity/Bcg* genes on chromosome 1. *Nature* **353**: 262–265.

Crews, D. E., Kamboh, M. I., Bindon, J. R. and Ferrell, R. E. (1991). Genetic studies of human apolipoproteins. 17. Population genetics of apolipoprotein polymorphisms in American Samoa. *American Journal of Physical Anthropology* **84**: 165–170.

Cropp, C. S., Lidereau, R., Campbell, C., Champene, M. H. and Callahan, R. (1990). Loss of heterozygosity on chromosomes 17 and 18 in breast carcinoma: two additional regions identified. *Proceedings of the National Academy of Sciences, USA* **87**: 7737–7741.

Crow, J. F. (1986). *Basic Concepts in Population, Quantitative, and Evolutionary Genetics*, New York, W. H. Freeman.

Crow, J. F. and Kimura, M. (1970). *Introduction to Population Genetics Theory*, New York, Harper & Row.

Cupples, L. A., Risch, N., Farrer, L. A. and Myers, R. H. (1991). Estimation of morbid risk and age at onset with missing information. *American Journal of Human Genetics* **49**: 76–87.

Cupples, L. A., Terrin, N. C., Myers, R. H. and d'Agostino, R. B. (1989). Using survival methods to estimate age-at-onset distributions for genetic diseases with an application to Huntington disease. *Genetic Epidemiology* **6**: 361–371.

Cutler, R. G. (1978). Evolutionary biology of senescence. In Behnke, J., Finch, C. and Moment, G., editors, *The Biology of Aging*, New York, Plenum, pp. 311–360.

Dadone, M. M., Hasstedt, S. J., Hunt, S. C., Smith, J. B., Ash, K. O. and Williams, R. R. (1984). Genetic analysis of sodium–lithium countertransport in 10 hypertension-prone kindreds. *American Journal of Medical Genetics* **17**: 565–577.

Daiger, S. P., Chakraborty, R., Reed, L., Fekete, G., Schuler, D., Berenssi, G. *et al.* (1989a). Polymorphic DNA haplotypes at the phenylalanine hydroxylase (PAH) locus in European families with phenylketonuria (PKU). *American Journal of Human Genetics* **45**: 310–318.

Daiger, S. P., Reed, L., Huang, S.-S., Zeng, Y.-T, Wang, T., Lo, W. H. Y. *et al.* (1989b). Polymorphic DNA haplotypes at the phenylalanine hydroxylase (PAH) locus in Asian families with phenylketonuria (PKU). *American Journal of Human Genetics* **45**: 319–324.

Darlu, P., Saquier, P. P. and Bois, E. (1990). Genealogical and genetical African admixture estimations, blood pressure and hypertension in a Caribbean community. *Annals of Human Biology* **17**: 387–400.

Dausset, J., Cann, H., Cohen, D., Lathrop, M., Lalouel, J.-M. and White, R. (1990). Centre d'Etude du Polymorphisme Humain (CEPH): collaborative genetic mapping of the human genome. *Genomics* **6**: 575–577.

Davidoff, A. M., Humphrey, P. A., Iglehart, J. D. and Marks, J. R. (1991). Genetic basis for p53 overexpression in human breast cancer. *Proceedings of the National Academy of Sciences, USA* **88**: 5006–5010.

Davies, K. (1991). The essence of inactivity. *Nature* **349**: 15–16.

Davignon, J., Gregg, R. R. and Sing, C. F. (1988). ApoE polymorphism and atherosclerosis. *Arteriosclerosis* **8**: 1–21.

Dawson, D. V., Kaplan, E. B. and Elston R. C. (1990). Extensions to sib-pair linkage tests applicable to disorders characterized by delayed onset. *Genetic Epidemiology* **7**: 453–466.

De Duve, C. (1991). *Blueprint for a Cell: The Nature and Origin of Life*, Burlington, North Carolina, Neil Patterson/Carolina Biological.

Dean, A., Dykhuizen, D. and Hartl, D. (1988). Theories of metabolic control in quantitative genetics. In Weir, B. S., Eisen, E. J., Goodman, M. M. and Namkoong, G., editors, *Proceedings of the Second International Conference on Quantitative Genetics*. Sunderland, Massachusetts, Sinauer Associates, pp. 536–548.

Deka, R. (1981). Fertility and haemoglobin genotypes: a population study in Assam (India). *Human Genetics* **59**: 172.

Deka, R., Gogol, B. C., Hundreiser, J. and Flatz, G. (1987). Hemoglobinopathies in northeast India. *Hemoglobin* **11**: 531–538.

Del Puente, A., Knowler, W. C., Pettitt, D. J. and Bennett, P. H. (1989). High incidence and prevalence of rheumatoid arthritis in Pima Indians. *American Journal of Epidemiology* **129**: 1170–1178.

Del Junco, D. J., Luthra, H. S., Annegers, J. F., Worthington, J. W. and Kurland, J. T. (1984). The familial aggregation of rheumatoid arthritis and its relationship to the HLA-DR4 association. *American Journal of Epidemiology* **119**: 813–829.

Demenais, F. M. and Amos, C. I. (1989). Power of the sib-pair and lod-score methods for linkage analysis of quantitative traits. In *Multipoint Mapping and Linkage Based on Affected Pedigree Members: Genetic Analysis Workshop* **6**, New York, Alan R. Liss, pp. 201–206.

Demenais, F. M. and Bonney, G. E. (1989). Equivalence of the mixed and regressive models for genetic analysis. 1. Continuous traits. *Genetic Epidemiology* **6**: 597–617.

Demenais, F. M. and Elston, R. C. (1981). A general transmission probability model for pedigree data. *Human Heredity* **31**: 93–99.

Devilee, P., Van Den Roek, M., Kuipers-Dijskhoorn, N., Kolluri, R., Khan, P. M., Pearson, P. L. and Cornelisse, C. J. (1989). At least four different chromosomal regions are involved in loss of heterozygosity in human breast carcinoma. *Genomics* **5**: 554–560.

DiLella, A. G. and Woo, S. L. C. (1987). Molecular basis of phenylketonuria and its clinical applications. *Molecular Biology and Medicine* **4**: 181–192.

Doebley, J. and Stec, A. (1991). Genetic analysis of the morphological differences between maize and teosinte. *Genetics* **129**: 285–295.

Doebley, J., Stec, A., Wendel, J. and Edwards, M. (1990). Genetic and morphological analysis of a maize-teosinte F_2 population: implications for the origin of maize. *Proceedings of the National Academy of Sciences, USA* **87**: 9888–9892.

Doll, R. and Peto, R. (1980). *The Causes of Cancer*, Oxford, Oxford University Press.

Dorit, R. L., Schoenbach, L. and Gilbert, W. (1990). How big is the universe of exons? *Science* **250**: 1377–1382.

Dryja, T. P., Mukai, S., Petersen, R., Rapaport, J. M., Walton, D. and Yandell, D. W. (1989). Parental origin of mutations of the retinoblastoma gene. *Nature* **339**: 556–558.

Easton, D. and Peto, J. (1990). The contribution of inherited predisposition to cancer incidence. *Cancer Surveys* **9**: 395–416.

Edwards, J. H. (1969). The simulation of mendelism. *Acta Genetica et Statistica Medica* **10**: 63–70.

Edwards, M. D., Stuber, C. W. and Wendel, J. F. (1987). Molecular-marker-facilitated investigations of quantitative-trait loci in maize. 1. Numbers, genomic distribution and types of gene action. *Genetics* **116**: 113–125.

Eisen, H. (1990). *General Immunology*. Philadelphia, J. B. Lippincott.

Elandt-Johnson, R. C (1971). *Probability Models and Statistical Methods in Genetics*, New York, J. Wiley.

Elandt-Johnson, R. C. and Johnson, N. L. (1980). *Survival Models and Data Analysis*. New York, J. Wiley.

Elston, R. C. (1973). Ascertainment and age of onset in pedigree analysis. *Human Heredity* **23**: 105–112.

Elston, R. C. (1981). Segregation analysis. In Harris, H. & Hirschorn, K., editors, *Advances in Human Genetics*, vol. 11, New York, Plenum, pp. 63–120.

Elston, R. C. (1986). Modern methods of segregation analysis. In Moolgavkar, S. H. and Prentice, R. L. editors, *Modern Statistical Methods in Chronic Disease Epidemiology*, New York, Wiley, pp. 213–224.

Elston, R. C. and Lange, K. (1975). The prior probability of autosomal linkage. *Annals of Human Genetics* **38**: 341–350.

Elston, R. C. and Sobel, E. (1979). Sampling considerations in the gathering and analysis of pedigree data. *American Journal of Human Genetics* **31**: 62–69.

Elston, R. C. and Stewart, J. (1971). A general model for the analysis of pedigree data. *Human Heredity* **21**: 523–542.

Emery, A. E. H. (1984). *An Introduction to Recombinant DNA*, New York, J. Wiley.

Ewens, W. J. and Clarke, C. P. (1984). Maximum likelihood estimation of genetic parameters of HLA-linked diseases using data from families of various sizes. *American Journal of Human Genetics* **36**: 858–872.

Ewens, W. J. and Shute, N. C. (1986). A resolution of the ascertainment sampling problem. 1. Theory. *Theoretical Population Biology* **30**: 388–412.

Falconer, D. S. (1989). *Introduction to Quantitative Genetics*, 3rd edn, London, Longman.

Faustman, D., Li, X., Lin, H. Y., Fu, Y., Eisenbarth, G., Avruch, J. and Guo, J. (1991). Linkage of faulty major histocompatibility complex class I to autoimmune diabetes. *Science* **254**: 1756–1761.

Felsenfeld, G. (1985). DNA. *Scientific American* **253**: 58–67.

Field, L. L. (1988). Insulin-dependent diabetes mellitus: a model for the study of multifactorial disorders. *American Journal of Human Genetics* **43**: 793–798.

Finch, C. E. (1990). *Longevity, Senescence, and the Genome*, Chicago, University of Chicago Press.

Fisher, R. A. (1918). The correlation between relatives on the supposition of Mendelian inheritance. *Transactions of the Royal Society of Edinburgh* **52**: 399–433.

Fisher, R. A. (1930). *The Genetical Theory of Natural Selection*, reprinted by Dover Books, New York, 1958.

Fleiss, J. L. (1981). *Statistical Methods for Rates and Proportions*, 2nd edn, New York, Wiley.

Flint, J., Hill, A. V. S., Bowden, D. K., Oppenheimer, S. J., Sill, P. R., Serjeantson, S. W. *et al*. (1986). High frequencies of α-thalassaemia are the result of natural selection by malaria. *Nature* **321**: 744–750.

Fountain, J. W., Bale, S. J., Houseman, D. E. and Dracopoli, N. C. (1990). Genetics of melanoma. *Cancer Surveys* **9**: 645–671.

Fumeron, F., Rigaud, D., Bertiere, M. C., Bardon, S., DeLy, C. and Apfelbaum, M. (1988). Association of apolipoprotein E4 isoform with hypertriglyceridemia. *Clinical Genetics* **34**: 1–7.

Gale, J. (1990). *Theoretical Population Genetics*, London, Alan Unwin.

Garchon, H.-H., Bedossa, P., Eloy, L. and Bach, J.-F. (1991). Identification and mapping to chromosome 1 of a susceptibility locus for periinsulitis in non-obese mice. *Nature* **353**: 260–262.

Garrod, A. E. (1908). The Croonian lectures on inborn errors of metabolism. *Lancet* **2**: 1–7, 73–79, 142–148, 214–220.

Gelehrter, T. and Collins, F. (1990). *Principles of Medical Genetics*, Baltimore, Maryland, Williams and Wilkins.

Gilbert, S. F. (1991). *Developmental Biology*, 3rd edn, Sunderland, Massachusetts, Sinauer Associates.

Gilbert, W. (1987). The exon theory of genes. *Cold Spring Harbor Symposia on Quantitative Biology*, LII: 901–905.

Gleick, J. (1987). *Chaos: Making A New Science*. New York, Viking Press.

Goate, A., Chartier-Harlin, M.-C., Mullam, M., Brown, J., Crawford, F., Fidani, L. and 14 others. (1991). Segregation of a missense mutation in the amyloid precursor protein gene with familial Alzheimer's disease. *Nature* **349**: 704–706.

Gojobori, T. and Nei, M. (1984). Concerted evolution of the immunoglobulin V_H gene family. *Molecular Biology and Evolution* **1**: 195–212.

Gojobori, T. and Nei, M. (1986) Relative contributions of germline gene variation and somatic mutation to imunoglobulin diversity in the mouse. *Molecular Biology and Evolution* **3**: 156–167.

Goldgar, D. E. (1990). Multipoint analysis of human quantitative genetic variation. *American Journal of Human Genetics* **47**: 957–967.

Goldschmidt, R. (1940). *The Material Basis of Evolution*. New Haven, Yale University Press.

Goldstein, J. L. and Brown, M. S. (1988). Familial hypercholesterolemia. In Scriver, C. R., Baudet, A. L., Sly, W. S. and Valle, D., editors, *The Metabolic Basis of Inherited Disease*, 6th edn, New York, McGraw-Hill, pp. 1215–1250.

Goodenough, U. W. (1991). Deception by pathogens. *American Scientist* **79**: 344–355.

Goodnight, C. J. (1988). Epistasis and the effect of founder events on the additive genetic variance. *Evolution* **42**: 441–454.

Grant, S. G. and Chapman, V. M. (1988). Mechanisms of X chromosome regulation. *Annual Review of Genetics* **22**: 199–233.

Grebner, E. E. and Tomczak, J. (1991). Distribution of three α-chain β-hexosaminidase A mutations among Tay-Sachs carriers. *American Journal of Human Genetics* **48**: 604–607.

Gueguen, R., Visvikis, S., Steinmetz, J., Siest, G and Boerwinkle, E. (1989). An analysis of genotype effects and their interactions by using the apolipoprotein E polymorphism and longitudinal data. *American Journal of Human Genetics* **45**: 793–802.

Gusella, J. F., Wexler, N. S., Conneally, P. M., Naylor, S. L., Anderson, M. A., Tanzi, M. A. and 11 others. (1983). A polymorphic DNA marker genetically linked to Huntington's disease. *Nature* **306**: 234–238.

Guttler, F., Ledley, F. D., Lidsky, A. S., DiLella, A. G., Sullivan, S. E. and Woo, S. L. C. (1987). Correlations between polymorphic DNA haplotypes at phenylalanine hydroxylase locus and clinical phenotypes of phenylketonuria. *Journal of Pediatrics* **110**: 68–71.

Hall, J. G. (1988). Somatic mosaicism: observations related to clinical genetics. *American Journal of Human Genetics* **43**: 355–363.

Hall, J. G. (1990a). Nontraditional inheritance. *Growth Genetics and Hormones* **6**: 1–4.

Hall, J. G. (1990b). Genomic imprinting: review and relevance to human disease. *American Journal of Human Genetics* **46**: 857–873.

Hall, J. M., Lee, M. K., Newman, B., Morrow, J. E., Anderson, L. E., Huey, B. and King, M.-C. (1990b). Linkage of early-onset familial breast cancer to chromosome 17q21. *Science* **250**: 1684–1689.

Hall, J. M., Zuppan, P. J., Anderson, L. A., Huey, B., Carter, C. and King, M.-C. (1990a). Oncogenes and human breast cancer. *American Journal of Human Genetics* **44**: 577–584.

Hall, P. A. and Watt, F. M. (1989). Stem cells: the generation and maintenance of cellular diversity. *Development* **106**: 619–633.

Hamilton, M, Pickering, G. S. W., Roberts, J. A. F. and Sowry, G. S. C. (1954). The etiology of essential hypertension. 4. The role of inheritance, *Clinical Science* **13**: 273–304.

Hanis, C. L., Chakraborty, R., Ferrell, R. E. and Schull, W. J. (1986). Individual admixture estimates: disease associations and individual risk of diabetes and gallbladder disease among Mexican Americans in Starr County, Texas. *American Journal of Physical Anthropology* **70**: 433–441.

Hanis, C. L., Hewitt-Emmet, D., Kabrusly, L. F., Maklad, M. N., Douglas, T. C., Mueller, W. H. *et al.* (1992). An ultrasound survey of gallbladder disease among Mexican-Americans in Starr County, Texas: frequencies and risk factors. *Ethnicity and Disease*, in press.

Hanks, S. K., Quinn, A. M. and Hunter, T. (1988). The protein kinase family: conserved features and deduced phylogeny of the catalytic domains. *Science* **241**: 42–52.

Hartl, D. L. and Clark, A. G. (1989). *Principles of Population Genetics*, 2nd edn, Sunderland, Massachusetts, Sinauer Associates.

Haseman, J. K. and Elston, R. C. (1972). The investigation of linkage between a quantitative trait and a marker locus. *Behavior Genetics* **2**: 3–19. (See important exchange of methodologic comments, *Behavior Genetics* **3**: 4.)

Hasstedt, S. J. (1982). A mixed model likelihood approximation for large pedigrees. *Computers in Biomedical Research* **15**: 295–307.

Hasstedt, S. J., Wu, L. L., Ash, K. O., Kuida, H. and Williams, R. R. (1988). Hypertension and sodium-lithium countertransport in Utah pedigrees: evidence for major locus inheritance. *American Journal of Human Genetics* **3**: 14–22.

Haviland, M. B., Kessling, A. M., Davignon, J. and Sing, C. F. (1991). Estimation of Hardy–Weinberg and pairwise disequilibrium in the apolipoprotein AI-CII-AIV gene cluster. *American Journal of Human Genetics* **49**: 350–365.

Hedrick, P. W. (1985). *Genetics of Populations*. Boston, M. A., Jones & Bartlett.

Hedrick, P. W., Thomson, G. & Klitz, W. (1987). Evolutionary genetics and HLA: another classic example. *Biological Journal of the Linnaean Society* **31**: 311–331.

Heitz, D., Rousseau, F., Devys, D., Saccone, S., Abderrahim, H., Le Paslier, D. *et al.* (1991). Isolation of sequences that span the fragile Xs and identification of a fragile X-related CpG island. *Science* **251**: 1236–1239.

Henry, I., Bonaiti-Pelle, C., Chehensse, V., Beldjord, C., Schwartz, C., Utermann, G. and Junien, C. (1991). Uniparental disomy in a genetic cancer-predisposing syndrome. *Nature* **351**: 665–667.

Herrlich, P. and Ponta, H. (1989). 'Nuclear' oncogenes convert extracellular stimuli into changes in the genetic program. *Trends in Genetics* **5**: 112–116.

Hertzberg, M., Jahromi, K., Ferguson, V., Dahl, H. H. M., Mercer, J., Mickleson, K. N. P. and Trent, R. J. (1989). Phenylalanine hydroxylase gene haplotypes in Polynesians: evolutionary origins and absence of alleles associated with severe phenylketonuria. *American Journal of Human Genetics* **44**: 382–387.

Higgs, D. R., Vickers, M. A., Wilkie, A. O. M., Pretorius, I.-M., Jarman, A. P. and Weatherall, D. J. (1989). A review of the molecular genetics of the human α-globin gene cluster. *Blood* **73**: 1081–1104.

Hilbert, P., Lindpaintner, K., Beckmann, J., Serikawa, T., Soubrier, F., Dubay, C. *et al.* (1991). Chromosomal mapping of two genetic loci associated with blood-pressure regulation in hereditary hypertensive rats. *Nature* **353**: 521–529.

Hobbs, H. H., Leitersdorf, E., Leffert, C. C., Cryer, D. R., Brown, M. S. and Goldstein, J. L. (1989). Evidence for a dominant gene that suppresses hypercholesterolemia in a family with defective low density lipoprotein receptors. *Journal of Clinical Investigation* **84**: 656–664.

Hobbs, H. H., Russell, D. W., Brown, M. S. and Goldstein, J. L. (1990). The LDL receptor locus in familial hypercholesterolemia. *Annual Review of Genetics* **24**: 133–170.

Hodge, S. E. (1981). Some epistatic two-locus models of disease. 1. Relative risks and identity-by-descent distributions in affected sib pairs. *American Journal of Human Genetics* **33**: 381–395.

Hodge, S. E. (1984). The information contained in multiple sibling pairs. *Genetic Epidemiology* **1**: 109–122.

Holliday, R. (1987). The inheritance of epigenetic defects. *Science* **238**: 163–170.

Holliday, R. (1989). A different kind of inheritance. *Scientific American*: 60–73.

Hollstein, M. C., Metcalf, R. A., Welsh, J. A., Montesano, R. and Harris, C. C. (1990). Frequent mutation of the p53 gene in human esophageal cancer. *Proceedings of the National Academy of Sciences, USA* **87**: 9958–9961.

Hollstein, M., Sidransky, D., Vogelstein, B. and Harris, C. C. (1991). p53 mutations in human cancers. *Science* **253**: 49–53.

Hopkins, P., Wu, L., Schumacher, M., Emi, M., Hegele, R., Hunt, S. *et al.* (1991). Type 3 dyslipoproteinemia in patients heterozygous for familial hypercholesterolemia and apolipoprotein E2: evidence for a gene–gene interaction. *Arteriosclerosis and Thrombosis* **11**: 1137–1146.

Hughes, A. L. and Nei, M. (1988). Pattern of nucleotide substitution at major histocompatability complex loci reveals overdominant selection. *Nature* **335**: 167–170.

Hughes, A. L. and Nei, M. (1989). Nucleotide substitution at major histocompatability complex class II loci: evidence for overdominant selection. *Proceedings of the National Academy of Sciences, USA* **86**: 958–962.

Hughes, A. L. and Nei, M. (1990). Evolutionary relationships of class II major-histocompatability-complex genes in mammals. *Molecular Biology and Evolution* **7**: 491–514.

Hughes, A. L., Ota, T. and Nei, M. (1990). Positive Darwinian selection promotes charge profile diversity in the antigen-binding cleft of class I major-histocompatibility-complex molecules. *Molecular Biology and Evolution* **7**: 515–524.

Humphries, S. E. (1988). DNA polymorphisms of the apolipoprotein genes to their use in the investigation of the genetic component of hyperlipidaemia and atherosclerosis. *Atherosclerosis* **72**: 89–108.

Humphries, S. E., Talmud, P. J. and Kessling, A. M. (1987). Use of DNA polymorphisms of the apolipoprotein genes to study the role of genetic variation in the determination of serum lipid levels. In Bock, G. and Collins, G. M., editors, *Molecular Approaches to Human Polygenic Disease*, CIBA Foundation Symposium No. 130, Chichester, J. Wiley, pp. 128–129.

Hutchinson, J. (1986). Relationship between African admixture and blood pressure variation in the Caribbean. *Human Heredity* **36**: 12–18.

Jacob, H., Lindpainter, K., Lincoln, S., Kusami, K., Bunker, R., Mao, Y.-P. *et al.* (1991). Genetic mapping of a gene causing hypertension in the stroke prone spontaneously hypertensive rat. *Cell* **67**: 213–224.

Jacquard, A. (1974). *The Genetic Structure of Populations*, New York, Springer.

John, S. W., Rozen, R., Scriver, C. R., Laframboise, R. and Laberge, C. (1990). Recurrent mutation, gene conversion, or recombination at the human phenylalanine hydroxylase locus: evidence in French-Canadians and a catalog of mutations. *American Journal of Human Genetics* **46**: 970–974.

Jorde, L. B. and Lathrop, G. M. (1988). A test of the heterozygote-advantage hypothesis in cystic fibrosis carriers. *American Journal of Human Genetics* **42**: 808–815.

Julier, C., Hyer, R. N., Davies, J., Merlin, F., Soularue, P., Briant, L. *et al.* (1991). Insulin-IGF2 region on chromosome 11p encodes a gene implicated in HLA-DR4-dependent diabetes susceptibility. *Nature* **354**: 155–157.

Kalbfleisch, J. D. and Prentice, R. L. (1980). *The Statistical Analysis of Failure-Time Data*, New York, Wiley.

Kamboh, M. I., Weiss, K. M. and Ferrell, R. E. (1991). Genetic studies of human apolipoproteins. XVI. APOE polymorphism and cholesterol levels in the Mayans of the Yucatan Peninsula, Mexico. *Clinical Genetics* **39**: 26–32.

Kan, Y. W. and Dozy, A. M. (1980). Evolution of the hemoglobin S and C genes in world populations. *Science* **209**: 388–390.

Kaprio, J., Ferrell, R. E., Kottke, B. A. and Sing, C. F. (1989). Smoking and reverse cholesterol transport: evidence for gene–environmental interaction. *Clinical Genetics* **36**: 266–268.

Kelly, A. P., Monaco, J. J., Cho, S. and Trowsdale, J. (1991). A new human HLA class II-related locus, *DM*. *Nature* **353**: 571–573.

Kerem, B., Rommens, J., Buchanan, J., Markiewicz, D., Cox, T., Chakravarti, A. *et al.* (1989). Identification of the cystic fibrosis gene: genetic analysis. *Science* **245**: 1073–1080.

Kerem, B., Rommens, J., Buchanan, B., Markiewicz, D., Cox, T. and Chakravarti, A. (1990). The relationship between genotype and phenotype in cystic fibrosis – analysis of the most common mutation ($\triangle$*F508*). *New England Journal of Medicine* **323**: 1517–1522.

Khoury, M. J., Beaty, T. H. and Cohen, B. H. (1986). The interface of genetics and epidemiology. *Journal of Chronic Disease* **39**: 963–978.

Khoury, M. J., Beaty, T. H. and Flanders, W. D. (1990). Epidemiological approaches to the use of DNA markers in the search for disease susceptibility genes. *Epidemiologic Reviews* **12**: 41–55.

Khoury, M. J., Beaty, T. H. and Liang K.-Y. (1988). Can familial aggregation of disease be explained by familial aggregation of environmental risk factors? *American Journal of Epidemiology* **127**: 674–683.

Kidd, K. K. (1992). Trials and tribulations in the search for genes causing neuropsychiatric disorders. *Social Biology*, in press.

Kimura, M. (1981). Possibility of extensive neutral evolution under stabilizing selection with special reference to nonrandom usage of synonymous codons. *Proceedings of the National Academy of Sciences, USA* **78**: 5773–5777.

Kimura, M. (1989). The neutral theory of molecular evolution and the world view of the neutralist. *Genome* **31**: 24–31.

Klag, M. J., Whelton, P. K., Coresh, J., Grim, C. E. and Kuller, L. H. (1991). The association of skin color with blood pressure in US Blacks with low socioeconomic status. *Journal of the American Medical Association* **265**: 599–602.

Klein, G. (1988). Oncogenes and tumor suppressor genes. *Reviews in Oncology* **1**: 427–437.

Knowler, W. C., Pettitt, D. J., Bennett, P. H. and Williams, R. C. (1983). Diabetes mellitus in the Pima Indians: genetic and evolutionary considerations. *American Journal of Physical Anthropology* **62**: 107–114.

Knowler, W. C., Williams, R. C., Pettitt, D. J. and Steinberg, A. G. (1988). $Gm^{3;4,13,14}$ and type 2 diabetes mellitus: an association in American Indians with genetic admixture. *American Journal of Human Genetics* **43**: 520–526.

Knudson, A. G. (1971). Mutation and cancer: statistical study of retinoblastoma. *Proceedings of the National Academy of Sciences, USA* **68**: 820–823.

Kondo, I., Berg, K., Drayna, D. and Lawn, R. (1989). DNA polymorphism at the locus for human cholesteryl ester transfer protein (CETP) is associated with high density lipoprotein cholesterol and apolipoprotein levels. *Clinical Genetics* **35**: 49–56.

Konecki, D. S. and Lichter-Konecki, U. (1991). The phenylketonuria locus: current knowledge about alleles and mutations of the phenylalanine hydroxylase gene in various populations. *Human Genetics* **87**: 377–388.

Korenberg, J. R. (1991). Down syndrome phenotypic mapping. In Epstein, C. J., editor, *The Morphogenesis of Down Syndrome*, New York, Wiley-Liss, pp. 43–52.

Krawczak, M. and Cooper, D. N. (1991). Gene deletions causing human genetic disease: mechanisms of mutagenesis and the role of the local DNA sequence environment. *Human Genetics* **86**: 425–441.

Krieger, H., Morton, N. E., Rao, D. C. and Azevedo, E. (1980). Familial determinants of blood pressure in northeastern Brazil. *Human Genetics* **53**: 415–418.

Kunkel, L. M. A., Mocaco, A. P., Middlesworth, W., Ochs, H. D. and Latt, S. A. (1985). Specific cloning of DNA fragments absent from the DNA of a male patient with an X chromosome deletion. *Proceedings of the National Academy of Sciences, USA* **82**: 4778–4782.

Laird, C. D. (1987). Proposed mechanism of inheritance and expression of the human fragile X syndrome of mental retardation. *Genetics* **117**: 587–599.

Lalouel, J. M., Le Mignon, L., Simon, M., Fauchet, R., Bourel, M., Rao, D. C. and Morton, N. E. (1985). A combined qualitative (disease) and quantitative (serum iron) genetic analysis of idiopathic hemochromatosis. *American Journal of Human Genetics* **37**: 700–718.

Lalouel, J. M., Rao, D. C., Morton, N. E. and Elston, R. C. (1983). A unified model for complex segregation analysis. *American Journal of Human Genetics*, **35**: 816–826.

Lande, R. (1987). Quantitative genetics and evolutionary theory. In Weir, B. S., Eisen, E. J., Goodman, M. M. and Namkoong, G., editors, *Proceedings of the Second International Conference on Quantitative Genetics*, Sunderland, Massachusetts, Sinauer Associates, pp. 71–84.

Lander, E. S. (1988). Mapping complex genetic traits in humans. In Davies, K. E., editor, *Genomic Analysis: A Practical Approach*, Oxford, IRL Press, pp. 171–189.

Lander, E. S. and Botstein, D. (1986a). Strategies for studying heterogeneous genetic traits in humans by using a linkage map of restriction fragment length polymorphisms. *Proceedings of the National Academy of Sciences, USA* **83**: 7353–7357.

Lander, E. S. and Botstein, D. (1986b). Mapping complex genetic traits in humans: new methods using a complete RFLP linkage map. *Cold Spring Harbor Symposia on Quantitative Biology* **L1**: 49–62.

Lander, E. S. and Botstein, D. (1989). Mapping Mendelian factors underlying quantitative traits using RFLP linkage maps. *Genetics* **121**: 185–199.

Lange, K. (1986a). The affected sib-pair method using identity by state relations. *American Journal of Human Genetics* **39**: 148–150.

Lange, K. (1986b). A test statistic for the affected-sib-set method. *Annals of Human Genetics* **50**: 283–290.

Lange, K. and Boehnke, M. (1983). Extensions to pedigree analysis. 4. Covariance components models for multivariate traits. *American Journal of Medical Genetics*. **14**: 513–524.

Lange, K. and Elston, R. C. (1975). Extensions to pedigree analysis. 1. Likelihood calculations for simple and complex pedigrees. *Human Heredity* **25**: 95–105.

Lange, K., Westlake, J. and Spence, M. A. (1976). Extensions to pedigree analysis. 3. Variance components by the scoring method. *Annals of Human Genetics* **39**: 485–491.

Lathrop, G. M. and Lalouel, J.-M. (1984). Easy calculations of LOD scores and genetic risks on small computers. *American Journal of Human Genetics* **36**: 460–465.

Lathrop, G. M. and Lalouel, J.-M. (1988). Efficient computations in multilocus linkage analysis. *American Journal of Human Genetics* **42**: 498–505.

Lathrop, G. M., Lalouel, J.-M., Julier, C. and Ott, J. (1984). Strategies for multilocus linkage analysis in humans. *Proceedings of the National Academy of Sciences, USA* **81**: 3443–3446.

Lathrop, G. M., Lalouel, J.-M., Julier, C. and Ott, J. (1985). Multilocus linkage analysis in humans: detection of linkage and estimation of recombination. *American Journal of Human Genetics* **37**: 482–498.

Lathrop, G. M., Lalouel, J. M. and White, R. L. (1986). Construction of human linkage maps: likelihood calculations for multilocus linkage analysis. *Genetic Epidemiology* **3**: 39–52.

Leppert, M., Breslow, J. L., Wu, L., Hasstedt, S., O'Connell, P., Lathrop, G. M. *et al*. (1988). Inference of a molecular defect of apolipoprotein B in hypobetalipoproteinemia by linkage analysis in a large kindred. *Journal of Clinical Investigations* **82**: 847–851.

Lewin, B. (1990). *Genes IV*, New York, Cell Press/Oxford University Press.

Levitan, M. (1988). *Textbook of Human Genetics*, 3rd edn, New York, Oxford University Press.

Li, C. C. (1961). *Human Genetics*, New York: McGraw-Hill.

Li, C. C. (1976). *First Course in Population Genetics*, Pacific Grove, California, Boxwood Press.

Li, C. C. (1987). A genetical model for emergencies: in memory of Laurence H. Snyder, 1901–86. *American Journal of Human Genetics* **41**: 517–523.

Li, W.-H. and Graur, D. (1990). *Fundamentals of Molecular Evolution*. Sunderland, Massachusetts, Sinauer Associates.

Li, W.-H., Tanimura, M., Luo, C.-C., Datta, S. and Chan, L. (1988). The apolipoprotein multigene family: biosynthesis, structure, structure–function relationships, and evolution. *Journal of Lipid Research* **29**: 245–271.

Lipkin, M. (1988). Biomarkers of increased susceptibility to gastrointestinal cancer: new application studies of cancer prevention in human subjects. *Cancer Research* **48**: 235–245.

Lipkin, M., Scharf, S. and Schechter, L. (1980). Memorial Hospital registry of population groups at high risk for cancer of the large intestine: age of onset of neoplasms. *Preventive Medicine* **9**: 335–345.

Livingstone, F. B. (1986). Anthropological aspects of the distributions of the human hemogloblin variants. In Winter, W. P., editor, *Hemoglobin Variants in Human Populations*, Boca Raton, Florida, CRC Press, pp. 17–28.

Livingstone, F. B. (1989). Who gave whom hemoglobin S: the use of restriction site haplotype variation for the interpretation of the evolution of the β^s-globin gene. *American Journal of Human Biology* **1**: 289–302.

Longini, I. M., Higgins, M. W., Hinton, P. C., Moll, P. P. and Keller, J. B. (1984) Environmental and genetic sources of familial aggregation of blood pressure in Tecumseh, Michigan. *American Journal of Epidemiology* **120**: 131–144.

Lowenfels, A. B., Lindstrom, C. G., Conway, M. J. and Hastings, P. R. (1985). Gallstones and risk of gallbladder cancer. *Journal of the National Cancer Institute* **75**: 77–80.

Luo, C.-C., Li, W.-H., Moore, M. N. and Chan, L. (1986). Structure and evolution of the apolipoprotein multigene family. *Journal of Molecular Biology* **187**: 325–340.

Lusis, A. J. (1988). Genetic factors affecting blood lipoproteins: the candidate gene approach. *Journal of Lipid Research* **29**: 397–429.

Lyonnet, S., Caillaud, C., Rey, F., Berthelon, M., Frezal, J., Rey, J. and Munnich, A. (1989). Molecular genetics of phenylketonuria in Mediterranean countries. A mutation associated with partial phenylalanine hydroxylase deficiency. *American Journal of Human Genetics* **44**: 511–517.

Lynch, H. T., Fain, P. R. and Goldgar, D. E. (1981). Familial breast cancer and its recognition in an oncology clinic. *Cancer* **47**: 2730–2739.

Lynch, H. T., Kimberling, W. J., Biscone, K. A., Lynch, J. F., Wagner, C. A., Brennan, K. *et al*. (1986). Familial heterogeneity of colon cancer risk. *Cancer* **57**: 2089–2096.

Lynch, M. (1988). The rate of polygenic mutation. *Genetical Research* **51**: 137–148.

Magenis, E. R., Brown, M. G., Lacy, D. A., Budden, S. and LaFranchi, S. (1987). Is Angelman syndrome an alternate result of del(15)(q11q13)? *American Journal of Medical Genetics* **28**: 829–838.

Majumder, P. P., Chakraborty, R. and Weiss, K. M. (1983). Relative risk of disease in the presence of incomplete penetrance and sporadics. *Statistics in Medicine* **2**: 13–24.

Majumder, P. P., Das, S. K. and Li, C. C. (1988). A genetical model for vitiligo. *American Journal of Human Genetics* **43**: 119–125.

Majumder, P. P., Ramesh, A. and Chinnappan, D. (1989). On the genetics of prelingual deafness. *American Journal of Human Genetics* **44**: 86–99.

Malkin, D., Jolly, K., Barbier, N., Look, A., Friend, S., Gebhardt, M. *et al.* (1992). Germline mutations of the p53 tumor-suppressor gene in children and young adults with second malignant neoplasms. *New England Journal of Medicine* **326**: 1309–1315.

Malkin, D., Li, F. P., Strong, L. C., Fraumeni, J. L., Nelson, C. E., Kim, D. H. *et al.* (1990) Germ line p53 mutations in a familial syndrome of breast cancer, sarcomas, and other neoplasms. *Science* **250**: 1233–1238.

Manton, K. G. and Stallard, E. (1988). *Chronic Disease Modelling*, London, Charles Griffin.

Martinez, M. and Goldin, L. R. (1991). Detection of linkage for heterogeneous disorders by using multipoint linkage analysis. *American Journal of Human Genetics* **49**: 1300–1305.

McNeill, W. H. (1976). *Plagues and Peoples*, New York, Doubleday.

Meindl, R. (1978). Hypothesis: a selective advantage for cystic fibrosis heterozygotes. *American Journal of Physical Anthropology* **74**: 39–45.

Meyer, J. M. and Eaves, L. J. (1988). Estimating genetic parameters of survival distributions: a multifactoral model. *Genetic Epidemiology* **5**: 265–275.

Meyerowitz, R. (1988). Splice junction mutation in some Ashkenazi Jews with Tay-Sachs disease: evidence against a single defect within this ethnic group. *Proceedings of the National Academy of Science, USA* **85**: 3955–3959.

Moll, P. P., Harburg, E., Burns, T. L., Schork, M. A. and Ozgoren, F. (1983). Heredity, stress and blood pressure, a family set approach: the Detroit project revisited. *Journal of Chronic Diseases* **36**: 317.

Moll, P. P., Michels, V. V., Weidman, W. H. and Kotke, B. A. (1989). Genetic determination of plasma apolipoprotein AI in a population-based sample. *American Journal of Human Genetics* **44**: 124–139.

Moll, P. P., Sing, C. F., Lussier-Cacan, S. and Davignon, J. (1984). An application of a model for a genotype-dependent relationship between a concomitant (age) and a quantitative trait (LDL cholesterol) in pedigree data. *Genetic Epidemiology* **1**: 301–314.

Monaco, A. P., Bertelson, C. J., Middlesworth, W., Colletti, C.-A., Aldridge, J., Fischbeck, K. H. *et al.* (1985). Detection of deletions spanning the Duchenne muscular dystrophy locus using a tightly linked DNA segment. *Nature* **316**: 842–845.

Monk, M. (1990). Variation in epigenetic inheritance. *Trends in Genetics* **6**: 110–114.

Moolgavkar, S. H. (1992). Stochastic models of carcinogenesis. In Rao, C. R. and Chakraborty, R., editors, *Handbook of Statistics*, vol. 8, pp. 373–393.

Morel, P. A., Dorman, J. S., Todd, J. A., McDevitt, H. O. and Trucco, M. (1988). Aspartic acid at position 57 of the DQβ chain protects against type 1 diabetes: a family study. *Proceedings of the National Academy of Sciences, USA* **85**: 8111–8115.

Morita, R., Ishikawa, J., Tsutsumi, M., Hikiji, K., Tsukada, Y., Kamidono, S. *et al.* (1991). Allelotype of renal cell carcinoma. *Cancer Research* **51**: 820–823.

Morrell, D., Cromartie, E. and Swift, M. (1986). Mortality and cancer incidence in 263 patients with ataxia-telangiectasia. *Journal of the National Cancer Institute* **77**: 89–92.

Morton, N. E. (1982). *Outline of Genetic Epidemiology*, Basel, S. Kriger.

Morton, N. E., Gulbrandsen, C. L., Rao, D. C., Rhoads, G. C. and Kagon, A. (1980). Determinants of blood pressure in Japanese-American families. *Human Genetics* **53**: 261–266.

Morton, N. E. and Maclean, C. J. (1974). Analysis of family resemblance. 3. Complex segregation of quantitative traits. *American Journal of Human Genetics* **26**: 489–503.

Motulsky, A. G., Burke, W., Billings, P. R. and Ward, R. H. (1987). Hypertension and the genetics of red cell membrane abnormalities. In Bock, G. and Collins, G. M., editors, *Molecular Approaches to Human Polygenic Disease*, CIBA Foundation Symposium, No. 130, Chichester, J. Wiley, pp. 150–166.

Mousseau, T. A. and Roff, D. A. (1987). Natural selection and the heritability of fitness components. *Heredity* **59**: 181–197.

Mulligan, L. M., Matlashewski, G. J., Scrable, H. J. and Cavenee, W. K. (1990). Mechanisms of p53 loss in human sarcomas. *Proceedings of the National Academy of Sciences, USA* **87**: 5863–5867.

Murphy, E. A. and Chase, G. A. (1975). *Principles of Genetic Counseling*, Chicago, Yearbook Medical Publishers.

Navon, R., Kolodny, E. H., Mitsumoto, H., Thomas, G. H. and Proia, R. L. (1990). Ashkenazi-Jewish and non-Jewish adult G_{M2} gangliosidosis patients share a common genetic defect. *American Journal of Human Genetics* **46**: 817–821.

Navon, R., and Proia, R. L. (1989). The mutations in Ashkenazi Jews with adult G_{M2} gangliosidosis, the adult form of Tay-Sachs disease. *Science* **243**: 1471–1473.

Neel, J. V. (1962). Diabetes mellitus: a "thrifty genotype" rendered detrimental by "progress". *American Journal of Human Genetics* **14**: 353–362.

Neel, J. V., Centerwall, W. R., Chagnon, H. A. and Casey, H. L. (1970). Notes on the effect of measles and measles vaccine in a virgin-soil population of South American Indians. *American Journal of Epidemiology* **91**: 418–429.

Nei, M. (1987). *Molecular Evolutionary Genetics*, New York, Columbia University Press.

Nei, M. and Hughes, A. L. (1991). Polymorphism and evolution of the major histocompatibility complex loci in mammals. In Selander, R. K., Clark, A. G. and Whittam, T. S., editors, *Evolution at the Molecular Level*, Sunderland, Massachusetts, Sinauer Associates, pp. 222–247.

Nichols, R. D., Knoll, J. H. M., Butler, M. G., Karam, S. and Lalande, M. (1989). Genetic imprinting suggested by maternal hetero-disomy in non-deletion Prader-Willi syndrome. *Nature* **342**: 281–285.

Nigro, J. M., Baker, S. J., Preisinger, A. C., Jessup, J. M., Hostetter, R., Clearly, K. *et al.* (1989). Mutations in the p53 gene occur in diverse human tumor types. *Nature* **342**: 705–708.

Nutels, N. (1968). Medical problems of newly contacted Indian groups. In Pan American Health Organization Report, *Biomedical Challenges Presented by the American Indian*, Washington, pp. 68–76.

Oberle, I., Rousseau, F., Heitz, D., Kretz, C., Devys, D., Hanauer, A. *et al.* (1991). Instability of a 550-base pair DNA segment and abnormal methylation in fragile X syndrome. *Science* **252**: 1097–1102.

Ohno, S. (1970). *Evolution by Gene Duplication*, New York, Springer.

Ohno, S. (1985). Perspectives on immortal genes. *Trends in Genetics* **1**: 196–200.

Okano, Y., Eisensmith, R. C., Guttler, F., Lichter-Konecki, U., Konecki, D. S., Trefz, F. K. *et al.* (1991). Molecular basis of phenotypic heterogeneity in phenylketonuria. *New England Journal of Medicine* **324**: 1232–1238.

Okano, Y., Wang, T., Eisensmith, R. C., Guttler, F. and Woo, S. L. C. (1990). Recurrent mutation in the human phenylalanine hydroxylase gene. *American Journal of Human Genetics* **46**: 919–924.

Ott, J. (1991). *Analysis of Human Genetic Linkage*, 2nd edn, Baltimore, Maryland, Johns Hopkins Press.

Paterson, A. H., Damon, S., Hewitt, J. D., Zamir, D., Rabinowitch, H. D., Lincoln, S. E. *et al.* (1991). Mendelian factors underlying quantitative traits in tomato: comparison across species, generations, and environments. *Genetics* **127**: 181–197.

Paterson, A. H., DeVerna, J. W., Lanini, B. and Tanksley, S. D. (1990). Fine mapping of quantitative trait loci using selected overlapping recombinant chromosomes, in an interspecies cross of tomato. *Genetics* **124**: 735–742.

Paterson, A. H., Lander, E. S., Hewitt, J. D., Peterson, S., Lincoln, S. E. and Tanksley, S. D. (1988). Resolution of quantitative traits into Mendelian factors by using a complete linkage map of restriction fragment length polymorphisms. *Nature* **335**: 721–726.

Paw, B. H., Tieu, P. T., Kaback, M. M., Lim, J. and Neufeld, E. F. (1990). Frequency of three *HexA* mutant alleles among Jewish and non-Jewish carriers identified in a Tay-Sachs screening program. *American Journal of Human Genetics* **47**: 698–705.

Penrose, L. S. (1935). The detection of autosomal linkage in data which consists of pairs of brothers and sisters of unspecified parentage. *Annals of Eugenics* **6**: 133–138.

Pericak-Vance, M. A., Bebout, J. L., Gaskell, P. C., Yamaoka, L. H., Hung, W.-Y., Alberts, M. J. *et al.* (1991). Linkage studies in familial Alzheimer disease: evidence of chromosome 19 linkage. *American Journal of Human Genetics* **48**: 1034–1050.

Perucca, D., Szepetowski, P., Simon, M.-P. and Gaudray, P. (1990). Molecular genetics of human bladder carcinomas. *Cancer Genetics and Cytogenetics* **49**: 143–156.

Perusse, L., Moll, P. P. and Sing, C. F. (1991). Evidence that a single gene with gender- and age-dependent effects influences systolic blood pressure determination in a population-based sample. *American Journal of Human Genetics* **49**: 94–105.

Peto, R. (1978). Epidemiology, multistage models, and short-term mutagenicity tests. In Hiatt, H. H., Watson, J. D. and Winsten, J. A., editors, *Origins of Human Cancer* (4 vols.), vol. C, Cold Spring Harbor, New York, Cold Spring Harbor Laboratory Press, pp. 1403–1414.

Plomin, R., DeFries, J. C. and McClearn, G. E. (1990). *Behavioral Genetics*, New York, Freeman.

Plomin, R., McClearn, G. E., Gora-Maslak, G. and Neiderhiser, J. M. (1991). Use of recombinant inbred strains to detect quantitative trait loci associated with behavior. *Behavioral Genetics* **21**: 99–116.

Ploughman, L. M. and Boehnke, M. (1989). Estimating the power of a proposed linkage study for a complex genetic trait. *American Journal of Human Genetics* **44**: 543–551.

Rajavashisth, T. B., Lusis, A. J. and Kaptein, J. S. (1990). A model for the evolution of apolipoprotein genes from oligonucleotide repeats. In Dutta, S. K. and Winter, W. P., editors, *DNA Systematics* vol. 3 *Human and Higher Primates*, Boca-Raton, Florida, CRC Press.

Rao, D. C. (1985). Applications of path analysis in human genetics. In Krishnaian, P. R., editor, *Multivariate Analysis - IV*, New York, Elsevier, pp. 467–484.

Rao, D. C., Keats, B. J. B., Morton, N. E., Yee, S. and Lew, R. (1978). Variability in human linkage data. *American Journal of Human Genetics* **30**: 516–529.

Rao, D. C., McGue, M., Wette, R. and Glueck, C. J. (1984). Path analysis in genetic epidemiology. In Chakravarti, A., editor, *Human Population Genetics: The Pittsburgh Symposium*, New York, Van Nostrand Reinhold, pp. 35–81.

Rebbeck, T. R., Turner, S. T., Michels, V. V. and Moll, P. P. (1991). Genetic and environmental explanations for the distribution of sodium-lithium countertransport in pedigrees from Rochester, MN. *American Journal of Human Genetics* **48**: 1092–1104.

Reilly, S. L., Ferrell, R. E., Kottke, B. A., Kamboh, M. I. and Sing, C. F. (1991). The gender-specific apolipoprotein E genotype influence on the distribution of lipids and apolipoproteins in the population of Rochester, MN. 1. Pleiotropic effects on means and variances. *American Journal of Human Genetics* **49**: 1155–1167.

Rey, F., Berthelon, M., Caillaud, C., Lyonnet, S., Abadie, V., Blandin-Savoja, F. *et al.* (1988). Clinical and molecular heterogeneity of phenylalanine hydroxylase deficiencies in France. *American Journal of Human Genetics* **43**: 914–921.

Rice, J. P. (1986). Genetic epidemiology: models of multifactorial inheritance and path analysis applied to quantitative traits. In Moolgavkar, S. H. and Prentice, R. L., editors, *Modern Statistical Methods in Chronic Disease Epidemiology*, New York, Wiley, pp. 225–243.

Rice, T., Bouchard, C., Borecki, I. B. and Rao, D. C. (1990). Commingling and segregation analysis of blood pressure in a French-Canadian population. *American Journal of Human Genetics* **46**: 37–44.

Riordan, J., Rommens, J., Kerem, B., Aon, N., Rozmahel, R., Grzelczak, Z. *et al*. (1989). Identification of the cystic fibrosis gene: cloning and characterization of complementary DNA. *Science* **245**: 1066–1073.

Risch, N. (1990a). Linkage strategies for genetically complex traits. 1. Multilocus models. *American Journal of Human Genetics* **46**: 222–228.

Risch, N. (1990b). Linkage strategies for genetically complex traits. 2. The power of affected relative pairs. *American Journal of Human Genetics* **46**: 229–241.

Risch, N. (1990c). Linkage strategies for genetically complex traits. 3. The effect of marker polymorphism on analysis of affected relative pairs. *American Journal of Human Genetics* **46**: 242–253.

Risch, N. (1991). A note on multiple testing procedures in linkage analysis. *American Journal of Human Genetics* **48**: 1058–1065.

Robertson, A. (1967). The nature of quantitative genetic variation. In Brink, R. A., editor, *Heritage from Mendel*, Madison, Wisconsin, University of Wisconsin Press, pp. 265–280.

Roitt, I., Brostoff, J. and Male, D. (1989). *Immunology*, 2nd edn, London, Wower Medical Publishing.

Romeo, G., Devoto, M. and Galietta, J. L. V. (1989). Why is the cystic fibrosis gene so frequent? *Human Genetics* **84**: 1–5.

Rommens, J., Iannuzzi, M., Kerem, B., Drumm, M., Melmer, G., Dean, M. *et al*. (1989). Identification of the cystic fibrosis gene: chromosome walking and jumping. *Science* **245**: 1059–1065.

Ron, E., Modan, B. and Boice, J. D. (1988). Mortality after radiotherapy for ringworm of the scalp. *American Journal of Epidemiology* **127**: 713–725.

Rose, M. R. (1991). *Evolutionary Biology of Aging*, New York, Oxford University Press.

Rothman, K. J. (1990). No adjustments are needed for multiple comparisons. *Epidemiology* **1**: 43–46.

Rotter, J. I. and Diamond, J. M. (1987). What maintains the frequencies of human genetic diseases? *Nature* **329**: 289–290.

Roy, M., Sing, C. F., Betard, C. and Davignon, J. (1991). Impact of a >10kb deletion in the low density lipoprotein receptor (LDLR) gene on means, variances and correlations of measures of lipid metabolism in French-Canadians. *American Journal of Human Genetics* **49**(Suppl): 106.

Roychoudhury, A. K. and Nei, M. (1988). *Human Polymorphic Genes: World Distribution*, New York, Oxford University Press.

Runnegar, B. (1985). Collagen gene construction and evolution. *Journal of Molecular Evolution* **22**: 141–149.

Rushforth, N. B., Bennet, P. H., Steinberg, A. G., Burch, T. A. and Miller, M. (1971). Diabetes in the Pima Indians: evidence of bimodality in glucose tolerance distributions. *Diabetes* **20**: 756–765.

Sambrook, J., Fritsch, E. and Maniatis, T. (1989). *Molecular Cloning: A Laboratory Manual* (3 vols.). Cold Spring Harbor, New York, Cold Spring Harbor Laboratory Press.

Saunders, J., and O'Malley, Y. C. (1950). The illustrations from the works of Andreas Vesalius of Brussels. New York, Dover Books.

Scheffer, H., te Meerman, G., Kruize, Y., van den Berg, A., Penninga, D., Tan, K. *et al.* (1989). Linkage analysis of families with hereditary retinoblastoma: nonpenetrance of mutation revealed by combined use of markers within and flanking the RB1-gene. *American Journal of Human Genetics* **45**: 252–260.

Schlessinger, D. (1990). Yeast artificial chromosomes: tools for mapping and analysis of complex genomes. *Trends in Genetics* **6**: 248–258.

Schork, N. and Schork, M. A. (1989). Testing separate families of segregation hypotheses: bootstrap methods. *American Journal of Human Genetics* **45**: 803–813.

Schwartz, A. G., Boehnke, M. and Moll, P. P. (1988). Family risk index as a measure of familial heterogeneity of cancer risk: a population-based study in metropolitan Detroit. *American Journal of Epidemiology* **128**: 524–535.

Scrable, H. J., Sapienza, C. and Cavenee, W. K. (1990). Genetic and epigenetic losses of heterozygosity in cancer predisposition and progression. *Advances in Cancer Research* **54**: 25–62.

Scriver, C. R., Kaufman, S. and Woo, S. L. C. (1988). Mendelian hyperphenylalanemia. *Annual Review of Genetics* **22**: 301–321.

Searle, A. G., Peters, J., Lyon, M. F., Hall, J. G., Evans, E. P., Edwards, J. H. and Buckle, V. J. (1989). Chromosome maps of man and mouse. 4. *Annals of Human Genetics* **53**: 89–140.

Serjeantson, S. W. (1983). HLA and susceptibility to leprosy. *Immunological Reviews* **70**: 89–112.

Shashikant, C., Utset, M., Violette, S., Wise, T., Einat, P., Einat, M. *et al.* (1991). Homeobox genes in mouse development. *Critical Reviews in Eukaryotic Gene Expression* **1**: 207–245.

Sheinfeld, J., Schaeffer, A. J., Cordon-Cardo, C., Rogatko, A. and Fair, W. R. (1989). Association of the Lewis blood-group phenotype with recurrent urinary tract infections in women. *New England Journal of Medicine* **320**: 773–777.

Shrimpton, A. E. and Robertson, A. (1988a). The isolation of polygenic factors controlling bristle score in *Drosophila melanogaster*. 1. Allocation of third chromosome sternopleural bristle effects to chromosome sections. *Genetics* **118**: 437–443.

Shrimpton, A. E. and Robertson, A. (1988b). The isolation of polygenic factors controlling bristle score in *Drosophila melanogaster*. 2. Distribution of third chromosome bristle effects within chromosome sections. *Genetics* **118**: 445–459.

Shute, N. C. E. and Ewens, W. J. (1988a). A resolution of the ascertainment sampling problem. 2. Generalizations and numerical results. *American Journal of Human Genetics* **43**: 364–373.

Shute, N. C. E. and Ewens, W. J. (1988b). A resolution of the ascertainment sampling problem. 3. Pedigrees. *American Journal of Human Genetics* **43**: 374–386.

Siber, G. R., Santosham, M., Reid, R., Thompson, C., Almeido-Hill, J., Morell, A. *et al.* (1990). Impaired antibody response to *Haemophilus influenzae* type b polysaccharide and low IgG2 and IgG4 concentrations in Apache children. *New England Journal of Medicine* **323**: 1387–1392.

Sidransky, D., Von Eschenbach, A. V., Tsai, Y. C., Jones, P., Summerhayes, I., Marshall, F. *et al.* (1991). Identification of p53 gene mutations in bladder cancers and urine samples. *Science* **252**: 706–709.

Sing, C. F. and Boerwinkle, E. (1987). Genetic architecture of inter-individual variability in apolipoprotein, lipoprotein and lipid phenotypes. In Bock, G. and Collins, G. M., editors, *Molecular Approaches to Human Polygenic Disease*, CIBA Foundation Symposium No. 130, Chichester, Wiley, pp. 99–121.

Sing, C. F., Boerwinkle, E., Moll, P. P. and Templeton, A. R. (1988). Characterization of genes affecting quantitative traits in humans. In Weir, B. S., Eisen, E. J., Goodman, M. M. and Namkoong, G., editors, *Proceedings of the 2nd International Conference on Quantitative Genetics*, Sunderland, Massachusetts: Sinauer, pp. 250–269.

Sing, C. F., Boerwinkle, E. and Turner, S. T. (1986). Genetics of primary hypertension. *Clinical and Experimental Hypertension*, Part A: *Theory and Practice* **A8**(4&5): 623–651.

Sing, C. F. and Moll, P. P. (1989). Genetics of variability of CHD risk. *International Journal of Epidemiology* **18**(Suppl.): S183–S195.

Sing, C. F. and Moll, P. P. (1990). Genetics of atherosclerosis. *Annual Review of Genetics* **24**: 171–187.

Sing, C. F. and Orr, J. D. (1976). Analysis of genetic and environmental sources of variation in serum cholesterol in Tecumseh, Michigan. 3. Identification of genetic effects using 12 polymorphic genetic blood marker systems. *American Journal of Human Genetics* **28**: 453–464.

Skolnick, M., Bishop, D., Carmelli, D. L., Gardner, E., Hadley, R., Hasstedt, S. *et al.* (1981). A population-based assessment of familial cancer risk in Utah Mormon genealogies. In Arrighi, F., Rao, P. and Stubblefield, E., editors, *Genes, Chromosomes, and Neoplasia*, New York, Raven Press, pp. 477–500.

Skolnick, M. H., Cannon-Albright, L. A., Goldgar, E. E., Ward, J. H., Marshall, C. J., Schumann, G. B. *et al.* (1990). Inheritance of proliferative breast disease in breast cancer kindreds. *Science* **250**: 1715–1720.

Sokal, R. R. and Rohlf, F. J. (1981). *Biometry: The Principles and Practice of Statistics in Biological Research*, 2nd edn, San Francisco, W. H. Freeman.

Solomon, E. (1990). Colorectal cancer genes. *Nature* **343**: 412–414.

Solter, D. (1988). Differential imprinting and expression of maternal and paternal genomes. *Annual Review of Genetics* **22**: 127–146.

Squire, W. (1982). On measles in Fiji. *Transactions of the Epidemiological Society of London* **4**: 72–74.

Srivastava, S., Zou, Z., Pirollo, K., Blattner, W. and Chang, E. H. (1990). Germ-line transmission of a mutated p53 gene in a cancer-prone family with Li-Fraumeni syndrome. *Nature* **348**: 747–749.

Stern, M. P. (1988). Type II diabetes mellitus: interface between clinic and epidemiological investigation. *Diabetes Care* **11**: 119–126.

Stryer, L. (1988). *Biochemistry*, 3rd edn, New York, W. H. Freeman.

Stuber, C. W., Edwards, M. D. and Wendel, J. F. (1987). Molecular marker-facilitated investigations of quantitative trait loci in maize. 2. Factors influencing yield and its component traits. *Crop Science* **27**: 639–648.

Suarez, B. K. and Hodge, S. E. (1979). A simple method to detect linkage for rare recessive diseases: an application to juvenile diabetes. *Clinical Genetics* **15**: 126–136.

Suarez, B., O'Rourke, D. and Van Eerdewegh, P. (1982). Power of the affected-sib-pair method to detect disease susceptibility loci of small effect: an application to multiple sclerosis. *American Journal of Medical Genetics* **12**: 309–326.

Suarez, B. K., Rice, J. P. and Reich, T. (1978). The generalized sib pair IBD distribution: its use in the detection of linkage. *Annals of Human Genetics* **42**: 87–94.

Suarez, B. K. and Van Eerdewegh, P. (1984). A comparison of three affected-sib-pair scoring methods to detect HLA-linked disease susceptibility genes. *American Journal of Medical Genetics* **18**: 135–146.

Sudhof, T. C., Goldstein, J. L., Brown, W. S. and Russell, D. W. (1985). The LDL receptor gene: a mosaic of exons shared with different proteins. *Science* **228**: 815–822.

Susser, E. and Susser, M. (1989) Familial aggregation studies: a note on their epidemiologic properties. *American Journal of Epidemiology* **129**: 23–30.

Suzuki, D. T., Griffiths, A. J. F., Miller, J. H. and Lewontin, R. C. (1989). *An Introduction to Genetic Analysis*, 4th edn. New York, W. H. Freeman.

Swift, M., Morrell, D., Massey, R. and Chase, C. (1991). Incidence of cancer in 161 families affected by ataxia-telangiectasia. *New England Journal of Medicine* **325**: 1831–1836.

Swift, M., Reitnauer, P. J., Morrell, D. and Chase, C. L. (1987). Breast and other cancers in families with ataxia-telangiectasia. *New England Journal of Medicine* **316**: 1289–1294.

Szathmary, E. J. E. (1990). Diabetes in Amerindian populations: the Dogrib studies. In Armelagos, G. and Swedlund, A., editors, *Health and Disease of Populations in Transition*, New York, Bergin & Garvey, pp. 75–104.

Tanaka, T. and Nei, M. (1989). Positive darwinian selection observed at the variable-region genes of immunoglobulin. *Molecular Biology and Evolution* **6**: 447–459.

Taylor, J. M., Lauer, S., Eslhourbagy, N., Reardon, C., Tacman, E., Walker, D. *et al.* (1987). Structure and evolution of human apolipoprotein genes: identification of regulatory elements of the human apolipoprotein E gene. In Bock, G. and Collins, G. M., editors, *Molecular Approaches to Human Polygenic Disease*, CIBA Foundation Symposium No. 130, Chichester, Wiley, pp. 70–86.

Templeton, A. R., Boerwinkle, E. and Sing, C. F. (1987). A cladistic analysis of phenotypic associations with haplotypes inferred from restriction endonuclease mapping. 1. Basic theory and an analysis of alcohol dehydrogenase activity in *Drosophila*. *Genetics* **117**: 343–351.

Templeton, A. R., Sing, C. F., Kessling, A. and Humphries, S. (1988). A cladistic analysis of phenotype associations with haplotypes inferred from restriction endonuclease mapping. 2. The analysis of natural populations. *Genetics* **120**: 1145–1154.

Thomson, G. (1988). HLA disease associations: models for insulin dependent diabetes mellitus and the study of complex human disorders. *Annual Review of Genetics* **22**: 31–50.

Thomson, G., Robinson, W. P., Kuhner, M. K. *et al.* (1989). Genetic heterogeneity, modes of inheritance, and risk estimates for a joint study of Caucasians with insulin-dependent diabetes mellitus. *American Journal of Human Genetics* **43**: 799–816.

Thompson, E. A. (1986). *Pedigree Analysis in Human Genetics*, Baltimore, Maryland, Johns Hopkins Press.

Thompson, J. N. and Mascie-Taylor, C. G. N. (1985). Detection of simple polygenic segregations in a natural population. *Proceedings of the National Academy of Sciences, USA* **82**: 8552–8556.

Thompson, J. N. and Thoday, J. M. (1979). *Quantitative Genetic Variation*, New York, Academic Press.

Thompson, J. S. and Thompson, M. W. (1986). *Genetics in Medicine*, Philadelphia, W. B. Saunders.

Todd, J. A., Acha-Orbea, H., Bell, J. I., Chao, N., Fronek, Z., Jacob, C. O. *et al.* (1988a). A molecular basis for MHC class II associated autoimmunity. *Science* **240**: 1003–1009.

Todd, J. A., Aitman, T., Cornall, R., Ghosh, S., Hall, J., Hearne, C. *et al.* (1991). Genetic analysis of autoimmune type I diabetes mellitus in mice. *Nature* **351**: 542–547.

Todd, J. A., Bell, J. I. and McDevitt, H. O. (1988b). A molecular basis for genetic susceptibility to insulin-dependent diabetes mellitus. *Trends in Genetics* **4**: 129–134.

Toguchida, J., Ishizaki, K., Sasaki, M. S., Nakamura, Y., Ikenaga, M., Kato, M. *et al.* (1989). Preferential mutation of paternally derived RB gene as the initial event in sporadic osteosarcoma. *Nature* **338**: 156–158.

Toguchida, J., Yamaguchi, T., Dayton, S., Beacha, R., Herrera, G., Ishizaki, K. *et al.* (1992). Prevalence and spectrum of germline mutations of the p53 gene among patients with sarcoma. *New England Journal of Medicine* **326**: 1301–1308.

Trabuchet, G., Elion, J., Baudot, G., Pagnier, J., Bouhass, R., Nigon, V. M. *et al.* (1991). Origin and spread of β-globin gene mutations in India, Africa, and Mediterranea: analysis of the 5′ flanking and intragenic sequences of β^S and β^C genes. *Human Biology* **63**: 241–252.

Tsui, L.-C. (1992). The spectrum of cystic fibrosis mutations. *Trends in Genetics* **8**: 392–398.

Tsui, L.-C., Buchwald, M., Barker, D., Braman, J. and Knowlton, R. (1985). Cystic fibrosis locus defined by a genetically linked polymorphic marker. *Science* **230**: 1054–1057.

Turelli, M. (1987). Population genetic models for polygenic variation and evolution. In Weir, B., Goodman, M. M., Eisen, E. J. and Namkoong, G., editors. In *Proceedings of the 2nd International Conference on Quantitative Genetics*, Sunderland, Massachusetts, Sinauer Associates, pp. 601–618.

Turner, S. T., Weidman, W. H., Michels, V. V., Reed, T. J., Ormson, C. L., Fuller, T. and Sing, C. F. (1989). Distribution of sodium–lithium countertransport and blood pressure in caucasians five to eighty-nine years of age. *Hypertension* **13**: 378–391.

Utermann, G. (1987). Apolipoproteins, quantitative lipoprotein traits and multifactorial hyperlipidemia. In Bock, G. and Collins, G. M., editors, *Molecular Approaches to Human Polygenic Disease*, CIBA Foundation Symposium No. 130, Chichester, Wiley, pp. 52–69.

van Mazijk, J., Pinheir, F. P. and Black, F. L. (1982). Measles and measles vaccine in isolated Amerindian tribes. 1. The 1971 Trio (Tiriyo) epidemic. *Tropical and Geographical Medicine* **34**: 3–6.

Vanheim, C. M., Rotter, J. I., Maclaren, N. K., Riley, W. J. and Anderson, C. E. (1986). Preferential transmission of diabetic alleles with the HLA gene complex. *New England Journal of Medicine* **315**: 1314–1318.

Vaupel, J. W., Manton, K. G. and Stallard, E. (1979). The impact of heterogeneity in individual frailty on the dynamics of mortality. *Demography* **16**: 439–454.

Vincent, A., Heitz, D., Petit, C., Kretz, C., Oberle, I. and Mandel, J.-L. (1991). Abnormal pattern detected in fragile-X patients by pulsed-field gel electrophoresis. *Nature* **349**: 624–626.

Vogel, F. and Motulsky, A. V. (1986). *Human Genetics*, 2nd edn, New York, Springer.

Vogelstein, B., Fearon, E. R., Hamilton, S. R., Kern, S. E., Preisinger, A. C., Leppert, M. *et al.* (1988). Genetic alterations during colorectal-tumor development. *New England Journal of Medicine* **319**: 525–532.

Vogelstein, B., Fearon, E. R., Kern, S. E., Hamilton, S. R., Preisinger, A. C., Nakamura, Y. and White, R. (1989). Allelotype of colorectal carcinomas. *Science* **244**: 207–211.

Wagener, D., Cavalli-Sforza, L. L. and Barakat, R. (1978). Ethnic variation of genetic disease: roles of drift for recessive lethal genes. *American Journal of Human Genetics* **30**: 262–270.

Waksman, B. H. (1989). Multiple sclerosis: relationship to a retrovirus? *Nature* **337**: 599.

Wang, T., Okano, Y., Eisensmith, R., Huang, S.-Z., Zeng, Y.-T., Lo, W. H. Y. and Woo, S. L. C. (1989). Molecular genetics of phenylketonuria in Orientals: linkage disequilibrium between a termination mutation and haplotype 4 of the phenylalanine hydroxylase gene. *American Journal of Human Genetics* **45**: 675–680.

Wang, T., Okano, Y., Eisensmith, R. C., Lo. W. H. Y., Huang, S.-H., Zeng, Y.-T, and Woo, S. L. C. (1991a). Identification of a novel phenylketonuria (PKU) mutation in the Chinese: further evidence for multiple origins of PKU in Asia. *American Journal of Human Genetics* **48**: 628–630.

Wang, T., Okano, Y., Eisensmith, R., Harvey, M., Lo, W., Huang, S. *et al.* (1991b). Founder effect of a prevalent phenylketonuria mutation in the Oriental population. *Proceedings of the National Academy of Sciences, USA* **88**: 2146–2150.

Ward, R. H. (1990). Familial aggregation and genetic epidemiology of blood pressure. In Laragh, J. H. and Brenner, B. M., editors, *Hypertension:*

Pathophysiology, Diagnosis, and Management, New York, Raven Press, pp. 81–100.

Ward, R. H., Chin, P. G. and Prior, I. A. M. (1980). The effect of migration on the familial aggregation of blood pressure. *Hypertension* **2**: 143–154.

Warram, J. H., Krolewski, A. S., Gottlieb, M. S. and Kahn, C. R. (1984). Difference in risk of insulin-dependent diabetes in offspring of diabetic mothers and diabetic fathers. *New England Journal of Medicine* **311**: 149–152.

Weatherall, D. J. (1991). *The New Genetics and Clinical Practice*, 3rd edn, Oxford, Oxford University Press.

Weatherall, D. J., Bell, J. I., Clegg, J. B., Flint, J., Higgs, D. R., Hill, A.V.S. *et al.* (1988). Genetic factors as determinants of infectious disease transmission in human communities. *Philosophical Transactions of the Royal Society of London, B* **321**: 327–348.

Weder, A. B. and Schork, N. J. (1989). Mixture analysis of erythrocyte lithium-sodium countertransport and blood pressure. *Hyptertension* **13**: 145–150.

Weeks, D.E. and Lange, K. (1988). The affected-pedigree-member method of linkage analysis. *American Journal of Human Genetics* **42**: 315–326.

Weinberg, R. A. (1991). Tumor suppressor genes. *Science* **254**: 1138–1146.

Weir, B. S. (1990). *Genetic Data Analysis*, Sunderland, Massachusetts: Sinauer.

Weiss, K. M. (1984). The number of members of the genus *Homo* who have ever lived, and its evolutionary implications. *Human Biology* **56**: 637–649.

Weiss, K. M. (1985). 'Phenotype amplification' as illustrated by cancer of the gallbladder in New World peoples. In Chakraborty, P. and Szathmary, E., editors, *Etiology of Complex Diseases in Small Populations: Ethnic Differences and Research Approaches*, New York, Alan R Liss, pp. 179–198.

Weiss, K. M. (1989). Are the known chronic diseases related to the human lifespan and its evolution? *American Journal of Human Biology* **1**: 307–319.

Weiss, K. M. (1990a). Duplication with variation: metameric logic in evolution from genes to morphology. *Yearbook of Physical Anthropology* **33**: 1–23.

Weiss, K. M. (1990b). Cancer models and cancer genetics. *Epidemiology* **1**: 486–490.

Weiss, K. M. (1990c). The biodemography of variation in human frailty. *Demography* **27**: 185–206.

Weiss, K. M. (1990d). Biology, homology, and epidemiology. In Adams, J., Lam, D. A., Hermalin, A. I. and Smouse, P. E., editors, *Convergent Questions in Genetics and Demography*, New York, Oxford University Press, pp. 189–206.

Weiss, K. M. (1992). Medieval *mappaemundi* and the conceptual map of genetics. In Sing, C. F. and Hanis, C. L., editors, *Genetic Variability in Human Diseases: Cells, Individuals, Families and Populations*, Oxford, Oxford University Press, in press.

Weiss, K. M., Buchanan, A. V., Valdez, R., Moore, J. H. and Campbell, J. (1992). Amerindians and the price of modernization. In Smith, M., Bilsborough, A., Schell, L. and Watts, E., editors, *Urban Health and Ecology in the Third World*, Cambridge, Cambridge University Press, in press.

Weiss, K. M. and Chakraborty, R. (1984). Multistage risk models and the age pattern in familial polyposis coli. *Cancer Investigation* **2**: 443–448.

Weiss, K. M. and Chakraborty, R. (1990). Multistage models and the age-patterns of cancer: does the statistical analogy imply genetic homology? In Herrera, L., editor, *Familial Adenomatous Polyposis*, New York, Alan R. Liss, pp. 79–89.

Weiss, K. M., Chakraborty, R., Majumder, P. P. and Smouse, P. E. (1982). Problems in the assessment of relative risk of chronic disease among biological relatives of affected individuals. *Journal of Chronic Diseases* **35**: 539–551.

Weiss, K. M., Chakraborty, R., Smouse, P. E., Buchanan, A. V. and Strong, L. C. (1986). Familial aggregation of cancer in Laredo, Texas, a generally low-risk Mexican-American population. *Genetic Epidemiology* **3**: 121–143.

Weiss, K. M., Ferrell, R. E. and Hanis, C. L. (1984a). A New World syndrome of metabolic diseases with a genetic and evolutionary basis. *Yearbook of Physical Anthropology* **27**: 153–178.

Weiss, K. M., Ferrell, R. E., Hanis, C. L. and Styne, P. (1984b). Genetics and epidemiology of gallbladder disease in New World native people. *American Journal of Human Genetics* **36**: 1259–1278.

Weller, J. I. (1986). Maximum likelihood techniques for the mapping and analysis of quantitative trait loci with the aid of genetic markers. *Biometrics* **42**: 627–640.

Weller, J. I., Soller, M. and Brody, T. (1988). Linkage analysis of quantitative traits in an interspecific cross of tomato (*Lycopersicon esculentum* × *Lycopersicon pimpinellifolium*) by means of genetic markers. *Genetics* **118**: 329–339.

Weston, A., Willey, J. C., Modali, R., Sugimura, H., McDowell, E. M., Resau, J. *et al.* (1989). Differential DNA sequence deletions from chromosomes 3, 11, 13, and 17 in squamous-cell carcinoma, large-cell carcinoma, and adenocarcinoma of the human lung. *Proceedings of the National Academy of Sciences, USA* **86**: 5099–5103.

White, R. and Caskey, T. (1988). The human as an experimental system in molecular genetics. *Science* **240**: 1483–1488.

White, R. and Lalouel, J.-M. (1987). Investigation of genetic linkage in human families. *Advances in Human Genetics* **16**: 121–228.

White, R. and Lalouel, J.-M. (1988). Sets of linked genetic markers for human chromosomes. *Annual Review of Genetics* **22**: 259–279.

Whorton, R. G. and Thompson, M. W. (1988). Genetics of Duchenne muscular dystrophy. *Annual Review of Genetics* **22**: 601–630.

Williams, R. C. (1985). HLA 2: the emergence of the molecular model for the human major histocompatibility complex. *Yearbook of Physical Anthropology* **28**: 79–96.

Wright, A. F., Goedert, M. and Hastle, N. D. (1991). Beta amyloid resurrected. *Nature* **349**: 653–654.

Wright, S. (1968–1978). *Evolution and the Genetics of Populations*; vol. 1 *Genetics and Biometric Foundations* (1968); vol. 2 *The Theory of Gene Frequencies* (1969); vol 3 *Experimental Results and Evolutionary Deductions* (1977); vol. 4 *Variability Within and Among Natural Populations* (1978) Chicago, University of Chicago Press.

Yu, S., Pritchard, M., Kremer, E., Lynch, M., Nancarrow, J., Baker, E. *et al.* (1991). Fragile X genotype characterized by an unstable region of DNA. *Science* **252**: 1179–1181.

Zerba, K. E., Kessling, A. M., Davignon, J. and Sing, C. F. (1991). Genetic structure and the search for genotype–phenotype relationships: an example from disequilibrium in the *ApoB* gene region. *Genetics* **129**: 525–533.

Zhu, X., Dunn, J. M., Phillips, R. A., Goddard, A. D., Paton, K. E., Becker, A. and Gallie, B. L. (1989). Preferential germline mutation of the paternal allele in retinoblastoma. *Nature* **340**: 312–313.

Index